INNOVATION AND CONSOLIDATION IN AVIATION

Innovation and Consolidation in Aviation

Selected Contributions to the Australian Aviation Psychology Symposium 2000

Edited by
GRAHAM EDKINS
PETER PFISTER

LONDON AND NEW YORK

First published 2003 by Ashgate Publishing

Published 2016 by Routledge
2 Park Square, Milton Park, Abingdon, Oxfordshire OX14 4RN
711 Third Avenue, New York, NY 10017, USA

First issued in paperback 2016

Routledge is an imprint of the Taylor & Francis Group, an informa business

British Library Cataloguing in Publication Data
Australian Aviation Psychology Symposium (2000)
Innovation and consolidation in aviation : selected contributions to the Australian Aviation Psychology Symposium 2000
1.Aeronautics - Human factors - Congresses 2.Airlines - Safety measures - Congresses 3. Airlines - Employees - Training of - Congresses
I.Title II.Edkins, Graham III.Pfister, Peter
387.7'06

Library of Congress Cataloging-in-Publication Data
Australian Aviation Psychology Symposium (5th : 2000 Sydney, N.S.W.)
Innovation and consolidation in aviation : selected contributions to the Australian Aviation Psychology Symposium 2000 / edited by Graham Edkins and Peter Pfister.
p. cm.
Fifth conference held in Manley Beach, Sydney, Australia in the year 2000.
Includes bibliographical references.
1. Aeronautics--Human factors--Congresses. 2. Aviation psychology--Congresses. I. Edkins, Graham. II. Pfister, Peter. III. Title.
TL553.6.A87 2003
629.132'52'019--dc21

2003055320

ISBN 13: 978-1-138-27744-1 (pbk)
ISBN 13: 978-0-7546-1999-4 (hbk)

Contents

PART 3: TRAINING, LICENSING AND REGULATIONS

PART 4: AUTOMATION AND HUMAN COMPUTER INTERACTION

PART 5: HUMAN FACTORS APPLICATIONS WITHIN THE ATS ENVIRONMENT

PART 6: OTHER HUMAN FACTORS ISSUES

Foreword

The aviation industry has matured to a point where there is now a far more sophisticated understanding and application of Human Factors concepts and knowledge than there has been in the past, in the design, construction, operation and maintenance functions of the aviation system, in both civil and military aviation. The fact that error is a normal characteristic of human behaviour, and that its occurrence is inevitable in any system in which humans are components, has generally been acknowledged and accepted by the aviation industry. Human error is now viewed as normal, and a phenomenon to be expected. There is a growing realisation that aviation systems must therefore be designed to be error tolerant, and that systems and operational procedures which rely upon error free human performance are fundamentally flawed.

In aviation, the traditional 'blame, punish and re-train' solutions to the 'problem' of human error have generally been replaced by informed, multidisciplinary systemic analysis, particularly in the area of flight operations. There is an increasingly proactive focus on attempting to understand the nature of, and the reasons for, the occurrence of human error in normal operations, rather than basing such knowledge primarily upon information derived from accident and incident investigation.

Perhaps the next major challenge for aviation psychologists will be to scientifically evaluate whether the changes in operational philosophy and strategies within civil and military aviation which have arisen from the application of Human Factors concepts and knowledge have made significant measurable contributions to the efficiency, safety, profitability, and cost effectiveness of civil and military aviation operations.

This book comprises selected papers from the Fifth Australian Aviation Psychology Symposium that was conducted at Manly, Sydney, in November 2000. The theme of the Symposium was *Innovation and Consolidation*. At the symposium, practitioners, academics, aviators, management and regulators from a diversity of cultures and companies, came together to discuss and debate not only innovative programs and approaches, but also the need to consolidate established and successful practical applications of aviation psychology and Human Factors concepts, knowledge, and expertise.

Innovation and Consolidation contains the work of 60 researchers and practitioners, in 31 chapters. When one examines the diversity of the contributions and the sophistication of some of the chapters, it is clear that this is a handbook that will provide a rich resource for the practitioner looking for practical applications, as well as the researcher seeking inspiration and peer based information. Some of the chapters in this book also explore the application of mature aviation Human Factors strategies, principles and approaches to other industries such as health care and nuclear power.

The editors, Graham Edkins and Peter Pfister have expended a great deal of effort in the preparation of this publication. All of the symposium papers selected for inclusion in this book have been peer reviewed by a well-qualified team of colleagues from around the globe, representing both practitioners in the industry, as well as research academics from leading Universities. A particularly pleasing development is the increasing amount of collaborative work being reported, with teams made up of specialists from diverse multidisciplinary backgrounds. This trend was already remarked upon in the publication *Aviation Resource Management*, edited by Lowe and Hayward, which resulted from the Fourth AAvPA conference, held in 1998. It has been further expanded in the present book.

This welcome growth in collaborative research is a sign that aviation is receiving increasing attention from other high technology industries, and is seen as a leader in the field of Human Factors and systems safety. It is also proof that many aviation Human Factors concepts can be applied to other industries, and, conversely, that aviation has much to learn from such industries. This exciting development bodes well for the future of Human Factors.

The common factor linking the human performance dimension across a wide range of high-technology industries is systems psychology. All of these industries constitute complex sociotechnical systems, in which the principles of Human Factors and system safety apply equally. Fundamentally, we are systems psychologists who work in many different areas, one of which is aviation. However, as psychologists, we all speak a common language, and share a common body of knowledge. In close consultation with operational specialists from many other areas in our particular industry, such as engineering, line operations, aviation medicine, maintenance, cabin safety, and occupational health and safety, we can apply our Human Factors knowledge, skills, and expertise to the practical solution of operational challenges. Most importantly, by pooling our Human Factors expertise and knowledge, and by seeking the opportunity to work in different industries, we can ensure that mistakes made in one industry are not repeated in others. It also means that industries do not have to go through the same learning curves. They are able to take mature products and concepts researched and developed in one industry, and immediately tailor them to the operating environments of their own.

Each member of a multidisciplinary team brings to it in-depth specialist knowledge from his or her particular field. This is a very important consideration that must be taken into account when assembling such a team. It is vitally important that each member of the team should be appropriately qualified in his or her particular field of expertise. I believe that those of us working as specialists in the fields of psychology, systems safety and Human Factors in aviation have achieved a great deal in making specialists from other areas within the aviation industry aware of some of the basic concepts of Human Factors and systems safety. This has been done by conducting short courses, seminars, and conferences, together with an ongoing process of education and cross training in the workplace. In addition, professional organisations such as the Australian Aviation Psychology Association and the European Association for Aviation Psychology have made considerable and productive efforts to involve operational personnel in the activities of these associations.

However, while this success has been commendable, it is important not to lose sight of the fact that only academically qualified and experienced psychologists possess the necessary in-depth knowledge required to properly and rigorously address behavioural Human Factors issues. As in all other scientific fields, such advanced knowledge can only be gained by years of tertiary study, preferably at the postgraduate level.

Psychology covers many diverse areas, including such fields as perception, memory, human-computer interaction, human information processing, cognition, organisational behaviour, behavioural research techniques, scientific methodology, advanced statistical analysis, and many others. Comprehensive knowledge and understanding of such areas is necessary to ensure that Human Factors and systems safety issues are not treated at the superficial level, and are soundly based in psychological knowledge, scientific theory, and rigorous methodology. To illustrate this point, it is no more acceptable for someone with only a very basic working knowledge of, for example, aircraft engines to have the primary responsibility within a team involved in an advanced engine design and maintenance project, than it is for a person with only a basic knowledge of Human Factors to be the primary Human Factors specialist in a major multidisciplinary research project, or the implementation of an operational program, that has a critical behavioural dimension. It is desirable that all members of such a team possess a basic awareness of some of the fundamental concepts of Human Factors into order to facilitate communication and exchange of information. However, they should also understand what an academically qualified Human Factors specialist is able to offer their particular endeavour.

The many excellent papers presented in this book illustrate the benefits of collaborative multidisciplinary endeavours, and the need to ensure that the application of Human Factors concepts and knowledge are soundly based from a theoretical and scientific point of view. This will ensure that the outcomes of such programs are both scientifically and operationally valid.

Dr Robert Lee
Canberra, Australia

Preface

The Fifth Australian Aviation Psychology Symposium was staged by the Australian Aviation Psychology Association (AAvPA), a not for profit association established with the objective of increasing the contribution of aviation psychology and Human Factors to the safety and efficiency of the aviation industry. The Symposium was held in the aftermath of the 2000 Sydney Olympic games and like the sporting event was judged by those in attendance as the "best ever" Symposium. The Symposium attracted over 300 delegates from the international aviation community with representation ranging from Argentina to the United States.

The success of the Symposium was partly due to the commitment of key local and international experts to return and use "Manly" as a "corroboree" or meeting place in which to exchange, debate and advance contemporary thinking in Human Factors and aviation psychology. Returning industry thought leaders included Professor James Reason (University of Manchester), Professor Bob Helmreich (University of Texas), Captain Dan Maurino (ICAO), Dr Robert Lee (Australian Transport Safety Bureau), Rene Almerberti (IMASAA, France), Bert Ruitenberg (IFATCA), Dr David O'Hare (University of Otago, NZ), Dr Anne Isaac (Eurocontrol), Dr Nick McDonald (Trinity College, Dublin) and Professor Penny Sanderson (Swinburne University). In addition, a number of distinguished individuals accepted our invitation for the first time to be part of the Manly experience, including: Professor Erik Hollnagel (Linkoping University, Sweden), Professor Rhona Flin (Aberdeen University, Scotland), Dr Gary Klein (Klein Associates, USA), Professor Ron Westrum (Eastern Michigan University) and Mick Toller (CASA, Australia).

Since the Symposium, a number of significant international events have had a profound effect on the financial stability of the airline industry. As a result of the terrorist attacks on the World Trade Centre in New York on September 11, 2001, there has been an unprecedented level of security changes on the ground and in the air. Increased in-flight security programs such as sky marshals, locked cockpit doors and combat/self defence training for aircrew has resulted in fundamental changes to established Crew Resource Management practices on board. Recently IATA in their 2001 Safety Report published a word of caution for operators not to let the increased security activities detract from established safety and Human Factors initiatives. The terrorist attacks in New York and more recently Bali has placed enormous financial pressure on airlines as the result of declining international passenger loads. Recently two of the largest air carriers in the United States filed for Chapter 11 bankruptcy and many other carriers are shedding staff to survive.

In Australia, the collapse of Ansett Australia has changed the dynamics of the local airline industry and forced many to look for alternative industry employment. Since the staging of the Fifth AAvPA Symposium and the publication of these selected papers, much has changed but the relevance of the material contained in this book has not.

The chapters contained herein are divided into six parts. Part 1 consists of contemporary system safety and accident prevention applications. In particular, the concept of normal operations safety monitoring is presented as a means by which to expand the envelope of our operational understanding of error management.

Part 2 contrasts Human Factors applications in various high risk industries such as aviation, healthcare and nuclear power. The papers outline different approaches to threat and error management in aviation and medicine, describe a corporate Human Factors program within a major international airline, and discuss the management of human performance in the nuclear power industry.

Part 3 consists of various papers outlining approaches to training, licensing and regulations with a focus on decision skills training.

Part 4 focuses on automation and human computer interaction, dealing specifically with auditory warnings, automation reliance and pilot attitudes towards GPS.

Part 5 deals with the ATS environment, describing innovative Human Factors applications from Europe, Australia and the United States.

Finally Part 6 presents other Human Factors issues including fatigue countermeasures for long haul aircrew and the causal factors behind inadvertent aircraft evacuation slide activation's.

Our sincere gratitude to the academics and practitioners who took time to peer review the papers and ensure a high quality output. Special thanks to our Assistant Editors, Joanne De Landre from CASA and Rebecca Atkins from The University of Newcastle for the many hours spent proof reading and refining the chapters.

Final thanks to John Hindley from Ashgate for his patience and vigilance in waiting for the final product.

Dr Graham Edkins and Dr Peter Pfister
Sydney, Australia

Editorial Team

Joint Editors

Dr H. Peter Pfister
Associate Professor
School of Behavioural Sciences
The University of Newcastle, Australia

Dr Graham Edkins
Manager Human Factors
Qantas Airways Ltd
Sydney, Australia

Assistant Editors

Rebecca Atkins
School of Behavioural Sciences
The University of Newcastle, Australia

Joanne DeLandre
Civic Aviation Safety Authority
Human Factors Team
Canberra, Australia

Review Team

Rebecca Atkins, The University of Newcastle
Christine Boag, Australian Transport Safety Board
Kirstie Carrick, The University of Newcastle
Jo DeLandre, CASA, Australia
Graham Edkins, Qantas Airways Ltd
Gerhart Fahnenbruck, Lufthansa
Mark Fleming, St Mary's University, Halifax, Canada
Adam Fletcher, University of South Australia
Steven Ginpil, CASA, Australia
Steve Hampton, Emery-Riddle Aeronautical University, USA
Andrew Heathcote, The University of Newcastle
Alan Hobbs, NASA Ames Research Center, USA
Erik Hollnagel, Linkpoingen University, Sweden
Tim Horberry, MURAC, University of Melbourne
Graham Hunt, Massey University, New Zealand
Robert Lee, Canberra, Australia
Claire Marrison, AirServices Australia
Andrew Neal, The University of Queensland, Australia
Mike Nendick, CASA, Australia
Dave O'Hare, University of Otago, New Zealand
Peter Pfister, The University of Newcastle
David Post, Lufthansa
Michael Rodgers, CASA, Australia
Peter Simpson, Qantas Airways Ltd
Mike Skinner, DSTO, Australia
Steve Smith, St Mary's University, Halifax, Canada
Allan Spigelman, John Hunter Hospital and The University of Newcastle
Karen Stephan, DSTO, Australia
Ross St George, CAA, New Zealand
Mike Walker, Australian Transport Safety Board
Mark Wiggins, University of Western Sydney, Australia

List of Tables

List of Figures

Selected Contributors' Biographies

John Allin Brown is a Senior Analyst, Operations and Human Factors Air Traffic Management, in the Boeing Company. He joined the RAF in 1963 as trainee radar technician then as a pilot. Flew Victor tanker operationally for 9 years, becoming a refueling instructor. Graduated from Empire Test Pilot School in 1978; spent 7 years at A&AEE, Boscombe Down, flight testing the Hercules, Nimrod, VC-10, Jetstream and Victor. He received Queen's Commendation for Valuable Service in the Air for flight test work in support of the Falklands crisis. Retired from the RAF in 1985 and then emigrated to the USA.

From 1985 to 1988 he worked as a Senior Test Pilot Instructor at the National Test Pilot School in Mojave, California where he developed and taught a crew station evaluation and design course to test pilot and flight test aircrew students. From 1988 to 1996, he worked for Northwest Airlines, first as a flight simulator development and approval test pilot. From 1992 to 1994 he managed the pilots' *crew resource management* training course, developing materials and delivering classroom training. John joined Boeing in 1996 as a Human Factors specialist working on the design of controller-pilot data link communication interfaces in the cockpit. He moved to current position in 1999. John is a SETP Associate Fellow; MRAeS.

Dr. Jan M. Davies is a consultant anesthesiologist in the Department of Anesthesia, Foothills Medical Centre, Calgary Health Region and a Professor of Anesthesia at the University of Calgary.

For the past 20 years she has worked and undertaken research in the area of system safety. She has acted as a consultant for various Canadian provincial medical examiners and coroners, the Canadian Medical Protective Association, plaintiff's lawyers, a major Canadian airline and an aviation safety organisation. She also served as an expert in quality assurance, human error and Human Factors to the Pediatric Cardiac Surgery Inquiry in Winnipeg, Manitoba (www.pediatriccardiacinquest.mb.ca). She is the author of a book about anesthesia written for the public and more than 120 articles and chapters, and has given more than 200 presentations.

She received a BSc (Honours) in Zoology from the University of British Columbia, an MSc in Experimental Surgery from the University of Alberta, and an MD from the University of Calgary. Her specialty training in anesthesia (FRCPC) was undertaken at Dalhousie University.

Dr. Graham Edkins has over 15 years experience in the application of system safety and Human Factors principles within the transport industry. Currently he is the Director Public Transport Safety for the Department of Infrastructure, responsible for the safety accreditation and regulation of rail, tram and bus services within the State of Victoria. Previously, he has held a number of senior management positions within Qantas Airways, including responsibility for the strategic development of a corporate wide Human Factors program within Qantas and its subsidiary companies. Prior to Qantas, he was a Senior Air Safety Investigator with the former Bureau of Air Safety Investigation (now ATSB), and involved in investigating general aviation, helicopter and passenger aircraft accidents in both Australia and overseas. Additionally, he has worked in the railway industry as a System Safety Consultant and Accident Investigator. He was awarded a Masters Degree in Organisational Psychology and also has a PhD degree in Organisational Psychology that was applied within the aviation domain, and involved the development and evaluation of an airline safety management program called INDICATE, recommended by the International Air Transport Association (IATA) as a world standard aviation safety program. He is the current President of the Australian Aviation Psychology Association, past Vice-Chairman and Australian Representative of the IATA Human Factors Working Group, founding member of the Aviation Medicine and Human Factors Centre; member of the Australian Psychological Society, and previous industry advisor to CASA on the Human Factors Advisory Group (HFAG) and Regulatory Reform Program (RRP).

Professor Rhona Flin (BSc, PhD Psychology) is Professor of Applied Psychology in the Department of Psychology at the University of Aberdeen. She is a Chartered Psychologist, a Fellow of the British Psychological Society and a Fellow of the Royal Society of Edinburgh.

For the last 14 years she has directed a team of psychologists working with high reliability industries (especially energy sector) on research and consultancy projects concerned with the management of safety and emergency response. The group has recently been working on projects relating to aviation safety (funded by EC and CAA), leadership and safety in offshore management (funded by HSE and the oil industry), health management and safety in the offshore oil industry (HSE), team skills and emergency management (funded by the nuclear industry), anaesthetists' and surgeons' non-clinical skills (funded by NHSES/ RCS, Edinburgh). Her recent books include *Incident Command: Tales from the Hot Seat* (ed. & Arbuthnot, Ashgate, 2002); *Sitting in the Hot Seat: Leaders and Teams for Critical Incident Management* (Wiley, 1996), *Managing the Offshore Installation Workforce* (ed. & Slaven, PennWell Books, 1996) and *Decision Making under Stress* (ed. & Salas, Strub & Martin, Ashgate, 1997). Details of projects and publications can be found on: www.psyc.abdn.ac.uk/serv02.htm

Dr. Ross St.George is a Safety Adviser with the Civil Aviation Authority of New Zealand. The position involves extensive safety promotion in the field, the development and delivery of safety programs to the aviation industry and Human Factors research input into a wide range of safety and operational issues.

Prior to taking up this with position with CAANZ in 1996 Ross held academic positions in both the Departments of Education and Psychology at Massey University (1974-1996). Specialist teaching and research areas included educational and psychological measurement, and Human Factors . At Massey Ross developed programs aviation Human Factors teaching and research. Aviation focussed publications have been in the areas of flight training, instruction and GPS technology.

Ross is a Registered Psychologist in New Zealand, holds Flight Crew Licenses for New Zealand and the United States of America, and trained in air accident investigation at the Cranfield College of Aeronautics. He is an aircraft owner and an active pilot.

Dr. Robert L. Helmreich is professor of psychology at The University of Texas at Austin. He received BS, MS, and Ph.D. degrees from Yale University. He is director of University of Texas Human Factors Research Project which investigates human performance and threat and error management in aviation and medicine.

Helmreich received the Flight Safety Foundation Distinguished Service Award in 1994. He is Chair of the Foundation's Icarus Committee. He received Laurels from *Aviation Week and Space Technology* in 1994 and 2002 for his research into Human Factors in aviation. He is a fellow of the Royal Aeronautical Society, the American Psychological Association, and the American Psychological Society. He research has been supported by the National Science Foundation, NASA, the Agency for Healthcare Research and Quality, and the Daimler-Benz Foundation. He has more than two hundred publications and is author, with Ashleigh Merritt, of *Culture at Work in Aviation and Medicine: National, Organizational, and Professional Influences.*

Dr. Erik Hollnagel is Full Professor of Human-Machine Interaction at Linköping University (Sweden) and an internationally recognised expert in the fields of system safety, cognitive systems engineering, cognitive ergonomics and intelligent human-machine systems. He has worked at universities and in various industries since 1971, including nuclear power, aerospace, software engineering, and vehicles. He has published widely including eight books and is, together with Pietro C. Cacciabue, Editor-in-Chief of the *International Journal of Cognition, Technology & Work.*

Dr. Graham J. F. Hunt is the founding professor and Head of the School of Aviation, at Massey University, New Zealand. He joined the Royal New Zealand Air Force in 1968 as an aviation psychologist and held the first three-service appointment for the development of "systems approaches" to curriculum development and performance assessment. He also became significantly involved in flight crew selection and retention studies research.

In 1972 he was awarded a New Zealand Defence Studies Doctoral Scholarship. This was only the second time since 1945 that such an award has been offered. He studied at the Wright-Patterson Air Force base (Human Factors laboratory), and the University of Pittsburgh's Learning Research and Development Centre.

After returning to New Zealand and working for a brief time at the New Zealand Council for Educational Research, Graham joined Massey University in 1977. His growing research interest in Human Factors and professional competency development led to his direction of a number of national and international programs including the New Zealand Civil Aviation *Authority's Human Resource Development in Aviation (HURDA)* study. Professor Hunt is a Fellow of the Royal Aeronautical Society and was recently awarded a Bronze Medal by the Royal Society of New Zealand for services to aviation. He is currently an advisor to the International Civil Aviation Organisation's (ICAO) Flight Crew Licensing and Training Program in Montreal where Annex 1 and Annex 6 of the Chicago Convention are being reviewed as a means for further improving ab initio air transport pilot training.

Dr. Kurt M. Joseph is a Human Factors Engineer at SBC Laboratories, Inc. in Austin, Texas, USA. Prior to joining SBC Labs, he worked for four and a half years as an Engineering Research Psychologist in the Human Factors Laboratory at the Federal Aviation Administration Civil Aerospace Medical Institute in Oklahoma City, Oklahoma, USA. While there, Kurt completed Human Factors research in support of FAA certification and standards programs for GPS receivers, and enhanced cockpit surveillance displays (e.g., Cockpit Display of Traffic Information) that rely on ADS-B technology. Kurt received his Ph.D. in applied experimental psychology, with an emphasis in Human Factors, from Kansas State University in 2000.

Dr. Shayne Loft is a postdoctoral research fellow at the Key Centre for Human Factors and Applied Cognitive Psychology, at the University of Queensland. Shayne loft is an early career researcher and his PhD involved an empirical examination of the psychological processes underlying conflict detection in simulated air traffic control environments. Shayne Loft led the development of the air traffic control task (ATC-lab) that simulates the dynamic and practical aspects of air traffic control. ATC-lab is currently being used for Human Factors research by senior academics at the University of Queensland and defence scientists in Canada. The findings from the PhD program made a significant contribution to our theoretical understanding of human performance in complex task environments. The resulting papers are currently in press or are under review in international journals. Shayne Loft regularly consults to Air Services Australia on air traffic control related Human Factors issues, such as the incidence and management of human error in the acceptance and transmission of Search and Rescue Times (SARTIMES). Shayne Loft is currently working with Air Services Australia to assess the way in which air traffic controllers evaluate longitudinal separation between aircraft and resolve conflicts in non-radar environments.

Oliver Lods is an Adjunct Lecturer in Aviation Human Factors at the University of Western Sydney. He has a Bachelor of Aviation Studies and has been a recipient of the Air Transport Services Prize for Academic Excellence. He has published course material for tertiary education and assisted in research projects for the Federal Aviation Administration and the National Aeronautics and Space

Administration. Oliver is a member of the Royal Aeronautical Society and a private pilot.

Captain Ian Lucas (FRAeS) Captain Ian Lucas' career in aviation has spanned 35 years, during which he has held a diverse range of roles within Qantas Airways Ltd.

Born in Queensland in 1948, Captain Lucas graduated from Qantas No 3 Cadet Course in July 1968 before being drafted for National Service in the Australian Army. Following military discharge in 1970, he resumed his duties with Qantas and qualified as a Second Officer on the B707. Throughout the next 20 years he flew B747 and B767 aircraft, earning his Command on the B767 in 1987 and taking on several training supervisory roles before being appointed Manager Training B767.

Throughout the 1990s he moved through senior management in such roles as Fleet Operations Director and Deputy Chief Pilot, General Manager B744 Operations & Deputy Chief Pilot, and spent three years in Melbourne as General Manager B737 Operations & Deputy Chief Pilot.

In November 1999 he was appointed Group General Manager Flight Operations & Chief Pilot.

He has held this role for the past four years, in perhaps the most challenging and turbulent times the international aviation industry has seen.

Captain Daniel E. Maurino is the Co-ordinator of the Flight Safety and Human Factors Program with ICAO. A retired airline captain, Dan is a member of the International Society of Air Safety Investigators (ISASI) and the Human Factors and Ergonomics Society (HFES). He represents ICAO at the IATA Human Factors Working Group and the Safety Advisory Committee. He is a member of the Flight Safety Foundation's ICARUS Committee, and a Fellow of the Royal Aeronautical Society.

Captain Maurino is associate editor of the *International Journal of Aviation Psychology* (IJAP), and a member of the editorial board of the *Human Factors and Aerospace Safety Journal*. He is co-author of *Beyond Aviation Human Factors*. His awards include the First Flight Safety Foundation/Airbus Industry Award for Achievement in Human Factors and Aviation Safety (1999), the Flight Safety Foundation/Aviation Week & Space Technology Distinguished Service Award (2000), the Royal Aeronautical Society Roger Green Medal (2001) and the Captain A G Vette Flight Safety Research Trust Award (2002).

Dr. Ashleigh Merritt now teaches at the University of Texas, where she focuses on courses on cross-cultural differences and organisational culture. In addition, she consults world-wide in the field of organisational safety, particularly in the fields of aviation and power generation. She has been extensively involved in the development of the Crew Resource Management concepts, including the current concepts of Line Operational Safety Audits (LOSAs). Prior to her teaching appointment, she was a consultant with John Wreathall & Co., and previously, Dedale, S.A. She co-authored "Culture at Work in Aviation and Medicine"

(Ashgate Press) with Prof. Robert Helmreich. She obtained her Ph.D. in psychology from the University of Texas, and her B.A. in Organisational Psychology from the University of Queensland.

Michael D. Nendick has been a Human Factors specialist with the Human Factors and System Safety section, Aviation Safety Promotion, at the Civil Aviation Safety Authority (CASA) Australia, since February 2001. His role includes providing specialist advice, developing standards for Human Factors and Safety Management Systems, auditing aviation safety systems, and running Human Factors training programs. Prior to this he was the undergraduate Aviation Program Director at the University of Newcastle, where he lectured for six years in Human Factors and advanced navigation systems, including Global Navigation Satellite Systems (GNSS) and GPS. His 25 years in aviation includes experience as a Royal New Zealand Air Force transport navigator on Andovers and C130s, air traffic controller, Human Factors consultant, academic, researcher, and private pilot. Mike has a MSc(Hons) Psychology from Massey, and a Diploma of Management (NZIM). He is currently enrolled as a Ph.D. candidate at the University of Newcastle, Australia, researching the development and validation of Human Factors competency-based training and assessment methodologies in aviation.

Dr. David O'Hare graduated with a first-class honours degree in psychology from the University of Exeter (England) in 1974 and a PhD in 1978. He obtained a lectureship in psychology at the University of Lancaster (England) before moving to New Zealand in 1982. He is currently associate professor of psychology at the University of Otago. He is co-author (with Stan Roscoe) of *Flightdeck Performance: The Human Factor* (Iowa State University Press) and has recently edited *Human Performance in General Aviation* (Ashgate). He is currently the principal investigator for two major grant-funded research projects on aeronautical decision making (US Federal Aviation Administration) and case-based learning for aviation safety (NASA). He is Associate Editor of *The International Journal of Aviation Psychology* and a consulting editor for *Aviation, Space and Environmental Medicine*.

Dr. Peter Pfister is Associate Professor in the School of Behavioural Sciences at The University of Newcastle and Head of the Discipline of Aviation. He is a member of ARC Key Centre for Human Factors and Applied Cognitive Psychology. He is also a Registered Psychologist in the State of New South Wales, and is a Fellow of the Australian Psychological Society. He is primarily a researcher in Human Factors and Human Factors related areas and is the principle PhD supervisor in this area at The University of Newcastle. He and his students work on projects in the Transport Industry, Mining, and Manufacturing Industry as well as the Health Profession having recently commenced a multi country study into Human Factors in Surgery.

Dr. Stephen C. Provost has been an academic in the School of Psychology, Southern Cross University at Coffs Harbour in NSW, Australia since 2001. His previous appointment was at The University of Newcastle in the School of Behavioural Sciences, where he learned everything he knows about aviation Human Factors. His main interest in psychology is in learning, from basic theories to applications, and he is a closet behaviorist. Other interests include the relationship between cortical activity, as measured by EEG and ERP changes, and learning, emotion, and psychopharmacology.

Dr. Penelope M. Sanderson is Professor of Cognitive Engineering and Human Factors in the ARC Key Centre for Human Factors at The University of Queensland, with appointments in the Schools of Psychology and of ITEE. She received a BA (Hons I) from University of WA and an MA and PhD from University of Toronto and then spent 11 years on the faculty of University of Illinois at Urbana-Champaign in the Engineering Psychology Joint Program. In 1997 she returned to Australia to take up an appointment to the first university chair in Australia named for Human-Computer Interaction. She was also the inaugural Director of the Swinburne Computer-Human Interaction Laboratory. Since her move to The University of Queensland in 2001 she has established the UQ Usability Laboratory, where research, consulting and teaching activities are combined. She also leads the Cognitive Engineering Research Group (CERG). She is an Associate Editor of International Journal of Human Computer Studies and is on the Editorial Board of Cognition, Technology, and Work. She retains an adjunct appointment at University of Illinois at Urbana-Champaign. Sanderson's research interests include visual and auditory display design to support effective human work in complex socio-technical systems.

Susannah J. Tiller was an Honours student at The University of Newcastle (Australia) when she presented at the Australian Aviation Psychology Symposium. Her Honours thesis, from which her chapter is derived, won the University's W. H. Ward prize for best honours-level thesis in applied cognitive psychology. She is currently undertaking PhD studies at the ARC Key Centre for Human Factors and Applied Cognitive Psychology at The University of Queensland.

Ron Westrum is Professor of Sociology and Interdisciplinary Technology at Eastern Michigan University, in Ypsilanti, Michigan. His areas of expertise are organisational dynamics associated with safety and technical innovation. He is the author of many papers and three books, the most recent being Sidewinder: Creative Missile Design at China Lake, published by the Naval Institute Press. Dr. Westrum is a frequent speaker on organisational safety culture at national and international meetings in the aviation, medicine, and nuclear power industries. He is the originator of the distinction between pathological, bureaucratic, and generative organisations, and is writing a book about the origins and dynamics of these three environments.

Dr. Mark Wiggins is a Senior Lecturer in Psychology and Head of the Human Factors and Performance Research Group within the MARCS Auditory Research Centre at the University of Western Sydney. He has a PhD from the University of

Otago, is a Registered Psychologist in New South Wales, and is a member of the Human Factors and Ergonomics Society and the Australian Psychological Society. Mark is a member of the Human Factors Advisory Group for the Civil Aviation Safety Authority and is a qualified pilot. He has been a consultant researcher for the Federal Aviation Administration, the National Aeronautics and Space Administration, the Australian Transportation Safety Bureau, Airservices Australia, and the State Rail Authority of New South Wales. He is the author of a number of refereed journal articles, book chapters, and books including *Aviation Social Science: Research Methods in Practice* published by Ashgate.

John Wreathall is a systems engineer, who specialises in analysing and developing solutions for management and organisational issues in industries such as nuclear power, rail transportation, healthcare, and aviation/space. He became interested in the role of human and organisational systems in managing safety in the 1970s as part of the British nuclear navy and commercial nuclear power programs. Mr. Wreathall moved to the USA in 1981, to develop new methods of human performance analysis in the nuclear power industry. He is now the CEO of a small consulting company, providing support in the development of new tools and methods to support the management of large industrial organisations world-wide. Mr. Wreathall obtained his B.Sc. (Eng) in nuclear engineering and his M.Sc. in systems engineering from the University of London.

Chapter 1

Heroic Compensations: The Benign Face of the Human Factor

James Reason
Department of Psychology, University of Manchester, UK

Introduction

The human factor gets a bad press. There are two approaches to studying human performance in high-technology hazardous systems: one involves the 'fly-on-the-wall' observation of normal activities; the other is triggered by the occurrence of an adverse event. An 'event' is something untoward that disrupts the flow of normal or intended activities and which may, and often does, have harmful consequences. In Human Factors research, at least, there can be little doubt that the dominant tradition is the event-dependent one. Such analyses focus upon the errors and violations that either constitute or contribute to an event. The worse the event, the more intensive the investigation of the preceding decisions and actions. As a result, we have learned a good deal about the varieties of unsafe acts and, to a lesser degree, we know something of the circumstances that can provoke and shape them (Hollnagel, 1993; Reason, 1990).

Unfortunately, this has established a very biased view of the human factor as something that is causally implicated in the large majority of bad events (Hollnagel, 1993). To compound the problem further, stating that people make errors is probably one of the least interesting observations about the human condition on a par with declaring that we breathe oxygen and will some day die. Such information is undoubtedly important, but hardly newsworthy. Nonetheless, errors are sufficiently uppermost in the minds of the managers of hazardous technologies that they often regard the main goal of safety management as the elimination of human fallibility rather than the avoidance of its damaging consequences (Amalberti & Wioland, 1997).

So, if human fallibility is a mere truism, what is there that is really interesting about human performance? The answer, I believe, lies on the reverse side of the coin. As operators of complex systems, people have an unmatched capacity to adapt and adjust to the surprises thrown up by a dynamic and uncertain world. This includes the often remarkable ability to compensate for their own errors. Making errors is a fact of life, but recovering from them particularly when these recoveries involve heroic improvisations is quite another matter. The story of the 'Gimli glider' will serve to illustrate this point.

The Gimli Glider

On July 23, 1983, a Boeing 767 aircraft en route to Edmonton from Ottawa ran out of fuel over Red Lake, Ontario, about halfway to its destination. The reasons for this were a combination of inoperative fuel gauges, fuel loading errors and mistaken assumptions on the part of the flight crew. These errors and system failures were dealt with at length in the 104-page report of the Board of Inquiry (Lockwood, 1985). Only three paragraphs were devoted to the most extraordinary feature of this event: the forced landing at Gimli, a disused military airstrip, from which all 61 passengers and eight flight crew walked away unharmed and the aircraft was fit for service after relatively minor repairs.

When the second engine stopped, the aircraft was at 35,000 feet and 65 miles from Winnipeg. All the electronic gauges in the cockpit had ceased to function, leaving only stand-by instruments operative. The First Officer, an ex-military pilot, recalled that he had flown training aircraft in and out of Gimli, some 45 miles away. When it became evident that they would not make it to Winnipeg, the Captain, in consultation with Air Traffic Control, redirected the aircraft to Gimli, now 12 miles away on the shores of Lake Winnipeg. The report continues as follows:

> Fortunately for all concerned, one of Captain Pearson's skills is gliding. He proved his skill as a glider pilot by using gliding techniques to fly the large aircraft to a safe landing. Without power, the aircraft had no flaps or slats to control the rate and speed of descent. There was only one chance of landing. By the time the aircraft reached the beginning of the runway, it had to be flying low enough and slowly enough to land within the length of the 7,200 foot runway. As they approached Gimli, Captain Pearson and First Officer Quintal discussed the possibility of executing a side-slip to lose height and speed close to the beginning of the runway. This the Captain did on the final approach and touched down within 800 feet of the threshold. (p. 29)

The last laconic sentence is a masterpiece of understatement. It is unlikely that either Boeing or Captain Pearson's employers had ever imagined the side-slip manoeuvre being applied to a wide-bodied jet airliner. As it turned out, however, it was almost certainly the only way that the aircraft could have made a safe landing under those circumstances. This was heroic improvisation at its most inspired.

Theoretical Framework

The literature provides relatively little in the way of theoretical guidance when it comes to understanding and facilitating these remarkable adaptations. Most safety-related studies have focussed upon identifying those factors that create moments of vulnerability rather than elucidating the nature of resilience. The two notable exceptions are, firstly, the observation-based analyses of high reliability organisations, or HROs (Weick, Sutcliffe, & Obstfeld, 1999); and, secondly, the more person-oriented work on mental readiness in the achievement of sporting and surgical excellence (Orlick, 1990). Although these two research areas derive from different disciplinary backgrounds, they reveal similar processes operating at both the organisational and individual levels.

There would appear to be at least two vital components underpinning both high reliability organisations and individual excellence: a mindset that expects unpleasant surprises and the flexibility to deploy different modes of adaptation in different circumstances. In short, there is a mental element and an action element. Of these, the former is at least as important as the latter. Effective contingency planning at both the organisational and the personal levels depends on the ability to anticipate a wide variety of crises. Both components are resource-limited. Any person or organisation can only foresee and prepare for a finite number of possible circumstances and crisis scenarios. Crises consume available coping resources very rapidly. Only those people or organisations that have invested a considerable amount of preparatory effort in the pre-crisis period will be able to deploy compensatory responses in a sufficiently timely and appropriate manner so as to maintain the necessary resilience.

These three concepts cognitive readiness, pre-prepared responses, and the restricted nature of coping resources proved to be extremely valuable in interpreting the data obtained from the surgical study described below.

The Surgical Compensation Study

Background

Over the past few years, we have been investigating the compensations carried out by UK paediatric cardio-thoracic surgical teams during the course of the neonatal arterial switch procedure (Carthey, de Leval, Reason, Leggatt, & Wright, 2000). We were fortunate in having a skilled Human Factors observer present at 165 of these procedures. In other words, it was an event-independent study. Twenty-one consultant surgeons in 16 institutions throughout the country performed these neonatal switch operations.

The Arterial Switch Operation

The arterial switch operation (ASO) involves correcting cardiovascular congenital defects in very young babies by transposing the great arteries the pulmonary artery and the aorta so as to permit the full circulation of oxygenated blood. Without such an intervention, the child would die. The children upon whom the ASO is performed are born with the great vessels of the heart connected to the wrong ventricles: the aorta is connected to the right ventricle and the pulmonary artery to the left ventricle. The operation may last for 5-6 hours and is highly demanding both technically and in human terms.

The most challenging part of the procedure involves relocating the coronary arteries, each comprising very thin friable tissue. The arterial switch procedure takes the surgical team and particularly the consultant surgeon close to the edges of the human performance envelope on a variety of parameters: psychomotor skills, naturalistic decision making, and in its claims upon knowledge, experience, leadership, management, and communication skills. Errors of one kind or another are almost inevitable under such conditions. What matters are not the errors *per se* but whether or not they are detected and recovered. In the surgical context, as we shall see, bad outcomes happen when major adverse events, usually the result of errors, go uncompensated; happy outcomes by far the majority are due in large part to effective compensation by the surgical team.

Events and their Compensation

On average, there were seven adverse events per procedure. One of these was life-threatening (a major event); the remaining six were relatively minor events that disrupted the surgical flow but did not immediately jeopardise the safety of the patient. Nearly all of these events arose as the result of errors on the part of the surgical team.

Over half the major events were successfully compensated. When this happened, there was no increase in the risk of death in that particular procedure. However, only 20 percent of the minor events were compensated. Surgical teams varied in their compensatory success. Good compensators had good outcomes. Compensation for minor events was far less important than their total number within a given operation. The larger the number of minor events, the less likely were the team to cope effectively with a major event. Minor events appeared to exert an additive effect by cumulatively eroding the limited compensatory resources of the surgical team.

The message from this study was clear. All surgeons make errors, but the best of them have the ability to compensate for any adverse effects. This ability depends on the skill and experience of the surgeon, as well as the extent to which they have mentally rehearsed the detection and recovery of their errors.

The Variability Paradox

The reduction or even elimination of human error has now become one of the primary objectives of system managers. Errors and violations are viewed, reasonably enough, as deviations from some desired or appropriate behaviour. Having mainly an engineering background, such managers attribute human unreliability to unwanted variability. And, as with technical unreliability, they see the solution as one of ensuring greater consistency of human action. They do this through procedures and by buying more automation. What they often fail to appreciate, however, is that human variability in the form of moment-to-moment adaptations and adjustments to changing events is also what preserves system safety in an uncertain and dynamic world. And therein lies the paradox. By striving to constrain human variability, they are also undermining one the system's most important safeguards.

The problem has been encapsulated by Weick's (1987) insightful observation that 'reliability is a dynamic non-event'. It is dynamic because processes remain under control due to compensations by human components. It is a non-event because safe outcomes claim little or no attention. The paradox is rooted in the fact that accidents are salient, while non-events, by definition, are not.

Recently, Weick et al. (1999) have challenged the received wisdom that an organisation's reliability depends upon the consistency, repeatability, and invariance of its routines and activities. Unvarying performance, they argue, cannot cope with the unexpected. To account for the success of high reliability organisations (HROs) in dealing with unanticipated events, they distinguish two aspects of organisational functioning: cognition and activity. The cognitive element relates to being alert to the possibility of unpleasant surprises and having the collective mindset necessary to detect, understand, and recover them before they bring about bad consequences. Traditional 'efficient' organisations strive for stable activity patterns yet possess variable cognitions; these differing cognitions are most obvious before and after a bad event. In HROs, on the other hand, flexibility is encouraged in their activity, but there is consistency in the organisational mindset relating to the operational hazards. This cognitive stability depends critically upon an informed culture or what Weick et al. (1999) have called 'collective mindfulness'.

Collective mindfulness allows an organisation to cope with the unanticipated in an optimal manner. 'Optimal' does not necessarily mean 'on every occasion', but the evidence suggests that the presence of such enduring cognitive processes is a critical component of organisational resilience. Since catastrophic failures are rare events, collectively mindful organisations work hard to extract the most value from what little data they have. They actively set out to create a reporting culture by commending, even rewarding, people for reporting their errors and near misses. They work on the assumption that what might seem to be an isolated failure is likely to come from the confluence of many 'upstream' causal chains. Instead of localising failures, they generalise them. Instead of applying local repairs, they strive for system reforms. They do not take the past as a guide to the future. Aware that system failures can take a wide variety of yet-to-be-encountered forms, they are continually on the lookout for 'sneak paths' or novel ways in which active failures and latent conditions can combine to defeat or by-pass the system defences.

Conclusions

1. Human variability is both a source of error and a vital system defence. How can we limit one while still promoting the other?

2. The key to resilience at both the individual and organisational levels lies in being mentally prepared for nasty surprises, and having the counter-measures in place to deal with them.
3. The ability to make effective compensations appears to be resource-limited, and is liable to be eroded by the cumulative effects of minor stressors.
4. Mental preparedness and flexibility of response help to limit the stress-related attrition of these crucial coping abilities.

References

Amalberti, R., & Wioland, L. (1997). Human error in aviation. In H. Soekkha (Ed.), *Aviation Safety* (pp. 91-108). Utrecht: VSP.

Carthey, J., de Leval, M. R., Reason, J., Leggatt, A. R., & Wright D. J. (2000). Adverse events in cardiac surgery: The role played by human and organisational factors. In C. Vincent & B. de Mol (Eds.), *Safety in Medicine* (pp. 117-138). Amsterdam: Pergamon Press.

Hollnagel, E. (1993). *Human Reliability Analysis: Context and Control.* London: Academic Press.

Lockwood, The Hon. G. H. (1985). *Final Report of the Board of Inquiry into Air Canada Boeing 767 C-GAUN Accident Gimli Manitoba July 23, 1983.* Ottawa: Government of Canada.

Orlick, T. D. (1990). *In Pursuit of Excellence* (2nd ed.). Ottawa: Human Kinetics.

Reason, J. (1990). *Human Error.* New York: Cambridge University Press.

Weick, K. E. (1987). Organizational culture as a source of high reliability. *California Management Review, 29*, 112-127.

Weick, K. E., Sutcliffe, K. M., & Obstfeld, D. (1999). Organizing for high reliability: Processes of collective mindfulness. *Research in Organizational Behavior, 21,* 23-81.

Chapter 2

Aviation Safety and Human Factors: The Years to Come

Daniel E. Maurino
ICAO, Montreal, Canada

Introduction

The retrospective analysis of actions and inactions by operational personnel involved in accidents and incidents has been the traditional method utilised by aviation to assess the impact of human performance in regard to safety. The established safety paradigm, and prevailing beliefs about what constitutes safe and unsafe acts, guide this analysis in such a way that it traces back an event under consideration until a point in which investigators find a behaviour that did not produce the results intended. At such point, human error is concluded. This conclusion is generally arrived at with limited consideration of the processes that could have led to the 'bad' outcome. Furthermore, when reviewing events, investigators know that the behaviours displayed by operational personnel were 'bad' or 'inappropriate', because the negative outcomes are a matter of record. This is, however, a benefit the operational personnel involved did not have when they selected what they thought at the time were good or appropriate behaviours, and which would lead to a good outcome. In this sense, it is suggested that investigators examining human performance in safety occurrences enjoy the benefit of hindsight. Furthermore, conventional safety wisdom holds that, in aviation, safety is first. Consequently, human behaviours and decision-making in aviation operations are considered to be one hundred percent safety oriented. This is not true, and a more realistic approach is to consider human behaviours and decision-making in operational contexts as a compromise between production-oriented behaviours and decisions, and safety-oriented behaviours and decisions. The optimum behaviours to achieve the actual production demands of the operational task at hand may not always be fully compatible with the optimum behaviours to achieve the theoretical safety demands. All production systems (and aviation is no exception) generate a migration of behaviours: under the imperative of economics and efficiency, people are forced to operate at the edges of the system's safety space. Consequently, human decision-making in operational contexts lies at the intersection of production and safety, and is therefore a compromise. In fact, it might be argued that the trademark of experts is not years of experience and exposure to aviation operations, but rather how effectively they manage the compromise between production and safety. Operational errors do not reside in the person, as conventional safety knowledge would have the aviation industry believe. Operational errors primarily reside in latency within task and situational factors in the context, and emerge as consequences of mis-managing compromises between safety and production goals, largely influenced by the shared attitudes across individuals (i.e., culture). This compromise between production and safety is a complex and delicate balance and humans are generally very effective in applying the right mechanisms to successfully achieve it, hence the extraordinary safety record of aviation. Humans do occasionally mis-manage tasks and/or situational factors and fail in balancing the compromise, thus contributing to safety breakdowns. However, since successful compromises far outnumber failures, in order to understand human performance in context the industry needs to capture, through systematic analyses, the mechanisms underlying successful compromises when operating at the edges of the system, rather than those that failed. It is suggested that understanding the human contribution to successes and failures in aviation can be better achieved by monitoring normal operations, rather than accidents and incidents. The Line Operational Safety Audit (LOSA), discussed in detail by Helmreich in his chapter, is the vehicle endorsed by the International Civil Aviation Organisation (ICAO) for this purpose.

Strategies to Understand Operational Human Performance

Accident Investigation

The most widely used tool to document and attempt to understand operational human performance in aviation, and define remedial strategies, is the investigation of accidents. However, in terms of human performance, accidents yield data mostly about behaviours that failed to achieve the balance compromise between production and safety discussed in the previous section. It is suggested that positive outcomes provide a more sensible, supplemental foundation upon which to define remedial strategies and subsequently reshape them as necessary, than do negative ones. From the perspective of organisational interventions, there are limits to the lessons that may be extracted from accidents that might be applied to reshape remedial strategies. It might for example be possible to identify the type and frequency of external manifestations of errors in each of these generic accident scenarios, or discover specific training deficiencies that are particularly conspicuous in relation to identified errors. This, however, provides only a tip-of-the-iceberg perspective. Accident investigation, by definition, concentrates on failures, and in following the rationale advocated by LOSA, it is necessary to better understand the success stories to see if their mechanisms can somehow be bottled and exported. This can be better achieved through the monitoring of normal line operations and associated successful human performance. Nevertheless, there remains a clear role for accident investigation within the safety process. Accident investigation remains the appropriate tool to uncover unanticipated failures in technology or bizarre events, rare as they may be. More important, and in extreme terms, if only normal, daily operations were monitored, defining assumptions about safe/unsafe behaviours would prove to be a task without a frame of reference. Therefore, properly focussed accident investigation can reveal how specific behaviours, including errors and error management, can resonate with specific circumstances to generate an unstable and most likely catastrophic state of affairs. This requires a focussed and contemporary approach to the investigation. Should accident investigation restrict itself to retroactive analyses as discussed above, its only contribution in terms of human error would be to increase industry databases, the usefulness of which in regard to contemporary safety remains dubious. Even worse, it could provide the foundations for legal action, the allocation of blame and punishment.

Incident Investigation

Incidents are more telling markers than accidents, if not of operational human performance, at least on system safety, because they signal weaknesses within the overall system before the system breaks down. There are, nevertheless, limitations on the value of the information on operational human performance obtained from incident reporting systems.

Firstly, incidents are reported in the language of aviation and therefore capture only the external manifestations of errors. Furthermore, incidents are self-reported, and because of reporting biases, the processes and mechanisms underlying error as reported may or may not reflect reality. Secondly, and most important, incident reporting systems are vulnerable to what has been described as normalisation of deviance. Over time, operational personnel develop informal and spontaneous group practices and shortcuts to circumvent deficiencies in equipment design, clumsy procedures, or policies incompatible with operational realities, all of which complicate operational tasks. In most cases normalised deviance is effective, at least temporarily. However, since they are normal, it stands to reason that neither these practices nor their downsides will be reported to, nor captured by, incident reporting systems. Normalised deviance is further compounded by the fact that the most willing reporters may not be able to fully appreciate what are indeed reportable events. If operational personnel are continuously exposed to substandard managerial practices, poor working conditions, or flawed equipment, how could they recognise such factors as reportable problems? While these factors would arguably be reported if they generate incidents, there remains the difficult task of evaluating how they can create less than safe situations, and thus overcome the temptation to postulate that deviations explain incidents simply because they are deviations. Incident reporting systems are certainly better than accident investigations to begin understanding system and operational human performance, but the real challenge lies in taking the next step – understanding the processes underlying human error rather than taking errors at face value. It is essential to move beyond the visible manifestations of error when designing remedial strategies. If such interventions are to be successful in modifying system and individual behaviours, errors must be considered as symptoms that suggest where to look further. In order to understand the mechanisms underlying errors in operational environments, flaws in system and human performance captured by incident reporting systems should be considered as symptoms of mismatches at deeper layers of the system. The value of the information generated by incident reporting systems lies in the early warning about areas of concern, but it is suggested that such information does not capture the concerns themselves.

Training

The observation of training behaviours, such as for example during flight crew simulator training, is another tool to which aviation has ascribed inordinate value in helping to understand operational human performance. However, the production component of operational decision-making does not exist under training conditions. While operational behaviours during line operations are a compromise balance between safety and production objectives, training behaviours are heavily biased towards safety. In simpler terms, the compromise between production and safety is not a factor in decision-making, and operational behaviours exhibited are 'by the book'. Therefore, behaviours under monitored conditions can provide an approximation to the way operational personnel may behave during line operations, and such observation may contribute to flesh out major operational questions, such as for example significant procedural problems. However, it would not be correct – and it might lead an organisation into a risky path – to assume that observing personnel under training provides the key to understand human decision-making and error in unmonitored operational contexts.

Fight Data Recorder Information

Digital Flight Data Recorder (DFDR) and Quick Access Recorder (QAR) information from normal flights can also be a valuable diagnostic tool (although the expense may prohibit its use in many airlines). There are, however, considerations about the data acquired through these tools. DFDR/QAR readout does provide information on the frequency of exceedences and the locations where they occur, but these data cannot yield information on the human behaviours that were precursors of the event. While DFDR/QAR data tracks potential systemic problems, pilot reports are still necessary to provide the context within which to fully diagnose the problems.

Nevertheless, DFDR/QAR data hold high cost/efficiency ratio potential. Although probably under utilised because of both cultural and legal reasons, DFDR/QAR data can assist in filtering operational contexts within which migration of behaviours towards the edge of the system takes place.

Normal Operations Monitoring

The supplemental approach proposed in this chapter to uncover the mechanisms underlying the human contribution to failures and successes in aviation safety, and therefore to the design of countermeasures against human error and safety breakdowns, focuses on the monitoring of normal line operations. Any typical line flight – a normal process – involves inevitable, yet mostly inconsequential errors (selecting wrong frequencies, dialling wrong altitudes, acknowledging incorrect read-backs, mishandling switches and levers, and so forth). Some errors are due to flaws in human performance, others are fostered by systemic shortcomings; most are a concatenation of both. The majority of these errors have no damaging consequences because (a) operational personnel employ successful coping strategies, and (b) system defences act as containment net. It is about these successful strategies and defences that aviation must learn to shape remedial strategies, rather than continuing to focus on failures as the industry has historically done. Monitoring normal line flights, utilising a validated observation tool, allows the capture of these successful coping strategies. There is emerging consensus within aviation that the time has come to adopt a positive stance and anticipate the damaging consequences of human error in system safety, rather than regretting its consequences. This is a sensible objective and a cost-effective way (in terms of dollars and human life) to achieve it, is by pursuing a contemporary approach rather than updating or over-optimising methods of the past. After 50 years of investigating failures and monitoring accident statistics, the relentless prevalence of human error would seem to indicate – unless it is believed that the human condition is beyond hope – a somewhat misplaced safety emphasis in regard to operational human performance and error.

A Contemporary Approach to Operational Human Performance and Error

Progressing to normal operations monitoring, and thus to the implementation of LOSA, requires revisiting and adjusting prevailing views of human error. In the past, safety analyses in aviation have viewed human error as an undesirable and wrongful manifestation of human behaviour into which operational personnel somehow wilfully elect to engage. In recent years, a considerable body of practically oriented research, based on cognitive psychology, has provided a completely different perspective on operational errors. This research has substantiated in practical terms a fundamental concept of human cognition: error is a normal component of human behaviour. Regardless of the quantity and quality of regulations the industry might promulgate, regardless of the technology it might design, and of the training humans might receive, error will continue to be a factor in operational environments because it simply is the downside of human cognition. Error is the inevitable downside of human intelligence; it is the price human beings pay for being able to think on their feet. Error is a conservation mechanism afforded by human cognition to allow humans to flexibly operate under demanding conditions for prolonged periods without draining their mental batteries.

There is nothing inherently wrong or troublesome with error itself, as a manifestation of human behaviour. The trouble with error in aviation lies with the negative consequences it may generate in operational contexts. This is a fundamental point: in aviation, an error is inconsequential if the negative consequences of it are trapped before it produces damage. In operational contexts, errors that are caught in time do not produce damaging consequences and therefore, for practical purposes, do not exist. Countermeasures to error, including training interventions, should not be restricted to attempts to avoid errors, but rather to make them visible and trap them before they produce damaging consequences. This is the essence of error management: human error is unavoidable but manageable. Error management is at the heart of LOSA and reflects the previous argument. Under LOSA, flaws in human performance and the ubiquity of error are taken for granted, and rather than attempting to improve the human condition, the objective becomes improving the context within which humans perform. LOSA aims ultimately – through changes in design, certification, training, procedures, management and investigation – at defining operational contexts that introduce a buffer zone or time delay between the commission of errors and the point in which their consequences become a threat to safety. The buffer zone/time delay allows crews to recover the consequences of errors, and the better the quality of the buffer, or the longer the time delay, the stronger the intrinsic resistance and tolerance of the operational context to the negative consequences of human error. Operational contexts should be designed in such a way that allows front-line operators second chances to recover the consequences of errors. An approach to human error from the perspective of applied cognition furthers the case for LOSA. Accident and incident reports, and existing database analyses may provide some of the answers, but it is doubtful that they will answer the fundamental questions regarding the role of human error in aviation safety. To what extent do flight crews employ successful coping strategies? To what extent do successful remedial strategies avert incidents and accidents? These are the questions for which a systematic answer is imperative in order to ascertain the role of human error in aviation safety, prioritise the issues to be addressed by remedial strategies, and reshape remedial strategies as necessary.

Managing Change once LOSA Data is Collected

LOSA is but a data collection tool. LOSA data, when analysed, are used to support changes designed to improve safety. These may be changes to procedures, policies, or operational philosophy. The changes may affect multiple sectors of the organisation that support flight operations. It is essential that the organisation has a defined process to effectively use the analysed data, and to manage the change the data suggests. LOSA data should be presented to management in at least operations, training, standards and safety, with a clear analysis describing the problems related to each of these areas as captured by LOSA. It is important to emphasise that while the LOSA report should clearly describe the problem the analysed data suggest it should not attempt to provide solutions. These will be better provided through the expertise in each of the areas in question. LOSA directs organisational attention to the most important safety issues in daily operations, and it suggests what are the right questions to be asked; however, LOSA does not provide the solutions. The solutions lie in organisational strategies. The organisation must evaluate the data obtained through LOSA, extract the appropriate information, and then deploy the necessary interventions to address the problems thus identified. LOSA will only realise its full potential if the organisational willingness and commitment exist to act upon the data collected and the information such data support. Without this imperative step, LOSA data will join the vast amounts of untapped data already existing throughout the international civil aviation community.

Conclusion

There is no denying that monitoring normal operations through LOSA on a routine basis and world-wide scale poses significant challenges. Significant progress has been achieved in tackling some of these challenges. For example, from a methodological point of view, some early problems in defining, classifying, and standardising the data obtained have been solved; and consensus has been developed regarding what data should be collected. From an organisational perspective, there is a need to consider using and integrating multiple data collection tools, including line observations, more refined incident reporting and Flight Data Analysis (FDA) systems. This in turn poses a challenge to the research community, to assist airlines by developing analytic methods to integrate multiple and diverse data sources. But most importantly, the real challenge for the large-scale implementation of LOSA will be overcoming the obstacles presented by a blame-oriented industry. This will demand continued effort over time before normal operations monitoring is fully accepted by the operational personnel, whose support is essential.

References

Amalberti, R. (1996). *La Conduite de Systèmes a Risques.* Paris, Presses Universitaires de France.

Klinect, J. R., Wilhelm, J. A., & Helmreich, R. L. (in press). Event and error management: Data from line operations safety audits. In *Proceedings of the Tenth International Symposium on Aviation Psychology*, The Ohio State University.

Mauriño, D. E., Reason, J., Johnston, A. N., & Lee, R. (1995). *Beyond Aviation Human Factors.* Aldershot, England: Avebury Technical.

Pariès, J. (1996). Evolution of the aviation safety paradigm: Towards systemic causality and proactive actions (pp. 39-49). In B. J. Hayward & A. R. Lowe (Eds.), *Proceedings of the 1995 Australian Aviation Psychology Symposium.* Aldershot, England: Avebury Technical.

Reason, J. (1998). *Managing the Risks of Organizational Accidents.* Aldershot, England: Avebury Technical.

Vaughan, D. (1996). *The Challenger Launch Decision.* Chicago, USA: The University of Chicago Press.

Woods, D. D., Johannesen, L. J., Cook, R. I., & Sarter, N. B. (1994). *Behind Human Error: Cognitive Systems, Computers, and Hindsight*. Wright-Patterson Air Force Base, Ohio: Crew Systems Ergonomics Information Analysis Center (CSERIAC).

Chapter 3

Managing Threat and Error: Data from Line Operations[1]

Robert L. Helmreich, James R. Klinect, and John A. Wilhelm
University of Texas Human Factors Research Project
Department of Psychology,
The University of Texas at Austin, USA

Introduction

Achieving safety in any socio-technical endeavour requires accurate data about the way an organisation functions in normal operations and how it responds to threat and error. In aviation, a number of sources of operational data are available, but each provides a different and incomplete view of the organisation in action. Airlines have historically relied on data from their training departments (performance in training), assessment during scheduled flights (line checks), and formal proficiency checks conducted in simulators. These data provide accurate information on a pilot's technical competence and ability to respond to particular challenges. In Line Oriented Evaluation (LOE) the data demonstrate the ability of a full crew to respond to normal and abnormal situations during simulated flights. Data are also obtained from line checks in which an evaluator grades performance during normal flight operations. These sources provide valid information on pilots' ability to perform when being evaluated in a jeopardy situation (i.e., one where poor performance could result in disqualification). The limitation of data obtained in this fashion is that they do not inform organisations how crews behave when they are not under formal surveillance. In other words, are the practices taught and exhibited during formal evaluation routinely followed by crews not under scrutiny?

Another source of safety data comes from monitoring digital flight data recorders (Flight Operations Quality Assurance: FOQA). FOQA data provide a precise record of flight parameters and crew actions and have been hailed as a major contribution to safety. However, the limitation of FOQA data is that no information is recorded about why particular actions were taken. Finally, formal analyses of accidents and incidents provide insights, but are limited by their post

[1] Research reported here was supported by a grant from the US Federal Aviation Administration, Robert Helmreich, Principal Investigator.

hoc nature and the fact that they involve rare events that may not be reflect modal behaviour.

LOSA

To obtain a more comprehensive picture of crew behaviour and system performance and to validate the impact of safety initiatives such as Crew Resource Management (CRM), our research group at the University of Texas at Austin has developed a methodology we call the Line Operations Safety Audit (LOSA). The scientific background of LOSA is systematic coding of observable behaviour in naturalistic settings, a methodological approach that has been previously applied by the University of Texas group in natural settings such as undersea habitats (Radloff & Helmreich, 1968).

LOSA involves placing specially trained, expert observers in cockpits during normal line flights. The critical difference between a LOSA observation and a line check is LOSAs guarantee of anonymity. Data are entered into a de-identified database and no actions of specific crews are reported to management or regulatory agencies. Needless to say, one of the critical requirements for LOSA is trust. If crews do not believe the confidentiality assurances of observers and management, the process, in effect, becomes a line check. Based on the data obtained to date, we are confident that this level of trust has been achieved. Another important aspect of LOSA is the fact that it captures exemplary as well as deficient performance. It is important for organisations to know areas where they excel as well as those in need of improvement.

All LOSA observers complete a multi-day training program in the use of data forms and have practice using detailed scenarios to achieve acceptable reliability. They generate a text narrative of each phase of each flight. This text provides the context in which classification of observed conditions and behaviours is conducted. Coding begins with crew behaviour in the pre-flight phase and includes ratings of the behavioural markers of CRM practices developed by the University of Texas group (see Helmreich & Foushee, 1993 for history). When the foundations of LOSA were first developed, data collection was limited to these markers, ratings of the flight environment, and demographics of the crew. More than 3,000 flight segments in five airlines were observed using this protocol. However, since 1997, we have focussed explicitly on threat and error, which symbolises the beginning of what is known as LOSA today.

Threats and Errors

Threats are events or errors that occur outside the influence of the flight crew but require their active management to maintain adequate safety margins. Adverse weather, terrain, heavy traffic, and aircraft malfunctions are all considered threats beyond the flight crews' control and within their responsibility to manage. A crewmember's lack of technical proficiency is classified as a threat. Errors committed by outside agents such as ATC, ground, or dispatch are also threats.

Flight crew error is classified in LOSA as a deviation from organisational or crew expectations or intentions. Errors committed by the flight crew are described and coded along with actions (if any) taken to mitigate the consequences of the error. Drawing from observations of error, we have been able to classify all errors into four broad categories.

1. *Procedural errors:* These are what most think of as errors, crews intending to follow a procedure but doing it incorrectly. Procedural errors include the usual classification of slips, lapses, and mistakes. This type of error can only be committed when actions are covered by formal procedures. Errors in the physical handling of the aircraft are also subsumed in this category.

2. *Communications errors:* These involve failures in the transfer of information including misstatements, misunderstandings, and omissions. The identification of communications errors in accident investigations provided impetus for the development of CRM training.

3. *Decision errors:* When crews choose to follow a course of action that unnecessarily increases risk to the flight in a situation not governed by formal procedures, this action is classified as a decision error. For example, crews may choose not to deviate around weather on their flight path, resulting in encounter with turbulence.

4. *Intentional non-compliance:* When crews obviously and intentionally violate company or regulatory requirements, these are classified as intentional non-compliance errors. For example, failing to abort an unstable approach as required by company procedures would fit in this classification.

The outcomes of threats and errors can be inconsequential or consequential. In our methodology, consequential errors are those that result in an undesired aircraft state or lead to additional crew error. Undesired aircraft states include deviations from desired navigational path or altitude, unstable approaches, long or hard landings, being on the wrong runway or taxiway, and arriving at the wrong airport or wrong country. Undesired states put flights at increased risk.

Undesired Aircraft State

When an error (or threat) becomes consequential by leading to an undesired aircraft state, the crew is no longer managing threat and error, the task has become management of the undesired state. This is not a semantic distinction as different countermeasures may be required to manage threats, errors, and undesired states.

LOSA Results[2]

We have conducted 11 LOSAs in US and foreign air carriers. More than 1,900 flight segments have been observed with the threat and error management methodology. The data show that 88 percent of flights observed encountered one or more threats, with an average of 2 per flight and a range of 0 to 17. The most frequently encountered threats were adverse weather (34 percent), ATC actions or errors (34 percent), and aircraft malfunctions (15 percent). Airlines participating in threat and error LOSA are shown in Table 3.1.

Table 3.1 Airlines participating in LOSA

Air New Zealand	Delta
Cathay Pacific	EVA Air
Continental (1998/ 2000)	Frontier
Continental Express	Qantas
Continental Micronesia	US Airways
	Uni Air (Taiwan)

Errors

Errors were observed on 63 percent of flights, again with an average of 1.8 per segment and a range of 0 to 17. The most frequent source error was data entry into mode control panel or flight management computer. The second most frequent involved use and completion of checklists. Figure 3.1 shows the percentage of each error type as well as the percentage of each type classified as consequential.

As Figure 3.1 indicates, the most frequent type of error was procedural, constituting 56 percent of those observed. The second most frequent was intentional non-compliance, accounting for 26 percent. In contrast decision errors were the least frequent at 6 percent, followed by communications errors with 13 percent.

Outcomes of Error

Examination of the outcomes of error reveals that 72 percent were inconsequential, strong evidence for the error tolerance of the aviation system. However, 21 percent resulted in undesired aircraft states and 8 percent resulted in additional error (the error chain). When the error resulted in an undesired aircraft state, the crew detected the state and took action in two-thirds of the cases.

---

[2] The LOSA archive continues to grow. Results shown here are meant to exemplify findings, not to provide norms on threat, error, or human factors behaviors. Results shown are from six LOSAs conducted at airlines in three countries.

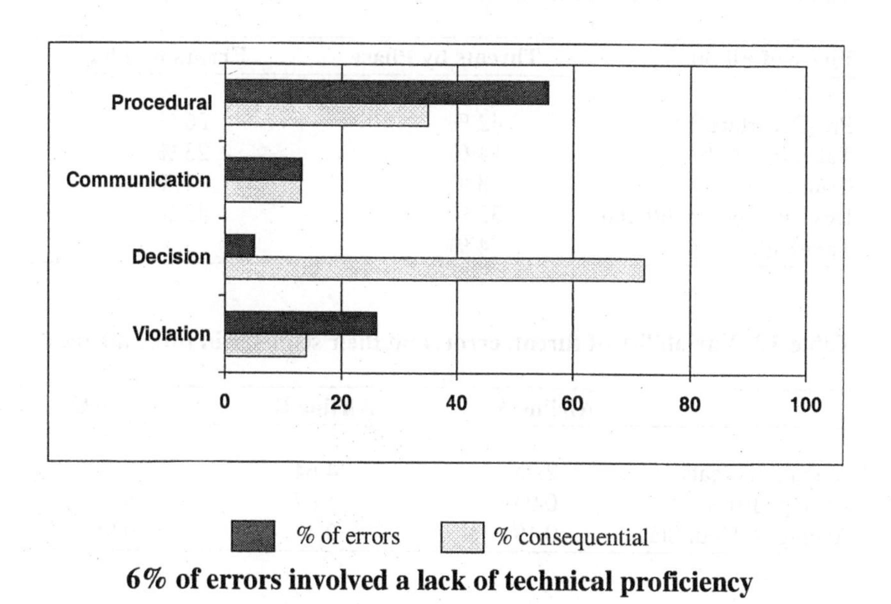

Figure 3.1 Percentage of each type of error and percentage classified consequential

However, in 26 percent of cases the state was undetected and in 8 percent it was detected and ignored. Again, the safety margins of the system are demonstrated by the fact that all flights observed were completed successfully. There was also substantial variability in the occurrence of threat and error across phase of flight. Table 3.2 shows the percentage of threats and errors observed in each phase. Consistent with the global accident rate, it is not surprising to find that the highest percentage of errors (42 percent) occurred in the approach and landing phase along with 32 percent of threats. What is surprising is the fact that 40 percent of threats were observed during the pre-departure/taxi phase. The latter finding again emphasises the importance of pre-flight and departure activities.

A high degree of variability was also found between airlines in the number of threats encountered, the number of errors committed, and the percentage leading to undesired aircraft states. Sample data from three airlines are shown in Table 3.3. This is hardly surprising as the airlines sampled differed in many ways. The implications of this are that the methodology can be used to sample any type of operation and can show norms and also variability in a highly complex system. The aggregate results presented are to show the methodology in action, not a veridical picture of how pilots worldwide deal with their operating environment.

Table 3.2 Percentage of threats and errors in each phase of flight

Phase of Flight	Threats by Phase	Errors by Phase
Pre-Departure/Taxi	42 %	26 %
Takeoff/Climb	14 %	23 %
Cruise	8 %	4 %
Descent/Approach/Land	32 %	42 %
Taxi/Park	4 %	5 %

Table 3.3 Variability of threat, error, and their sequelae in three airlines

	Airline A	Airline B	Airline C
Average Threats	2.25	4.64	3.27
Average Errors	0.93	1.97	4.32
Average # Und. State	0.19	0.42	0.95

Using LOSA Data

Organisations that have participated in LOSA have been enthusiastic about its value. The data provide management with information that assists in prioritising safety initiatives. Training departments can use the information to develop targeted training. The various types of error suggest different remedial strategies. For example, a high incidence of violations can point to poor procedures, weak leadership, and/or a culture of non-compliance. Procedural errors may suggest poor workload management or may be a reflection of inadequate procedures. Communications errors may reflect a need for more focus on CRM, especially interpersonal communications issues. Similarly, decision errors may suggest a need for further CRM concentration on expert decision making and risk assessment. Finally, proficiency involvement in errors may suggest a need to tighten standards for qualification and evaluation.

LOSA data can be used to answer very specific questions, for example is crew performance better with the captain or the first officer flying? (Based on the data from 3,800 flights, the answer is that it makes no difference if the environment is benign but effectiveness is significantly higher in complex, challenging environments with the first officer flying; Hines, 1997).

Another positive of the LOSA project is the fact that a database is being developed that allows organisations to contrast their results with those of other airlines. Such comparisons help in interpretation of the significance of, for example, the number of procedural and decision errors observed and the effectiveness of the organisation's threat and error countermeasures.

The International Civil Aviation Organisation (ICAO) has recognised the value of LOSA. Costa Pereira, the Secretary General of ICAO wrote to Jane Garvey, Administrator of the Federal Aviation Administration '[LOSA] acquires direct, first-hand data on the successful recovery from errors by flight crews during normal line flights. [It] is aimed at collecting data on successful human performance; and this is indeed a first in our industry, since aviation has traditionally collected data on failed human performance, such as an accident or incident investigation'. The proactive approach to collecting safety data exemplified by LOSA will become an ICAO requirement in 2005. The International Air Transport Association (IATA) and the International Federation of Airline Pilots Associations (IFALPA) and the US Airline Pilots Association (ALPA).

The University of Texas Threat and Error Management Model

The University of Texas Threat and Error Management Model is described in Chapter 10. The model is proving useful as a guide to the analysis and understanding of incidents and accidents and is being employed by the safety department of one major airline as the framework for interpreting confidential reports to an incident reporting system, the US Aviation Safety Action Programs (ASAP: FAA, 2000). Using the model as a template can aid in the identification and mitigation of threats before they have adverse consequences.

Conclusion

The non-jeopardy assessment of crew behaviour in normal flight operations through LOSA provides a valid picture of flight operations. The fact that many of the errors observed involved violations of procedures or regulations validates the conclusion that crews trust the methodology and do not perceive the observations as threatening their status. The data obtained can assist organisations in fostering a safety culture and targeting training to observed needs.

The challenge for industry is to make the process accessible to organisations with varying levels of resources and Human Factors expertise. Perhaps the greatest contribution to system safety will come from the development of a valid database of safety information that can be used to guide safety initiatives. Such a system is envisioned in the US Federal Aviation Administration's Global Aviation Information Network (GAIN), but many barriers to its growth and full utilisation exist, including a lack of trust in government maintained data archives.

References

Federal Aviation Administration. (2000). *Aviation Safety Action Programs: Advisory Circular 120-66A*. Washington, D.C. (Available at www.faa.gov)

Helmreich, R. L. (2000). On error management: Lessons from aviation. *British Medical Journal, 320*, 781-785.

Helmreich, R. L. (2000). Culture and error in space: Implications from analog environments. *Aviation, Space, and Environmental Medicine, 71*(9-II), 133-139.

Helmreich, R. L., & Foushee, H. C. (1993). Why crew resource management? Empirical and theoretical bases of human factors training in aviation. In E. Wiener, B. Kanki, & R. Helmreich (Eds.), *Cockpit Resource Management* (pp. 3-45). San Diego, CA: Academic Press.

Helmreich, R. L., & Musson, D. M. (2000). Threat and error management model: Components and examples. *British Medical Journal* [On-line], bmj.com. http://www.bmj.com/misc/bmj.320.7237.781/sld001.htm.

Helmreich, R. L., Wilhelm, J. A., Klinect, J. R., & Merritt, A. C. (2001). Culture, error, and crew resource management. In E. Salas, C. A. Bowers, & E. Edens (Eds.), *Applying Resource Management in Organizations: A Guide for Training Professionals*. Princeton, NJ: Erlbaum.

Hines, W. E. (1998). Teams and technology: Flight crew performance in standard and automated aircraft. Unpublished doctoral dissertation, The University of Texas at Austin.

Radloff, R. W., & Helmreich, R. L. (1968). *Groups Under Stress*. New York: Appleton-Century-Crofts.

University of Texas Human Factors Research Project Website: www.psy.utexas.edu/HumanFactors

Chapter 4

Management Influence on Safety Climate

Rhona Flin
Department of Psychology, University of Aberdeen, Scotland

Introduction

> A company may in many ways be likened to a human body. It has a brain and a nerve centre which controls what it does. It also has hands which hold the tools and act in accordance with directions from the centre. ... directors and managers ... represent the directing mind and will of the company and control what it does. The state of mind of these managers is the state of mind of the company and is treated by the law as such. (Lord Denning, 1957)

The concept of safety climate has now been incorporated into the parlance of safety management in high reliability organisations. This paper opens by examining the prime factors that influence the state of safety climate in an organisation, and then focuses on one of the key determinants – namely the role of management, especially the more senior managers.

Safety Climate

The term safety climate has tended to be used somewhat interchangeably with the broader concept of safety culture – a lack of differentiation that served to cloud an already obscure conceptual landscape. More recently an emerging consensus amongst industrial psychologists is beginning to differentiate safety climate as the surface features of an organisation's safety culture, discerned from the workforce's attitudes and perceptions at a given point in time – a 'snapshot of the state of safety' (Cox & Flin, 1998; Flin, Mearns, O'Connor, & Bryden, 2000).

Notwithstanding residual ambiguities in the culture vs. climate debate, a further problem in this domain has been the multiplicity of factors determined to be the key parameters of the safety climate. The lack of consensus echoed an earlier evolutionary stage of the theoretical and applied literature on personality prior to the emergence of the Big Five Factors (Costa & McCrae, 1992; De Raad, 1998) that while not receiving unanimous acceptance at least provided a sufficient consensus to allow the field to evolve. However when the common factors are extracted from a review of the principal themes measured in published accounts of safety climate research, then a more consensual position begins to emerge. Flin et al. (2000) reviewed 18 safety climate questionnaires from research teams working in different industries and found that there were five common factors; (i) Management and supervision, (ii) Safety system, (iii) Risk (e.g., risk taking behaviours), (iv) Work pressure, and (v) Competence. Similarly, Guldenmund (2000) scrutinised 15 measures of safety climate and found that Management's Safety Activity was one of the frequently appearing factors.

The UK Health and Safety Regulator (HSE, 1999) is apparently aware of this factor, and lists the organisational factors associated with a safety culture as:

- Senior *management* commitment
- *Management* style
- Visible *management*
- Good communication between all levels of employee (*management* action)
- A balance of health and safety and production goals (*management* prioritisation).

With a little judicious highlighting, it is not difficult to discern an emerging theme – managers, especially senior management are key influences on the safety culture.

Which Managers?

The occupational category 'management' has been used ambiguously within many of the studies of safety climate, describing various levels of management from CEO to first line supervisor. Consequently, it is often unclear which level of management is being assessed, but it is important that these distinctions are clarified, given that the various grades of managers play very different roles in the management of safety (Andriessen, 1978). There is no doubt that all managers must demonstrate their safety commitment but three of the principal strata of management – supervisors, site managers, and senior managers probably have key positions in a safety management system. I asked 200 managers and safety professionals attending the UK Electricity Association Conference (electricity generation and distribution) in May 2000, the following question. 'If you were to focus your attention on one of the following levels of management in order to improve safety, which one would you choose?' Their responses were:

- Senior Manager – 42 percent
- Site Managers – 11 percent
- Supervisors/Team Leaders – 47 percent

In fact none of these management strata have been studied very extensively by safety researchers.

Supervisors

Although 40 years ago Heinrich (1959, p. 22) advised, 'The supervisor or foreman is the key man in industrial accident prevention. His application of the art of supervision to the control of worker performance is the factor of greatest influence in successful accident prevention'. Only recently have researchers begun to examine the influence of the supervisor on safety and several recent studies have begun to show how safety is related to first line supervisors' leadership style and team management skills (Mearns, Flin, Fleming, & Gordon, 1997; Simard & Marchand, 1994; Zohar, 2001).

Site Managers

The influence of site managers on safety performance has received rather less attention in the literature (apart from some studies 20 years ago, e.g., Andriessen, 1978), perhaps due to the unfashionable nature of the leadership concept in 1980s. Notwithstanding the rather limited empirical evidence, the importance of the site manager on safety climate is acknowledged in regulatory guidance (e.g., ACSNI, 1993). Some new research is examining the effects of site managers on workplace safety, in relation to their attitudes and behaviours (O'Dea & Flin, 2001; O'Dea, Yule, & Flin, under review; Thompson, Hilton, & Witt, 1998). Conventional theories of leadership such as Bass's transformational model measured by the MLQ (Multifactor Leadership Questionnaire) appear to offer an insight into key leadership skills for influencing safety initiative and rule compliance at the worksite (O'Dea & Flin, 2000).

Senior Managers

Senior managers and directors who are often studied in relation to business successes, may be regarded as a neglected species when it comes to safety performance – a surprising omission, given their ultimate responsibility for safety losses and accidents. With a few notable exceptions (Hopkins, 1999, 2000) there is hardly anything written about the process by which they can achieve a strong safety culture in their organisations. Yet their responsibilities have not entirely gone unnoticed. Lawyers scrutinising company performance when investigating a series of major industrial catastrophes in the UK during the 1980s, had not taken long to realise that they should be focussing their attention on senior managers who ran the company rather than simply blaming operating staff who made errors.

Mr Justice Sheen (1987, p. 14) investigating the loss of hundreds of lives after the sinking of the car ferry Herald of Free Enterprise which sailed out of Zeebrugge with its bow doors open, concluded:

> But a full investigation into the circumstances of the disaster leads inexorably to the conclusion that the underlying or cardinal faults lay higher up in the company... From top to bottom the body corporate was infected with the disease of sloppiness.

Lord Cullen (1990, p. 301) who directed the Public Inquiry into the loss of the Piper Alpha oil platform operated by Occidental (with 167 fatalities), advised:

> No amount of detailed regulations for safety improvements could make up for the way that safety is managed by operators.

The UK safety regulator has started to endorse this position: 'Senior management commitment is crucial to a positive health and safety culture. It is best indicated by the proportion of resources (time, money, people) and support allocated to health and safety management and by the status given to health and safety' (HSE, 1999, p. 46; see also HSE, 2000). Moreover, new Guidance on corporate governance (the Turnbull Guidance) for directors of UK listed companies (issued by the Institute of Chartered Accountants in England and Wales) now requires them to develop a corporate-wide, risk management approach to internal control which includes health and safety management (McCrae & Balthazor, 2000). This development, coupled with new legislation on corporate manslaughter (Forlin, 2002; Slapper, 1999), has underlined the potential culpability of directors and senior managers if their safety management systems to not stand up to scrutiny following an accident.

Time for Safety

So what advice can be given to senior managers to enable them to develop and maintain an effective safety culture in their organisations? As in any other facet of management what is key is the behaviours that are demonstrated in relation to safety. Managers 'behave badly' when they send the wrong signals to more junior managers and to the workforce by their language and their actions, especially in relation to prioritisation and their time allocation. This last factor is probably the most crucial, as time is the most precious resource for senior managers well aware of Benjamin Franklin's maxim that 'Time is money'. The importance of time has recently been endorsed by the UK safety regulator when offering guidance on senior management commitment to safety.

> It is best indicated by the proportion of resources (time, money, people) and support allocated to health and safety management and by the status given to health and safety. (HSE, 1999, p. 46)

Thus the questions that should be posed to senior managers concerned about demonstrating their commitment to safety:

- Are you making time for safety?
- Do you allow your staff to take their time to do the job safely?
- Are they encouraged to stop the job and have a 'Time Out' for safety?
- Do you take the time to listen to safety concerns?
- Do you have time to spend at the worksite?

Why is time so crucial? Because it is the strongest signal of commitment from busy managers with little time to spare.

And what happens when they do not prioritise enough of their time to listening to safety concerns or warning signs? The late Barings International Bank is a good example. The Group Treasurer when asked why he had ignored the indications of a problem in Singapore with Nick Leason's trading activities, apparently replied 'But there always seemed to be something else more pressing' (quoted in Reason, 1997, p. 32).

In addition to demonstrating commitment by making time for safety, is there anything else managers can do? The nuclear power industry also emphasise that managers should frequently emphasise the importance of safety.

> On a personal basis, managers at the most senior level demonstrate their commitment by their attention to regular review of the processes that bear on nuclear safety, by taking direct interest in the more significant questions of nuclear safety or product quality as they arise, and by frequent citation of the importance of safety and quality in communications to staff. (IAEA, 1997, p. 10)

Leadership Style

There is a voluminous literature on the most appropriate leadership styles for effective management, at all levels, but this has little to say about which leadership style is most effective for safety management. Some nuclear industry guidance advises that the basic style required may be similar to that used for other business objectives, although they point out that because the workforce do not tend to find safety an inherently fascinating topic, then some additional emphasis may be required.

> There is no reason to suppose that leadership to improve safety is any different in principle from leadership to increase productivity or enhance job satisfaction. However it may require a distinctive blend of behaviours because of its relatively low intrinsic interest to the workforce. (ASCNI, 1993, p. 32)

The HSE (1999, p. 46) suggest 'A 'humanistic' approach to management involving more regard by managers for individuals' personal and work problems is likely to be effective'.

These appear to be sensible suggestions but as mentioned above there have been few if any empirical investigations of these important issues (O'Dea et al., under review). Understanding which styles are used effectively by senior managers to enhance the safety culture is extremely important, not only due to their indirect influence on the workforce but also because their behaviour patterns are often copied by ambitious junior managers who have a more direct influence on worksite activities.

Measuring Managers' Safety Commitment

What we do know is that the safety culture in an organisation is determined by perceptions of management commitment to safety, as judged by the workforce. Therefore in any effort to improve the safety culture by focussing on management factors, it is first necessary to measure how management commitment to safety is perceived. This should be done at different levels in the organisational hierarchy – at both a workforce level by the use of a safety climate survey, and at a managerial level using an upward appraisal technique.

Safety Climate Survey Following the *Piper Alpha* disaster in 1988 the operating companies extracting oil and gas from the North Sea made concerted efforts to improve their safety performance, including the introduction of safety climate measurement and benchmarking (Mearns et al., 1997; Mearns, Whitaker, & Flin, 2001). Research into safety climate on offshore installations in the North Sea has examined a number of factors, including how the workforce perceives senior management commitment. In one survey involving 13 companies 69 percent of the sample disagreed that 'Senior managers show a lack of commitment to health and safety', but only 57 percent agreed that 'Senior managers are genuinely concerned about the health and safety of their workforce'. In one of these companies the Managing Director was concerned that his commitment to safety was not being transmitted to his offshore workforce. He commissioned our research group to undertake an upward appraisal exercise to determine how effectively senior managers communicated their safety commitment to their immediate subordinates /direct reports (who were also senior managers).

Upward Appraisal on Safety There are no standard tools for measuring safety commitment and safety leadership in an upward appraisal exercise (i.e., perception of commitment from workforce). Therefore a special questionnaire was designed which included sections on safety commitment behaviours, prioritisation of safety, production, cost reduction, and reputation, as well as conventional leadership scales This questionnaire was given to 70 directors and senior managers from one oil company, as well as directors from their major contracting companies. Each manager was asked to complete the questionnaire describing his own safety attitudes, behaviours and leadership style and also to give a mirror-version of the questionnaire to five or six of his direct reports and to ask them to rate him in the same way.

These managers then attended a one-day safety workshop during which each manager was given a personal report describing his self-perception of his safety commitment contrasted against the view of his subordinates (shown as average and range data). Aggregate results were prepared for the group and presented at the workshop resulting in a frank discussion of whether senior managers were successfully communicating consistent messages about their safety commitment. The exercise produced a very positive response from the managers involved, with subsequent evidence of managers taking action to change their behaviour in relation to safety management (Bryden, Flin, & West, under review). Undertaking an upward appraisal survey of this type to scrutinise senior managers' safety behaviours is a powerful indicator of a strong commitment at the most senior level to improving the safety culture.

Conclusion

Managing an organisation's safety requires a long-term approach focussed on key determinants of the safety culture. One of the prime factors is the degree of management commitment to safety at all levels, from the first line supervisor to the managing director.

Managers must check whether their safety commitment is being transmitted to others. This can be achieved by the use of safety climate surveys to measure workforce perceptions of managers' attitudes and behaviours. In addition, the more immediate influence of managers on safety can be tested by using an upward appraisal survey. This shows the extent to which their commitment to safety is being communicated to their direct reports. Good safety management requires more than simply knowing 'the safety script'. It needs demonstrations of commitment shown by time allocation and prioritisation of safety, especially when managers are faced with conflicting safety and production goals.

References

ACSNI. (1993). *Organising for Safety.* Advisory Committee on the Safety of Nuclear Installations. Human Factors Study Group. Third Report. Suffolk: HSE Books.

Andriessen, J. (1978). Safe behaviour and safety motivation. *Journal of Occupational Accidents, 1,* 363-373.

Bryden, R., Flin, R., & West, P. (in submission). Upward appraisal of senior managers' safety commitment.

Costa, P., & McCrae, R. (1992). Four ways five factors are basic. *Personality and Individual Differences, 13,* 653-665.

Cox, S., & Flin, R. (1998). Safety culture: Philosopher's stone or man of straw. *Work & Stress, 12,* 189-201.

Cullen, Hon Lord. (1990). *The Public Inquiry into the Piper Alpha Disaster.* London: HMSO.

De Raad, B. (1998). Five big, big five issues: Rationale, content, structure, status and cross cultural assessment. *European Psychologist, 3,* 113-124.

Denning, L. J. (1957). In the case of H. L. Bolton (Engineering) Co Ltd v T J Graham & Sons Ltd 1 QB 159, England.
Flin, R., Mearns, K., O'Connor, P., & Bryden, R. (2000). Safety climate: Identifying the common features. *Safety Science, 34,* 177-192.
Forlin, G. (2002). *Corporate Manslaughter.* London: Butterworth.
Guldenmund, F. (2000). The nature of safety culture: A review of theory and research. *Safety Science, 34,* 215-257.
Heinrich, H. (1959). *Industrial Accident Prevention* (4th ed.). London: McGraw Hill.
HSE. (1999). *Reducing Error and Influencing Behaviour.* Suffolk: HSE Books.
HSE. (2000). *Revitalising our Potential.* Suffolk: HSE Books.
Hopkins, A. (1999). *Managing Major Hazards: The Lessons of the Moura Mine Disaster.* Sydney: Allen & Unwin.
Hopkins, A. (2000) *Lessons from Longford: The Esso Gas Plant Explosion.* Sydney: CCH.
IAEA. (1997). *Examples of Safety Culture Practices* (Safety Report Series No.1). Vienna: International Atomic Energy Authority.
McCrae, M., & Balthazor, L. (2000). Integrating risk management into corporate governance: The Turnbull guidance. *Risk Management,* 35-45.
Mearns, K., Flin, R., Fleming, M., & Gordon, R. (1997). *Human and Organisational Factors in Offshore Safety* (Report OTH 543). Suffolk: HSE Books.
Mearns, K., Whitaker, S., & Flin, R. (2001). Benchmarking safety climate in hazardous environments: A longitudinal, inter-organisational approach. *Risk Analysis,* 21, 771-786.
O'Dea, A., & Flin, R. (2000). *Safety leadership behaviours.* Paper presented at the Academy of Management Conference, Toronto, August.
O'Dea, A., & Flin, R. (2001). Site managers and safety leadership in the offshore oil and gas industry. *Safety Science,* 37, 39-57.
O'Dea, A., Yule, S., & Flin, R. (in submission). Managerial influence on workplace safety: A review of the literature.
Reason, J. (1997). *Managing the Risks of Organizational Accidents.* Aldershot: Ashgate.
Slapper, G. (1999). *Blood in the Bank.* Aldershot: Ashgate.
Sheen, Mr Justice. (1987). *mv Herald of Free Enterprise* (Report of Court 8074). London: Department of Transport.
Simard, M., & Marchand, A. (1994). The behaviour of first line supervisors in accident prevention and effectiveness in occupational safety. *Safety Science, 17,* 169-185.
Thompson, R., Hilton, T., & Witt, A. (1998). Where the rubber meets the shop floor: A confirmatory model of management influence on workplace safety. *Journal of Safety Research, 29,* 15-24.
Zohar, D. (2001). The effects of leadership dimensions, safety climate, and assigned priorities on minor injuries in work groups. *Journal of Organizational Behavior,* 23, 75-92.

Chapter 5

Models of Bureaucratic Failure

Ron Westrum
Department of Sociology, Eastern Michigan University
Ypsilanti, MI 48197, USA

Introduction

I want to discuss the way that bureaucracies allow accidents. While the Reason model is a good general guide to the nature of organisations' vulnerability to an accident, we must go further. I suggest that we think of accidents as falling into general types. In the process of classifying organisations into general types, we will become more aware of the differences in the accidents themselves and also in the ways that we can prevent them.

I will begin by proposing that the three cultures (pathological, bureaucratic, and generative) that I use to describe information flow have different susceptibilities to accidents. Pathological cultures have, I believe, the greatest susceptibility, and generative organisations the least. However, if we consider major accidents, we find that few originate in pathological cultures. Threat and intimidation play a relatively minor role in the universe of big accidents. The reason is, I believe, that bureaucratic organisations are far more common than pathological, and thus, in spite of supposedly better safety, they seem to fail more often. We need to address ourselves, then, to the ways that bureaucracies fail.

Dominating Features

I would argue that every accident possesses a dominating feature, and I will identify four of these: design flaw, task overload, persistent negligence, and violation. Let me explain a bit more about each of these accident features.

By 'design flaw' I mean that there are fundamental problems in hardware, software, staffing, or training. This kind of problem implies that the organisation has created a dangerous situation at the outset, either through failure to think through the design, or by not detecting flaws that are present. While a design flaw often illustrates a failure of what I have called 'requisite imagination', it may also be testimony to the observation that testing can never be perfect.

The dominating feature of the second type of accident is overload. For instance the system may be designed for routine operation, but because of some special circumstance, there is a workload increase. This can happen as the organisation scales up by increasing the scale of operations (as at Valuejet), or as it trims down through cutting back on resources (as at Bhopal), or as two organisations merge, as was the case at USAir. It can happen as a new threat to the system's integrity emerges, as with AIDS and the national blood systems. Or when the organisation is given a novel task in which it has little experience. But the key feature is that there are no longer sufficient resources to do the job well, whether this happens because the job increases or because the resources to do the job decrease. I will say more about this shortly.

The dominating feature of the third type of accident I call 'persistent neglect.' This occurs in situations where a problem is known to exist for some time, but nobody bothers to do anything about it. The organisation has a cavalier attitude about the latent pathogen as 'something that we just have to put up with'. And that is just what the organisation does; it puts up with the problem. When this persistent defect becomes implicated in an accident, many people say they 'knew all along' that there was a problem. And often they did. There may have been voices raised, formal reports about the problem, and even a 'dress rehearsal' for the accident, but the operation just plunges on ahead (One remembers the famous words of Adm. Farragut at Mobile Bay: 'Damn the torpedoes! Go ahead!').

The dominating feature of the fourth type of accident is wilful violation. This type is common in pathological cultures, where there is conflict and intimidation as a routine part of organisational life. A blatant contempt for rules and regulations goes hand in hand with risky decisions and a failure to cooperate. The 'cowboy' character of such operations is well expressed in Bob Helmreich's quotation that 'Checklists are for the lame and the weak.' Unlike organisations where persistently neglected issues are seen as unfortunate, this type of organisation feels no shame in its deviance. Now, of course, violation-produced accidents can happen in bureaucratic cultures as well, but they are more likely in pathological ones.

I would propose that the three cultural types have characteristically different make-ups of their accident mix. Of the four different dominating features, I believe that some are much more likely to be found in some contexts rather than others. While the two middle categories (persistent neglect and overload) are rather evenly distributed, in generative cultures design flaws are a more typical source of accidents. Violations, by contrast, are most common in pathological cultures. The diagram in Figure 5.1 represents my working hypothesis.

Now I would like to pursue the particular interest of this approach, and that is how accidents take place in organisations with bureaucratic cultures. The two types of failures that are most typical of such organisations are persistent neglect and overload.

Persistent neglect in many ways resembles overload, but the difference is that persistent neglect is the result of complacency, while overload is the result of pressure. When the dominating feature is neglect, the fundamental problems involved in the accident could probably have been fixed with the resources at hand, but the will to do this was absent. The attitude that 'there will always be this kind of problem' and that one just has to soldier on lasts right up until the accident. Then there is massive finger pointing and accusations. Then it is obvious to everyone that someone should have done something. But somehow this level of urgency cannot be mustered until the accident.

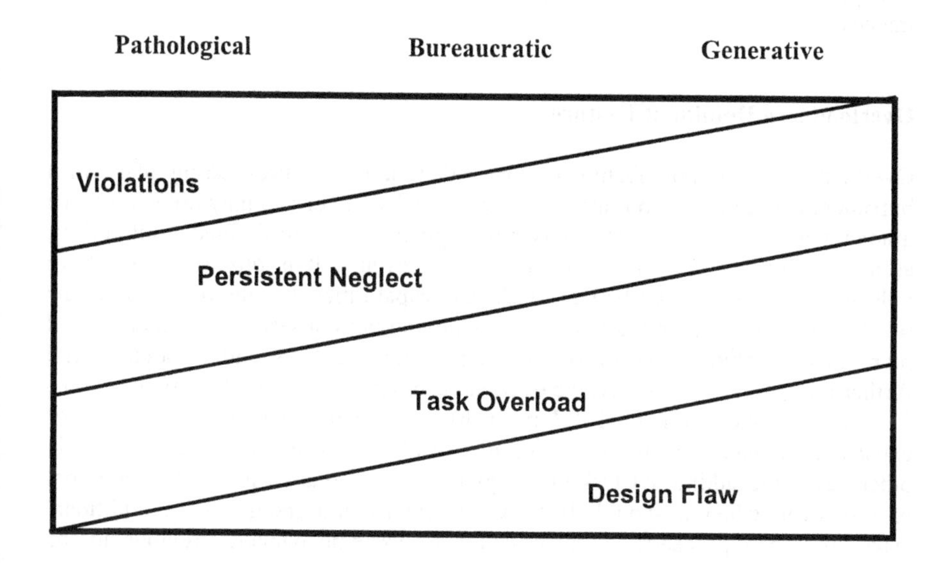

Figure 5.1 Neglect as a dominant feature

Persistent neglect is most stunning in the face of what I call the 'dress rehearsal' incident. This happens when the very same factors that will later produce an accident manifest themselves previously, but without a fatal result. The most curious thing is that the complacent organisation seems to miss entirely the implications of such events. There is nothing compelling for such organisations in the faint signals represented by the 'dress rehearsal'. The incident may pass unnoticed or unreported, and often the events are rationalised away with, 'well, things like this happen from time to time'. In other contexts, when there is a culture of conscious inquiry, such incidents will be transformed into serious warnings. The organisation then responds to the incident and takes action. We have seen this happen recently in the United States when several planned school massacres were stopped before they happened, because of a consciousness of what the result might be on the part of students and teachers. By contrast the incidents often get no response in a bureaucratic culture, leaving the path open for a tragedy.

Such an instance of a dress rehearsal accident took place in occupied northern Iraq, when two Air Force F-111's saw helicopters on the ground they assumed to be Iraqi. As the F-111's came back for a weapons pass, they realised that the helicopters were American army, and did not shoot. Later, the soldiers on the ground realised that they had come near to death, when a conversation in a bar revealed that the pilots had been about to shoot them. This chance meeting over drinks, however, did not result in either Air Force or Army picking up the phone and describing the incident. A year and a half later, in 1994, two Air Force F-15s shot down two US army helicopters full of military personnel, thinking that they were Iraqi. Apparently, no one in higher command was aware that such an accident was possible. Yet a different response to the incident could have averted the tragedy.

Overload as a Dominant Feature

Overload is a common feature of many devastating failures taking place in a bureaucratic culture. In contrast to persistent neglect, overload may not evolve in a gradual manner. Rather, overload usually originates at a well-defined point in time, as a result of a particular event or series of events. It is produced, as I have indicated, by a mis-match between tasks and capabilities. For the organisation the overload may seem temporary, and thus annoying but not serious. Yet an overload represents a natural window of opportunity for incidents and accidents. And furthermore, quite remarkably, the overload is often agreed to by the organisation.

Let me make explicit what I mean by 'agreement'. Obviously in cases of expansion or plant closure the organisation itself makes the key decisions. But in other cases the additional task(s) are given to the organisation and it does not protest, even when it should. Instead the organisation shoulders the additional tasks, and then proceeds in the direction of extreme danger. Previous to the Challenger explosion, for instance, NASA officials experienced a major cut in their safety resources. Yet the responsible officials did not threaten to resign, they just soldiered on.

In the awful Tenerife collision in the Canary Islands, the captain of the KLM aircraft was facing serious regulatory and customer problems if he did not complete the flight on time. And rather than protesting and calling for a 'time-out', he elected to take the serious risk of taking off, with horrific results. This willingness to endure the additional burden to get the job done is the hallmark of accidents dominated by overload problems. I call it 'the deal with the devil'.

This deal with the devil is even more obvious in the failure of the Canadian blood system to cope effectively with HIV contamination in the 1980s. I am in the early phases of an attempt to determine why the Canadian blood system failed so miserably. But some things are already clear. The Canadian Red Cross organisation effectively shouldered burdens that it could not handle with success (nor was it alone in this). But as it took on the burden of managing possible HIV contamination, it was not only overwhelmed by the task, it began to behave badly. So just as it hemmed and hawed, just as it delayed taking action, just as it failed to enforce blood safety, it also began to say things that were not so. It publicly indicated that the risk to the blood supply was minimal. It reassured haemophiliacs that there was little danger. It used contaminated blood even when safe blood was on its shelves. The Canadian Red Cross failed to coordinate the efforts of other agencies and even to check to see if its own initiatives were working. It failed to use the experience of the Americans as a guide to conducting its own work, and so forth. All this is detailed in the 3-volume Krevor Report (1997), which is a remarkable chronicle of what was on-going bureaucratic failure.

But it is all a result of a particular kind of crisis in bureaucratic organisations that comes from trying to do more than the organisation can accomplish. It reminds me of the famous drawing in the book of St.-Exupery, *The Little Prince*, of the boa constrictor that swallowed an elephant. But once the bureaucracy has swallowed the elephant, it does all these strange things. It deceives itself by believing that it can actually accomplish the task. It refuses help from the outside, even when that help is desperately needed. It falsifies records and ignores danger signs. It provides information that is demonstrably false to higher authority and to the public. It brands its critic 'Chicken Littles' who are running around and saying the sky is falling. But the sky is falling!

Irving Janis (1989), describing the symptoms of 'groupthink', came closest to understanding the behaviour of bureaucratic organisations that have swallowed an elephant. Yet often the cohesiveness he postulated as part of groupthink may well be absent in groups showing some of groupthink's symptoms. No such cohesiveness was present when NASA decided to launch the Challenger space shuttle on its fateful mission. Diane Vaughn (1996), in her analysis of the accident, is quite right to say that Janis's groupthink model did not fit the Challenger tragedy very well. In later work, however, Janis (1989) has suggested that a variety of constraints interfere with what he calls 'vigilant problem-solving'. Groupthink is merely one special case of such constraint. Clearly, the deal with the devil depends on the existence of such constraints on discussion and basic honesty. The existence of an unfulfillable contract by itself poses barriers to open discussion and the free examination of facts and alternatives. The deal with the devil sets in motion a series of nefarious processes that create or conceal latent pathogens as a matter of course. The organisation is trying to make bricks without straw and all too likely will fail in the project it has undertaken, because it does not have the resources to do the job.

Why Do They Say Yes?

The observations about overload lead to a host of questions. Why does the bureaucracy agree to do what is obviously impossible? Why does it refuse help? The main reason, I believe, is bureaucratic turf. The bureaucracy refuses to admit inadequacy, because this protects its right to its tasks. To admit it cannot do them risks the tasks being taken away and given to another agency. The organisation does not advertise its overload because such an admission would make it vulnerable. There are many reasons that I did not respect John Edgar Hoover, director of the Federal Bureau of Investigation for some 50 years. But I must admit that Hoover was always careful never to take on tasks that he did not think his agency could accomplish.

My friend and sometime co-author Tim Clark (1997) has written a remarkable book, *Avoiding Extinction*, about the management of the black-footed ferret species in Wyoming. Under the management of Wyoming Game and Fish department, the ferrets nearly disappeared forever. Clark has shown that this organisation regularly denied that the situation was deteriorating, refused help from outside experts and organisations, made faulty decisions, and yet in almost every case insisted that only Wyoming Game and Fish could manage the ferret population. The ferret story is a good example of how far a bureaucratic organisation will go to protect its organisational turf.

And if organisations can be pressured into accepting the impossible, how much easier it is to pressure individuals. Those assigned to do the job cannot say 'no' because their refusal will bring sanctions from above (think of the plot of the film 'Blade Runner'). Hence a desire for compliance, and fear of personal consequences, push toward agreeing to do something they do not have the time, resources, or skills to do. When subordinates agree to do the herculean task, managers move the burden from their shoulders to those lower down. If the task gets done, all good; if it does not get done, a scapegoat has been created.

Conclusion

What I have tried to suggest here is that organisational cultures, as measured by information flow types, tend to shape the vulnerabilities that organisations have to accidents. The worst flow cultures allow violations that put operations at risk. When something goes wrong in these organisations, it is often the result of ignoring the rules. By contrast, in bureaucratic organisations, more often it is ignoring problems that get organisations into trouble. Either by neglect or excessive zeal, bureaucratic organisations tend to take on more than they can handle or fail to use due diligence in pursuit of their goals. But even the high-performing generative organisations can slip up when they fail to see the implications of what they are doing, or fail to note a hidden flaw.

These speculations lend themselves to empirical testing, and it is to be hoped that my suggestions will receive a rigorous test in the not-too-distant future.

References

Clark, T. (1997). *Averting Extinction: Reconstructing Endangered Species Recovery*. New Haven: Yale University Press.

Janis, I. (1989). *Crucial Decisions: Leadership in Policy-Making and Crisis Management*. New York: Free Press.

Krevor, H. (1997). *Final Report on the Commission of Inquiry on the Blood System in Canada* (Vols. 1-3). Ottawa: Canadian Government Publishing.

Snook, S. (2000). *Friendly Fire: The Accidental Shootdown of U.S. Blackhawks over Northern Iraq*. Princeton: Princeton University Press.

Vaughn, D. (1996). *The Challenger Launch Decision: Risky Technology, Culture, and Deviance at NASA*. Chicago: University of Chicago Press.

Weick, K. (1993). The Vulnerable System: An Analysis of the Tenerife Air Disaster. In K. Roberts (Ed.), *New Challenges to Understanding Organizations*. New York: Macmillan.

Westrum, R. (1994). Cultures with requisite imagination. In J. Wise, V. D. Hopkin, & P. Stayer (Eds.), *Verification and Validation of Complex Systems: Human Factors*. New York: Springer.

Chapter 6

Culture, Systems and Change in Aircraft Maintenance Organisation

Nick McDonald
Department of Psychology, Trinity College, Dublin

Introduction

Increasingly the notion of safety culture is invoked to explain failures at an organisational level leading to major disasters. Thus, for example, a recent issue of the journal Safety Science was entirely devoted to this topic. Human Factors practitioners are aware that implementation of their prescriptions can require a culture shift in the organisations for which they work. What does this notion of 'Culture' explain? Sometimes it seems as if 'Culture' is simply a kind of bucket in which phenomena for which there is no adequate explanation are dumped. It is enough to use the label and we avoid explaining anything. When one begins to explore more thoroughly the territory of theories of culture it becomes apparent that one of their chief functions is to explain stability – how do societies, communities, or organisations maintain shared systems of meaning or norms of activity? Such theories are, almost by definition, not very good at explaining change. If they are not very good at explaining change, then they are probably not very good at supporting effective prescriptions for how change might be planned or managed. This then leads to a further problem – there is a temptation for analyses of culture to become prescriptive rather than descriptive. Thus Pidgeon and O'Leary's (1994) analysis seeks to expound the characteristics of a 'good safety culture'. Whilst this analysis is very compelling, it is difficult to use it as a practical guide to management action other than as an exhortatory statement – this is what you should do.

Over the last five years, in collaboration with several European aircraft maintenance organisations, an aircraft manufacturer, some leading aviation research institutes, and with the support of the European Commission, we have been studying how the maintenance system works, and developing and trying to implement various interventions designed to change some aspects of how it works. This has given us the opportunity both to compare different organisations and how they manage maintenance operations, and to look at processes of change (planned or otherwise) over time. This paper will summarise some of the conclusions we have reached and presents these in a framework which analyses stability and change within organisations. While the content of the paper is all about aircraft maintenance, the underlying framework can be applied to other domains.

The Formal Structure of Aircraft Maintenance

JAR 145

The framework of JAR 145 regulations is built around the philosophy of granting approval to maintenance organisations, which have an adequate management system to ensure safe operations. Thus the regulator only indirectly regulates the safety of the operation – the responsibility is on operational and quality management to ensure safety. In particular this devolution of responsibility for safety and airworthiness is expressed through the requirements to designate an accountable manager and to have an independent quality system. These are expressed thus:

> A senior person or group of persons acceptable to the Authority, whose responsibilities include ensuring that the JAR 145 approved maintenance organisation is in compliance with JAR 145 requirements, must be nominated. Such person(s) must ultimately be directly responsible to the accountable manager who must be acceptable to the Authority. JAR 145.30
> The JAR-145 approved maintenance organisation must establish an independent quality system to monitor compliance with and adequacy of the procedures to ensure good maintenance practices and airworthy aircraft and aircraft components. Compliance monitoring must include a feedback system to the person or group of persons specified in JAR 145.30(a) and ultimately to the accountable manager to ensure, as necessary, corrective action. Such system must be acceptable to the Authority. JAR 145.65

The Manufacturer

The Air Transport Association of America has set the industry standard for documentation. The principle specification relevant to maintenance documentation is ATA specification 100. It sets out guidelines for producing maintenance documentation for the manufacturer. Section 2-1-0 1. (Policy) advises that 'the Aircraft Maintenance Manual (AMM) shall provide the necessary procedures to enable a mechanic who is unfamiliar with the aircraft to maintain the aircraft properly, whether such action is required on the line or in the hangar/service centre'. Manufacturers are also particularly concerned to ensure that documentation is comprehensive, accurate, and up to date.

The Maintenance Organisation

The Maintenance Exposition Document is compiled in compliance with the regulations of the European Joint Aviation Authorities governing maintenance organisations (JAR 145). The document contains the company's formal information on Maintenance Management (i.e., roles, responsibilities, accountabilities), Maintenance Procedures (Line, Light, and Base), and Quality Systems Procedures.

The Maintenance Technician The Aircraft Engineer's International has promulgated a Mechanic's Creed that contains the following:

> I pledge myself never to undertake work or approve work which I feel to be beyond the limits of my knowledge, nor shall I allow any non-certified superior to persuade me to approve aircraft or equipment as airworthy against my better judgement.

How Does the Organisational System Work in Practice?

The ADAMS project provided the opportunity to study the functioning of four European aircraft maintenance organisations. Unless otherwise stated the research discussed below is reported in McDonald, Corrigan, Daly, and Cromie (2000).

Quality and Safety

Auditing practices vary widely between different maintenance organisations and national aviation authorities. While all organisations place emphasis on effective auditing of documentation, some audit fixed facilities and resources, but few attempt to audit how work is actually done. There are no common auditing standards.

Few quality-reporting systems work as they should, particularly in dealing with Human Factors information. Some organisations are only starting to implement a quality discrepancy reporting system, and either it only covers part of their operation or many technicians are not aware of its existence. For some it is not seen to be sufficiently independent of the disciplinary process to be fully trusted. For others there is a large volume of reports which give rise to a backlog and long delays on responding. Part of the problem is reported to be getting managers to take responsibility for dealing with reports when they have other more pressing matters to attend to (McDonald, 1999).

Organisations are not learning from their incidents. Following incidents or accidents, it is critical to future safety that organisations learn from what has happened and implement change to prevent similar incidents occurring. This is particularly true of the human and organisational factors that contribute to incidents. It is hard to find information on cases where learning and change has occurred. The Case Study on 'Organisational learning from incidents' (McDonald, 1999, Appendix 15) demonstrates that, often, despite focused efforts to solve the problem, incidents may have to occur several times before effective change happens. Organisations are rarely systematic in their follow-up to non-technical aspects of incidents, specifically the implementation of recommendations, the monitoring of their effectiveness in addressing the problems they are designed to change and ensuring that knock-on problems are avoided.

Planning and Organisation

For many companies, the traditional functional organisation with an established hierarchical structure and areas of specialisation dictates a top-down process of planning and organisation. For example, the Engineering department oversees the higher order and long term planning and produces the maintenance schedule (MS) for each aircraft. The Planning department receives the MS from the Engineering department and produces, certifies, and dispatches work-packs required to accomplish scheduled maintenance. The scheduling section in the Production area then receives the work-packs from the Planning Department and further breaks the packs down into the daily work. On completion of the checks on the aircraft the Planning department then audit and maintain the work-packs and any other records for the aircraft.

Process-based Planning

Some organisations are trying to move to a more process-based organisation that in effect calls for the break down of traditional departmental barriers in which the overall planning process is cross functional. Thus, while engineering, planning and material departments still exist, the planning process involves the integration of these to oversee the planning of long term and day-to-day maintenance activities. The planning and co-ordination of daily work take place within production control centres located in the hangars. The make up of the production control centres brings together functions previously carried out in planning, materials, and engineering.

Quality of Documentation

In most of the organisations studied, the quality of the documentation available to the maintenance technician (especially through microfilm readers and printers) fell well below basic ergonomic standards. Even where modern CD ROM systems were available they were not often used. Training was rarely provided in their use. On the other hand virtually every technician will admit to using 'black books' – unofficial documentation. This documentation is not available to scrutiny or inspection because of its illegal status.

Compliance with Procedures

Two hundred and eighty-six questionnaires were completed by maintenance engineers after they had completed a task. The questionnaire sought primarily to discover the normative level of deviation from task procedures, as well as inquiring into the reasons behind this non-conformance. Thiry-four percent of respondents reported not following the official procedure for the task. The most common reason given was that there was an easier way than the official method (45 percent) followed by 43 percent saying there was a quicker way. A number of factors, which were related to increased likelihood of non-conformance, were identified. Those individuals who consulted the manual but did not follow the official method were significantly more likely to report that:

1. The task card was unclear
2. The necessary steps to complete the task were unclear
3. To have employed guesswork or trial and error

To report that the maintenance history was desirable but unavailable.

Major Incidents

Increasingly evidence from major incident and accident enquiries, for example BAC111 (AAIB, 1992), Excalibur (AAIB, 1995), and Daventry (AAIB, 1996) is implicating failures at an organisational and regulatory level. Of particular concern are situations where there have been a series of incidents exhibiting similar underlying organisational problems, while the immediate characteristics of the incident might be quite different. For the official investigators and the authorities, it is difficult to know whether the recommendations from investigations have been implemented and, if so, whether they have been effective (Smart, 1997).

Inferences About Structure

What inferences about the normative system of aircraft maintenance can be made from these disparate observations about how it is practised? Several generalisations stand out:

1. For many organisations the top-down nature of their planning systems means that they tend to be relatively unresponsive to the short-term requirements of production. There is little opportunity for feedback and organisational boundaries inhibit effective co-ordination.
2. While virtually everyone admits using unofficial documentation, official documentation is not presented in a way that meets the needs of the user.

3. By common admission, work is routinely not done according to the requirements of the maintenance manual. This is particularly the case when task requirements are not clear.
4. Few organisations systematically attempt to monitor how work is actually carried out, being more concerned to ensure that the documentation is signed off in order.
5. Quality systems do not, in general, provide an effective method of gathering feedback, and, even when they do this, they do not provide a means of ensuring corrective action.
6. There is no transparent system (either within organisations or involving official national agencies) for demonstrating effective response to incidents through implementing recommendations designed to prevent similar incidents happening again.

The Double Standard

Putting this evidence together leads inexorably to the following conclusion:

1. There are two parallel systems of work in operation: an official one and the way in which work is actually being carried out. These two may overlap to some degree; however, there are considerable areas of divergence.
2. There are no currently effective mechanisms for reconciling discrepancies between these two systems, whether in terms of immediate feedback and adjustment, auditing how work is performed, quality reporting, or response to incidents.

This can only be described as a 'double standard'. The 'official standard' for task performance has a strong paper trail from the manufacturer to the maintenance organisation and back, through auditing, to the national authority. The 'actual standard' relies on unofficial documentation and informal work practices.

We should not infer that this system is inherently unsafe. It certainly lacks the transparency that might give an independent observer confidence in the system. Thus, technicians believe that there often are better and quicker ways of doing a task than those they understand to be officially sanctioned by the manual. It is not very difficult to find instances of where this is the case when one examines task performance and documentation in detail. However it is also possible to highlight instances where unofficial methods are pursued in apparently inappropriate circumstances, for example:

> In one case, national accident investigators, investigating an incident involving incorrect dismantling and assembly of a structure, were treated to a demonstration of the same pattern of violations, when observing the task done in the same location by the same team as involved in the incident.

Thus, in so far as unofficial methods may be inherently worse than official methods, which come with the authority of the designers, the system may be inherently unsafe. In so far as such methods may be better than the official methods, the system may be much safer through using the experience and judgement of professional technicians. Either way, it is clear that the system does not allow for effective learning so that the system as a whole can be made both more safe and efficient. We should caution, however, because even if we believe in the effectiveness of unofficial checks and balances in maintaining system safety, this may be true only for a stable system. When such a system is subject to externally produced changes (new technology, change in organisation, or personnel) the implications for safety will be very difficult to predict, and minor, apparently innocuous changes in technology, organisation, or personnel may have profound consequences for safety.

It is precisely this lack of transparency that makes this situation problematic. In order to elucidate the nature and dynamics of this 'double standard' it is useful to consider the beliefs that various agents have about the system. This is the realm of culture. In this context we will focus on beliefs about the role of documentation.

Analysing the Culture of Procedures

In order to understand the relationship between documented procedures and task performance it is useful to consider the function of such documentation as it is understood in relation to different roles in the maintenance process.

For the Manufacturer

Documentation has to be as far as possible legally watertight. This is expressed in terms of its being accurate, up-to-date, comprehensive, and covering sufficient detail to allow an inexperienced technician to complete the task. From this point of view, the necessity to follow task procedures as laid out appears self-evident.

For the Engineering and Planning Departments of the Maintenance Organisation

It is critical to ensure that the key technical requirements of the task are covered in a way that enables those responsible for certifying to sign to the effect that the task has been done according to the required procedures. If documented procedures are followed then safety should ensue.

For the Quality Department

It is essential that all the documentation is present and signed-off indicating conformance with the requirements. The extent to which what is represented in documentation is reconciled with what actually happens in practice varies greatly between organisations. Few have systematic procedures for sampling work as carried out.

For Maintenance Management

It is necessary to reconcile the technical requirements of the maintenance schedule, with the resources available, to satisfy the contract with the airline customer. While most operational managers know that following procedures to the letter is incompatible with achieving on-time performance, there is a strong belief in the importance of following procedures in all essential respects.

For the Certifying Technician

It is important to be satisfied that the work has been done, in all essentials, in conformance with the technical requirements. It is often not precisely clear what this means in practice.

For the Maintenance Technician

Documentation is rarely consulted except as a reminder of particular details (e.g., torque values) or when some procedure is unfamiliar. When help and support is needed to do the job it is normal to seek advice from a colleague, and only when this fails, to consult the manual. It is commonly held that sometimes it is necessary to go beyond what is formally required by the documentation, sometimes there are quicker or easier ways.

'Black Books'

Most technicians have their own unofficial documentation, which contains what they consider the most useful material for task support. The existence of this documentation cannot be officially admitted, as it is illegal.

A Differentiated Culture

If this analysis is correct, then it suggests a differentiated culture within the maintenance system, with particular beliefs being associated with particular roles and positions in the organisation. This idea is currently being empirically tested. It is sufficient to say at this stage that when people from a variety of roles in the maintenance system are asked whether they agree with a variety of questions about professionalism, most find it easy to agree with statements asserting the importance of doing what is necessary to ensure the airworthiness of the aircraft, or to get the job done safely. There is very little agreement about statements which suggest that technicians should blindly follow procedures or exercise their own judgement. If there is a differentiated culture, let us look at what this may mean at three levels: the technician, middle management, and senior or strategic management.

Professionalism

A strong professional culture of technicians is fostered by the apprenticeship system of training. This facilitates the transmission of an understanding of the core values of the job from one generation of technicians to the next. There are a number of features of this professional consciousness:

1. a strong sense of responsibility for the overall safety of the system, which goes beyond simply performing a technical task to a set standard;
2. a belief in professional judgement – that it is the role of the technician to use his or her own judgement, based on experience, knowledge, and skill in carrying out the work, rather than blindly following a set of procedures; and
3. a practical concern to get the job done, but to have it done safely.

This professional culture provides the flexibility to deal with situations, which are not fully anticipated or planned or where the task is not clear, and to make the judgement to do what it takes to get the job done. The maintenance technician has to reconcile the technical requirements, which have to be fulfilled, with the demands of production – getting the job done on time. Sometimes this requires compromise (the double standard). Those who manage this compromise best are highly valued by the company as being people on whom one can rely to get the job done (the 'right stuff'). However the nature of the compromise is never explicit or acknowledged. Occasionally it happens that one of these so-called 'best technicians' is implicated in a serious incident in which the nature of the compromise is laid bare to inspection. This is often a surprise to management. How this works can be illustrated in the following common scenario:

> An incident occurs in which there has been some clear violation of procedures. Management are surprised and disappointed because one of the best, most productive and reliable technicians was involved. If the incident is seriously investigated, it is discovered that there was a shortage of manpower, tools, or equipment at a critical point in the operation and the violation occurred in order to get the job done without additional delay.

At the end of the day it is the role of those at the point of production to do what it takes to get the job done effectively. The professionalism of a technician can easily lead to actions to compensate for deficiencies in the management system, particularly where the planning and organisation of work is not subject to routine validation for its adequacy in practice.

Middle Management

It is more difficult to characterise the culture of management as a homogeneous unity. For one thing the roles of management differ in different departments of the enterprise, as is suggested by the analysis of the 'culture of procedures' above. For another, organisations differ in their history and commercial environment, as well as in terms of their systems and policies. Within the ADAMS projects a number of issues were explored during management interviews that represent different aspects of 'management culture'. These include beliefs about how the organisation manages or should manage issues of discipline, blame following involvement in incidents, preventive safety strategies, openness and secrecy about safety events, accountability, the internal standard of safety, and the understanding of organisational goals. The questionnaire survey, which was administered to all levels of staff in the four companies, included sections on organisational safety climate. This gives an indirect measure of the management culture. The major factor in the safety climate scale, explaining 30 percent of the variance, concerns the perception of the company and its management: their commitment to safety standards and procedures, the importance placed on safe working conditions, funds and resources allocated to safety, and the priority put on safety in the face of other goals. It is suggested that this factor represents the way in which some of the normative beliefs and values of the most visible layer of management are perceived. Of course, these perceptions vary, to some extent, according to the perceiver, depending on their position in the organisation, and in this way they are also directly indicative of organisational sub-cultures.

Strategy and Leadership

When CEOs are asked how safety is prioritised compared to the other goals of aircraft maintenance operations (for example, low cost, on-time performance) the relationship of these goals is expressed differently in different companies. The following represent the views expressed in four different companies:

1. Safety is the most important element of the success of the company as a commercial enterprise.
2. Safety is embedded in the professional culture of the organisation. Safety is at the front of all technical decisions.
3. Commercial pressures create a potential conflict between safety, on the one hand and cost and punctuality on the other. Safety is however a prior requirement before addressing cost and punctuality goals.
4. The commitment to safety has to be seen in the context of the priority to ensure the financial survival of the company.

Thus the priority of safety is understood differently in different companies. These differences are probably due both to differences in the way in which the commercial, regulatory, and public environment is perceived by the CEO as well as internal values and priorities of the company.

Managing Culture – Maintaining Stability

If the above analysis is correct, then it implies that managing organisational change and transforming organisational cultures must be regarded as core issues in enhancing safety and reliability. In order to illustrate some of the difficulties of achieving this, two scenarios will be briefly discussed: a hypothetical example of an organisation's response to a series of incidents and an analysis of the difficulties encountered by some Human Factors interventions.

Response to Incidents

Take the following hypothetical example:

> An aircraft maintenance organisation has had a series of serious incidents over a relatively short period of time. While the immediate circumstances of these incidents are quite diverse, underlying them all is an apparent common pattern. This is exemplified by a lack of strict adherence to procedures (either organisational or maintenance manual procedures) and an habitual propensity to adopt 'go-arounds' (unofficial ways of doing things) where this allows highly pressured staff to accomplish the task with a minimum of inconvenience and disruption. Those directly involved in the incidents have their licences temporarily suspended. Investigations establish the basic facts of the incidents, relying on the co-operation of those directly and indirectly involved. These investigations lead to recommendations for an obligatory half-day retraining session, which emphasises the importance of following procedures. The aviation authorities, notified of these incidents, are concerned with what appear to be systematic failures at the level of the organisation and its management and their ability to ensure a safe system of work. Concern is expressed about the organisation's culture and it is firmly suggested that something be done to 'tweak the culture' in the direction of more active management of this type of problem. Mass meetings of the whole workforce are held in which the chief executive emphasises the fundamental importance of following procedures and stresses the risks to the company and its commercial future if the type of incidents recently experienced is repeated. Deviations from procedures will no longer be tolerated. A management team begin a crash program of rewriting organisational procedures, which have never been reviewed since the maintenance exposition document was first drafted.

Is it likely that this response will prevent such incidents happening in the future?

Can Human Factors Deliver Change?

Experience both in Europe and the United States has shown that Human Factors initiatives are often not always entirely successful. The following are the most common reasons:

1. *Marginalised*: Human Factors programs can become marginalised in a separate department or specific 'champion' who has little influence when decisions are taken. A lack of perceived effectiveness leads to the weakening and ultimate end of the program.
2. *One sided*: Many Human Factors programs have a single focus, often on training for example. When people then return to their previous work environment after training, disillusion occurs if that environment has not changed and the old ways of working are still reinforced.
3. *Focus on diagnosis not change*: Human Factors expertise has well-developed methods for diagnosing what went wrong. Often, there is too little focus on changing the situation to prevent it happening again.
4. *Lack of clear objectives*: Human Factors programs often have objectives that are not easily defined: for example, what is to be achieved by increasing awareness? What does the prevention of error mean? Such programs do not have a clear link between the focus of the intervention (usually people's attitudes or behaviour) and the outcomes that the organisation needs.
5. *No commitment to evaluation*: Few Human Factors interventions are accompanied by systematic evaluation of their effectiveness. Developing an effective Human Factors program involves a significant investment. It is appropriate to measure the effectiveness of this investment.

A key example of an initiative designed to elicit non-technical (Human Factors) information from technicians about the factors influencing their performance of maintenance operations is the MESH system devised by Reason for British Airways Engineering (Reason, 1995). This was a computer-based questionnaire for reviewing maintenance tasks and processes from a Human Factors point of view to anticipate potential sources of error or 'quality lapses'. The idea was that the routine use of this system would lead to the aggregation of data identifying weaknesses in the production system, which could then be addressed. Unfortunately this system never worked beyond a trial implementation period. Perhaps three flaws in the design of MESH were critical in undermining its potential for success:

1. The data it collected was not detailed and specific enough to identify particular parts, components, or processes which needed to be rectified. The level of specificity was that of the ATA chapter.
2. The information it collected about the type of problem encountered and what actually happened was not detailed enough to provide an understanding of the problems and what needed to be done about them. Thus the aggregated data could only identify the general source of the problem.
3. There was no organisational mechanism in place that could take the data, turn them into recommendations, and implement them to create more effective operations.

The result of the third weakness was that the system rapidly lost credibility because no effective action was perceived to have come about as a result of the effort of the technicians in inputting data to the system. Given the first two flaws, it is hard to see how the system could have responded without initiating further studies and investigations about the nature of the problems identified. There did not appear to be the organisational will to do this.

MESH was a highly innovative initiative which sought to shift our understanding of how to deal with error away from reactive approaches, which click into play following the occurrence of an incident, to a proactive approach, which attempts to identify potential error-causing situations in the normal running of the system, thus permitting preventive measures to be designed. What MESH demonstrates is that both the quality of information and the organisational use of that information are fundamental issues in making a system work.

The problems of organisational use of Human Factors information are well illustrated by the experience of reactive safety systems like MEDA and MEI. Maintenance Error Decision Aid (MEDA) is a structured investigation methodology for maintenance human error analysis. It was developed by Boeing, together with several airline companies during a three-year project (1993-95), with the intention of: (a) going beyond the classical approaches, which traditionally imply blame, training and punishment, and (b) providing maintenance organisations with a tailored and structured tool for human error analysis. There are a number of weaknesses in MEDA. Training in the investigation skills necessary to collect information about incidents is very cursory. The narrative description in MEDA is not stressed enough and information about the dynamics of the event is lost. However, looking at the experience of the industry indicates that the critical problems lie outside the system itself. In the US, it is reported that companies who have adopted the MEDA system will only conduct a MEDA investigation when the FAA is already investigating the incident in question. This is because the FAA has a mandate to prosecute infringements and for a company to initiate a MEDA investigation is seen simply to invite such prosecution (Rankin, 1999). Outside the US, the uptake of MEDA by aircraft maintenance organisations has been more extensive, but when one looks for evidence of the impact of the adoption of this technology, it is hard to find evidence that it has been effective (Hubener, 1999).

Maintenance Error Investigation (MEI) is a development of MEDA by British Airways Engineering. The intention was to develop a tool for maintenance error analysis, to be locally used by the maintenance managers and supervisors, for the thorough investigation of more serious incidents. The results of some of these investigations were reported in an exemplary company magazine, entitled 'Human Factors'. Unfortunately, it was hard to find evidence that this commendable exposition of the problems associated with particular incidents was not matched with a comparable initiative to change the organisational and Human Factors that had been identified as contributing to these incidents. The lesson is that, in order to achieve change, it may be necessary to gather evidence of deficiencies in the existing system, but this evidence in itself is not sufficient to ensure such change.

Some Theoretical Issues

Social System – Cultural System

Underlying the analysis outlined above are some theoretical distinctions that are essential to understanding the dynamics of stability and change in organisations. In the first place we have distinguished between the organisational system and the cultural system. This is essentially the distinction that was made by Kroeber and Parsons (1958) differentiating the fields of sociology and anthropology. Parsons, locates the basis of the social system in the norms and values of the community that in turn underlies the systematic expression of roles, obligations, and responsibilities. The cultural system on the other hand represents the meanings that are attached to social phenomena. For Archer (1988) the cultural system represents the universe of meanings that are potentially available to the members of the culture. Logic rather than cause underlies the relationships of the elements of the cultural system – most basically the relationships of consistency and contradiction. Reviewing different approaches to safety culture in organisations, Hale (2000) supports this distinction between safety culture and safety system.

Differentiating the Organisation

Depending upon one's standpoint and focus of interest, organisational cultures can be represented as being unified (homogeneous), differentiated, or fragmented (Martin, 1992). The same is also true of the social system – structural relationships do not always mesh and co-ordinate with each other in a functional unity without the clash of incompatible or ambiguous roles and responsibilities. These become significant when they are associated with different interests of subgroups and the exercise of power. We have suggested that there are in fact two organisational systems in aircraft maintenance – a formal, official one and an informal, unofficial one – and that these overlap imperfectly. Examining, at different levels of the organisation, the differentiation of cultural beliefs, which are held about maintenance tasks and operations, begins to explain the meanings that underlie this structural/functional disjuncture.

System and Process (Social/Cultural Interaction)

A further distinction will enable us to see how different elements of the organisational system and culture influence each other in ways that make for maintaining stability or permitting change. This distinction is between the system (organisational system, cultural system) and the patterns of interactions that are conditioned by, supported by, or constrained by the system (organisational or cultural). Following the distinction between organisational and cultural system we can describe these patterns as social interaction (social/organisational processes including leadership, team-working, investigating incidents, operating the disciplinary system, etc.), and socio-cultural interaction (the ways in which people make sense of the activities). The latter ranges from extended processes involved in learning the culture and becoming 'enculturated' (e.g., Kunda, 1993) to more short term deployment of these meanings in decision-making, assessing situations, making recommendations, holding people accountable, justifying actions according to the 'situational logic', blaming them etc. It is important to emphasise that these are two different aspects of the same process – no fundamental epistemological dualism is being proposed.

Two aspects of this are important here:

1. The system is what is produced by the past. It is the relatively stable set of relationships and beliefs that has been produced by the history (however recent) of the relevant community.
2. While the organisational system constrains what people can do, permitting certain avenues of activity and inhibiting others, it does not determine what they do. People choose how they act and this choice is again influenced by the cultural system of which they are a part – the knowledge, beliefs, and understanding that are potentially available to them.

This can be summarised in a very schematic table as follows:

Table 6.1 Systems and patterns of interaction

System	Patterns of interaction
Organisational system	*Social interaction*
(norms, values, social relationships, interests, power)	Social and organisational processes
	Causal relations
Conditioning influence – not determinant	
Cultural system	*Socio-cultural interaction*
(system of available beliefs)	Organisational sense-making
Logical relations	Enculturation
	'situational logic'

Cycles of Stability and Change

This analysis leads to a third proposition: These patterns of social and socio-cultural interaction can, in turn lead (or not lead) to changes in organisational and cultural systems. This, of course, is the crucial issue from the point of view of policy, regulation, and the management of operations – is it possible to achieve change in the ways in which systems operate and to overcome the barriers which prevent this achievement? Unless we can understand how this is possible, no intervention of policy, application or enforcement of rules and regulation, or management initiative, can be anything other than a shot in the dark.

Briefly, we can outline the elements of patterns of change and stability as follows:

Structural and Cultural Stability

When the organisational system and its associated cultural system are well integrated with no marginal groups with divergent beliefs, cultural forces work to produce a unified population by reproducing a stable corpus of ideas over time.

Structural Change, Cultural Stability

However, when the organisational system becomes more differentiated (as we have seen in this analysis of aircraft maintenance organisations) this is not necessarily accompanied by any productive activity to resolve emerging disfunctionalities in the system. Developing contradictions in the cultural system (the ways in which the system and the situations it produces are understood) are not resolved by generating new understanding but are avoided or suppressed, as illustrated in the above example.

Cultural Change, Structural Stability

Specialisation develops in the ways in which systems are understood, and this understanding becomes more differentiated and less unified throughout the organisation. However the organisational structure does not support processes that would enable the divergences and contradictions between these understandings to be worked out. People increasingly 'go around' the system constraints following the 'situational logic' appropriate to their circumstances. Human Factors programs, as we have seen, often encounter barriers arising from this pattern where the system cannot support the introduction of a new understanding of how the operations should be conducted.

Structural Change, Cultural Change

In this configuration, there is a productive conjunction between the organisational system facilitating organisational processes that can: generate new understandings that can be disseminated throughout the organisation and initiate and carry through change in the system itself. New ideas can be generated either within or outside the organisation and can support structural system change in order to accommodate new commercial, technical, or social realities. This is the model of the 'learning organisation'. This cycle of productive change is outlined in Figure 6.1 below. This indicates a time sequence from T1, representing the social and cultural system at the beginning of a period of analysis, through T2 and T3, representing ongoing organisational processes and organisational sense-making (Weick, 2000), leading to T4, the end of the period of analysis. Here, the outcome can be identified in terms of a change in the organisational and cultural systems. It is unlikely that the two cycles would manifest precise simultaneity. There is always tension and disjunction changes in system and practice and how these are understood.

Preconditions for Change

If this analysis has provided some insight into the factors that have inhibited effective change, what does it suggest about the preconditions for achieving change towards a better functioning system, a more integrated culture? The following brief points are informed both by our own experience and Pettigrew's (1985) comprehensive history of Organisational Development initiatives in ICI.

The most obvious precondition is the recognition that there is a problem or situation that requires change. This paper has been about fairly fundamental problems of the organisation. However when there is no effective organisational process for dealing with problems, even relatively simple problems can require considerable organisational effort to achieve a solution. This has been our experience in piloting a continuous improvement project involving a maintenance organisation and a manufacturer. Thus even simple problems can require fundamental change. However, in order to get the momentum behind change, it is very helpful to have a significant business crisis on the not too distant horizon. The problem, or situation should, of course, be seen to be relevant to this crisis. The role of the business environment and significant stakeholders may be quite important aspects of this crisis. Such stakeholders can include the regulatory authorities, and public opinion; the confidence of shareholders may also be important.

Culture and Understanding

Recognition that there is a problem is not enough. There needs to be an understanding of the problem, amongst the leadership of the organisation, and of possible ways in which it might be solved. The existence of models of successful change, best practice examples, and an accessible analytic framework can be instrumental in facilitating this. Do not underestimate the cultural barrier that may need to be overcome. Schoenberger (1997) has demonstrated how failings in the strategic management culture can compromise an entire industrial sector for a period of decades or more.

Competence is also essential in the appropriate depth and breadth throughout the organisation, which can enable the change processes to function effectively, and which can mediate the development of a new common understanding of the current situation and the emerging solution.

Organisational Structure

Change cannot happen without an organisational system that can support the change process and enable it to be managed in a coherent way. In many organisations, most of the people have not participated in any significant problem solving or change process. This may require the removal of institutional barriers to the change process and the creation of a framework of policy and procedure that is favourable to achieving the change objectives.

The Politics of Change

Achieving the change objectives will require the successful deployment of organisational power and politics, and occasionally some luck. However if the change is to be consolidated in a permanent development of the organisational system and culture, then it is important to pay attention to processes of stabilisation and embedding. Demonstrable success and effective participation may be amongst the most effective mechanisms for achieving this.

Acknowledgement

The work reported here was undertaken as part of the ADAMS project - Human Factors In Aircraft Dispatch and Maintenance Safety – which was undertaken within the European Commission program for Industrial Materials and Technology (Brite EuRam III), with a financial contribution from the European Commission Directorate General for Science, Research and Development.

References

Air Accident Investigation Branch (AAIB). (1992). Report on the incident to BAC 1-11, G-BJRT over Didcot, Oxfordshire on 10 June, 1990. London: HMSO.

Air Accident Investigation Branch (AAIB). (1995). Report on the incident to Airbus A320-212, G-KMAM at London Gatwick Airport on 26 August, 1993 (No. 2/95). London: HMSO.

Air Accident Investigation Branch (AAIB). (1996). Report on the incident to Boeing 737-400,G-OBMM, near Daventry, on 23 February, 1995. London: HMSO.

Archer, M. (1988). *Culture and Agency.* Cambridge: Cambridge UP.

Hale, A. R. (2000). Culture's confusions. *Safety Science, 34*, 1-14.

Hubener, S. (1999, April). Paper presented at the IQPC conference on Human Factors in Aircraft Maintenance. Amsterdam.

Kroeber, A. L., & Parsons, T. (1958): The concepts of culture and of social system. *American Sociological Review, 23,* 582-583.

Kunda, G. (1993). *Engineering Culture*. Temple University Press.

Martin, J. (1992). *Cultures in Organisations: Three Perspectives.* Oxford: Oxford University Press.

McDonald, N. (Ed.). (1999). Human Centred Management for Aircraft Maintenance (ADAMS project report). Department of Psychology, Trinity College Dublin.

Mc Donald, N., Corrigan, S., Daly, C., & Cromie, S. (2000). Safety management systems and safety culture in aircraft maintenance organisations. *Safety Science, 34*, 151-176.

Pettigrew, A. (1985). *The Awakening Giant*. Oxford: Blackwell.

Pidgeon, N., & O'Leary, M. (1994). Organisational safety culture: Implications for aviation practice. In N. Johnston, R. Fuller, & N. McDonald (Eds.), *Aviation Psychology in Practice* (pp. 21-43). Aldershot: Avebury Technical.

Rankin, W. (1999, April). Paper presented at the IQPC conference on Human Factors in Aircraft Maintenance. Amsterdam.

Reason, J. (1995). *Comprehensive Error Management in Aircraft Engineering*. London, Heathrow: British Airways Engineering.

Schoenberger, E. (1997). *The Cultural Crisis of the Firm.* Oxford: Blackwell.

Smart, K. (1997). Paper presented at the European Transport Safety Council conference on Accident Investigation. Brussels: ETSC.

Weick, K. (2000). *Making Sense of the Organisation.* Oxford: Blackwell.

Chapter 7

Barrier Analysis and Accident Prevention

Erik Hollnagel
Department of Computer and Information Science,
University of Linköping, 58183 Linköping, Sweden

Introduction

Accident – and incident – analyses are usually aimed at finding the set of causes that is necessary and sufficient to explain what happened. In many cases the determination of the cause, however, reflects the interests of the stakeholders as much as what actually happened. As Perrow (1986) noted:

> Formal accident investigations usually start with an assumption that the operator must have failed, and if this attribution can be made, that is the end of serious inquiry. Finding that faulty designs were responsible would entail enormous shutdown and retrofitting costs; finding that management was responsible would threaten those in charge, but finding that operators were responsible preserves the system, with some soporific injunctions about better training. (p. 146)

Even if the investigation into what caused the accident is more open-minded than Perrow assumes, focussing on finding the cause easily detracts interest from the other conditions that contributed to the accident. A proper accident analysis should not only look for possible specific causes, but also for the general system conditions at the time of the accident, specifically the barriers that may have failed. An understanding of the nature of barriers and defences, and a method for analysing and classifying their functions and failures, is important as a basis for being able effectively to prevent future accidents.

The Efficiency of Barriers

Large accidents invariably represent an unlikely combination of many individual factors or causes. The recurrence of an accident, or more likely of a similar accident, is therefore more effectively achieved by improving the barriers than by removing the causes. The reasoning behind this is simple. Since serious accidents are due to coincidences among multiple factors and conditions, there are no simple 'root causes'. Indeed, there may be a number of equivalent conditions that can turn an innocent event or a near miss into an accident. Removing just one or a few of these conditions – and moreover the ones that attracted most attention during the analysis – will therefore not prevent a recurrence. The solution to eliminate causes is only appropriate if it can be assumed that a sequential description of the accident is valid, e.g., a domino type of model described below.

Unlike the elimination of causes, barriers are effective because they can protect against a specific type of effect regardless of why it came about. To take a very simple example, an umbrella is effective against water in the air (precipitation) regardless of whether the source is rain, sleet, a fountain, a waterfall, a garden hose, etc. Similarly, a sprinkler system is effective against a fire regardless of the origin of the fire, and a parachute can be used to save lives regardless of the reason for needing to escape from an aircraft. Accident analyses should therefore not only look for causes but also try to find barriers that have failed or barriers that were missing, and in both cases go further to determine why they failed or were missing. As a response, introducing new or improved barriers is a more effective way of preventing a type of accident from occurring again than eliminating causes – provided, of course, that the barriers are effective and that they do not adversely affect the accomplishment of the task.

Accident models

Thinking about accidents involves a number of accident models, which are stereotypical ways of explaining how accidents occur. Although there are many individual models, they seem to fall into the three types summarised in Table 7.1. The simplest types of accident models describe the accident as the result of a sequence of events that occur in a specific order. The description of the sequence may either represent the scenario as a whole, or only the events that went wrong. The first type is illustrated by the so-called domino model (Heinrich, 1931), which depicts the accident as a set of dominos that tumble because of a unique initiating event. In this model the dominos that fall represent the action failures, while the dominos that remain standing represent the normal events. The outcome is a necessary consequence of one specific event, and it can therefore be considered a deterministic model. Another example is the Accident Evolution and Barrier model (Svenson, 1991), which only describes the sequence of events – or rather barriers – that failed. This puts the focus on what went wrong, but leaves out additional information that is potentially important. More generally, sequential models represent the event as a series of individual steps organised in their order of occurrence. Sequential models need not be limited to a single sequence of events but may be represented in the form of hierarchies such as the traditional event tree and networks such as Critical Path models (Program Evaluation and Review Technique or PERT) or Petri networks. A typical example is the generic representation known as the 'anatomy of an accident' (Green, 1988).

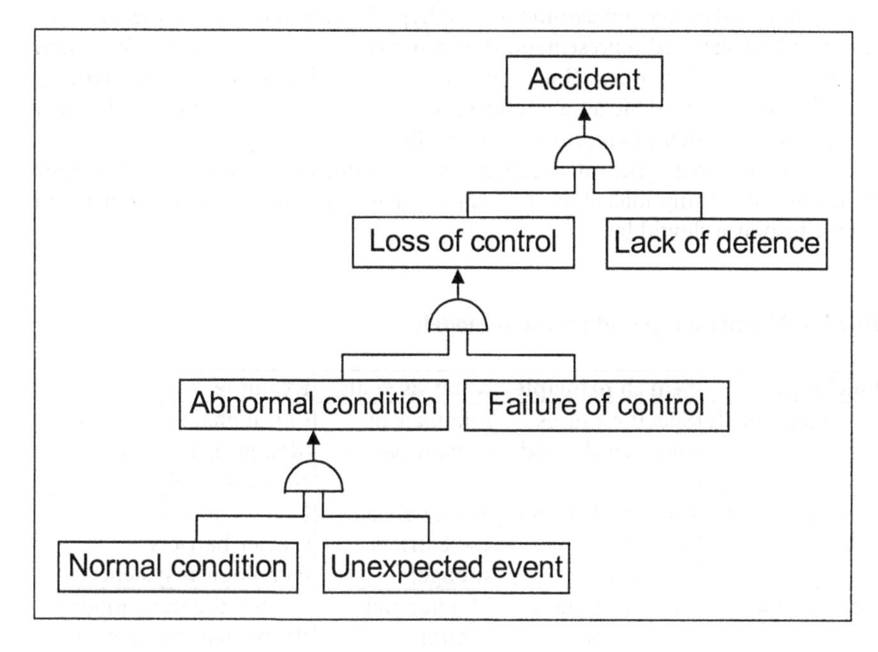

Figure 7.1 Anatomy of an accident

Sequential models are attractive because they are easy to understand and easy to represent graphically, but suffer from being oversimplified. In many cases epidemiological models provide a better solution. As the name implies, these models describe an accident in analogy with a disease, i.e., as the outcome of a combination of factors, some manifest and some latent, that happen to exist together in space and time. The classical example of that is the description of latent conditions proposed by Reason (1990). Other examples are models that consider barriers and carriers, models of 'sharp end'-'blunt end' interactions, and models of pathological system (organisation) states. Epidemiological models are valuable because they, at least, provide a basis for discussing the complexity of accidents that overcome the limitations of sequential models. Unfortunately, epidemiological models are rarely stronger than the analogy, i.e., they are difficult to specify in further detail, even though the concept of pathogens allows for a set of methods that can be used to characterise the general 'health' of a system (Reason, 1997).

A third type of model is the so-called systemic model. As the name denotes, these models endeavour to describe the characteristic performance on the level of the system as a whole, rather than on the level of specific cause-effect 'mechanisms'. Systemic models are found in control theory, in the analogical use of the Brownian movement, in chaos models, and in coincidence models. A token for the latter type is the Swiss cheese analogy (Reason, 1997), although this is not a model in the usual meaning of the term. In general, systemic models emphasise the need to base accident analysis on an understanding of the functional characteristics of the system, rather than on assumptions or hypotheses about internal mechanisms as provided by standard representations of process genotypes, such as information processing or decision trees. Since systemic models deliberately try to avoid a description of an accident as a sequence or ordered relation among individual events, they are difficult to represent graphically.

Each of the three types of accident models summarised in Table 7.1 carries with it a set of assumptions about how an accident analysis should take place and what the response should be.

Table 7.1 The main types of accident models

Model type	Search principle	Analysis goals	Example
Sequential models	Specific causes and well-defined links	Eliminate or contain causes	Linear chain of events (domino), Tree models Network models
Epidemiological models	Carriers, barriers, and latent conditions	Make defences and barriers stronger	Latent conditions Carrier-barriers Pathological systems
Systemic models	Tight couplings and complex interactions	Monitor and control performance variability	Control theoretic models 'Brownian' movement models, Coincidence models

1. For the sequential models the accident analysis becomes a search for recognisable, specific causes and well-defined cause-effect links. The underlying assumption is that causes, once they have been found, can be eliminated or encapsulated and that this will effectively prevent future accidents.
2. In the case of the epidemiological models, the accident analysis becomes a search for 'carriers' and latent conditions, as well as for reliable indications of general system 'health'. More generally, the search is for specific performance deviations, with the recognition that these can be complex phenomena themselves. In the case of the epidemiological models, the underlying assumption is that defences and barriers can be strengthened to prevent future accidents from taking place.
3. Finally, for the systemic models the analysis becomes a search for unusual dependencies and common conditions that turn into coincidences. This reflects the belief that the essential variability of a system can be detected and controlled. The variability is, however, not by itself seen as something negative that must be avoided at any cost. Quite to the contrary, it is acknowledged that performance variability is necessary for a system to learn, and that some variability indeed should be encouraged.

It is clearly not possible to analyse an accident without having some kind of underlying model. The quest should therefore not be for a 'model-free' analysis, but rather for an approach where the model is openly acknowledged so that the constraints can be made as light as possible. At the same time the model should be detailed enough to support an explicit and consistent method for analysis, as well as being a basis for recommending appropriate responses.

The distinction between the three classes of accident models proposed above does not imply that only systemic accident models should be used. Although it would clearly be inadvisable to rely exclusively on a sequential accident model, it should not be discarded outright. In some accidents there may clearly be easily distinguishable causes, and in such cases it makes sense to try to eliminate them. The same goes for epidemiological accident models. The three classes of accident models should therefore be used to complement each other, so that all facets of an accident are covered.

The Relativity of Causes

One of the fundamentals of Western thinking is the causality principle, which permeates both moral philosophy and scientific thinking. In the latter domain the paradigmatic example is Newton's laws. Since we generally 'know' that every cause has an effect, we automatically assume that every effect also has a cause, and furthermore that this cause can be found by deductive reasoning. According to this way of thinking an accident constitutes an effect, and it must therefore be possible to find the preceding cause – or set of causes. The common accident models as discussed above, especially the sequential models, reinforce this assumption.

A cause is, however, not an absolute condition or state of the system that is waiting to be discovered and which therefore can be determined with a high degree of certainty. As suggested by Perrow (1986), the search for a cause is often far from objective. Even if it is acceptably objective, there are always practical constraints of e.g., resources or time that limit the search. Every analysis must stop at some time, and interests that are quite remote from the scientific purpose of the accident investigation often determine the criterion. As Woods, Johnnesen, Cook, and Sater (1994) have pointed out, a cause is always the result of a judgement made in hindsight, and therefore benefits from the common malaise of *besserwissen*. More precisely, a cause can be defined as the *ex post facto* identification of a limited set of aspects of the situation that are seen as the necessary and sufficient conditions for the effect(s) to have occurred. A 'cause' usually has the following characteristics:

1. It can unequivocally be associated with a system structure or a function (people, components, procedures, etc.).
2. It is possible to do something to reduce or eliminate the cause within accepted limits of cost and time. This follows partly from the first characteristic, or rather, the first characteristic is a necessary condition for the second.

3. The cause conforms to the current 'norms' for explanations, as encapsulated by the theories that are part of the common lore. For instance, before the 1960s it was uncommon to use 'human error' as a cause, while it practically became *de rigueur* during the 1970s and 1980s. Later on, in the 1990s, the notion of organisational accidents became accepted, and the norm for explanations changed once more.

The reason for making these possibly obvious points is to emphasise that the determination of the 'cause' is a relative rather than an absolute process, and that it represents pragmatic rather than scientific reasoning. The outcome of an accident analysis should be treated with care, especially when it comes to thinking about responses to the accident. Knowing how systems have failed in the past is essential for predicting how they may fail in the future: design encapsulates experience. There is, however, an unfortunate disparity between accident analysis and accident prediction. In the case of analysis, it is by now commonly accepted that models should be of the epidemiological or systemic rather than of the sequential type. Yet in the case of design and prediction most models are of the sequential type. This can easily be demonstrated by referring to such widespread models as the event tree used in Probabilistic Safety Assessment, the family of fault tree or cause-consequence models and the failure mode and effect analysis methods. Since this disparity severely hampers our ability to anticipate failures that may occur, there is a need to develop accident models that are able to capture the complexity of coincidences and which can be applied to prediction.

What Are 'Errors'?

Causes are usually associated with actions, either directly or through a series of intermediate antecedent-consequent steps (Hollnagel, 1998). The notion of an action gone wrong or an 'error' has been widespread, but as several people have pointed out it is a potentially misleading oversimplification because it implies that an action can be either correct or incorrect (Hollnagel, 1993; Senders & Moray, 1991; Woods et al., 1994). The absolute binary distinction is both theoretically suspect and practically useless. The latter becomes obvious when considering the fact that actions can fall at least into the following categories (cf. Amalberti, 1996):

1. Actions where the actual outcome matches the intended outcome. Such actions are regarded as correctly performed actions, even though the outcome may have come about in other ways.
2. Incorrectly performed actions where the failure is detected and corrected. This can happen while the action is being carried out, e.g., mistakes in typing or data entry, or immediately after, as long as the system makes a recovery possible. In these cases the actual and intended outcomes may still match, and the action is therefore often considered as correct.

3. Incorrectly performed actions where the failure is detected but not corrected or recovered. Recovery can be impossible for several reasons, for instance that the system has entered an irreversible state, that there is insufficient time or resources, etc. In these cases the actual and intended outcomes do not match, and the action may therefore be characterised as an error.
4. Incorrectly performed actions where the failure is detected but ignored. This may happen because the expected consequences of the failure are seen as unimportant in an absolute or relative sense. This assessment may be correct or incorrect, depending, among other things, on the users' knowledge of the system in question. If it turns out that the consequences were not negligible, the action may in retrospect be classified as an error.
5. Incorrectly performed actions, which are not detected at the time, and therefore not recovered. These will as a rule lead to unwanted consequences, and hence can be classified as errors.

This description of five categories of action is clearly preferable to the two-way distinction between correct actions and errors. Furthermore, it is not necessarily bad if an action is incorrectly performed. So long as the outcome does not lead to a serious and irreversible condition, the incorrectly performed action provides an important opportunity to learn. Learning cannot take place if everything is done correctly, and if there are no unexpected outcomes (of course, one could argue that if everything is done correctly, then there is no need to learn either).

The extended classification of actions is consistent with the position that it is the outcome rather than the action in itself that is incorrect (indeed, the verdict of an incorrectly performed action is clearly a relative rather than an absolute judgement). It must be acknowledged that human performance individually and in groups, as well as the performance of technological artefacts, always is variable. Sometimes the variability remains within acceptable limits, but at other times it becomes so large that it leads to unexpected and unwanted consequences. In both cases, however, the basis for the performance variability is the same, and that which makes us classify one action as an 'error' and the other as not are the outcomes. It follows from this view that rather than trying to identify specific causes and eliminating them, we should try to detect the performance variability in order to control it, either by reducing it at the source or by protecting against the outcomes. In both cases managing the performance variability becomes more important than searching for and eradicating errors.

Barrier Systems and Barrier Functions

In relation to accidents, a barrier is an obstacle, an obstruction, or a hindrance that may either (a) prevent an action from being carried out or an event from taking place, or (b) prevent or lessen the impact of the consequences, for instance by slowing down uncontrolled releases of matter and energy.

In the practical work with barriers it is useful to make a distinction between barrier functions and barrier systems. A barrier function can be defined as the specific manner by which the barrier system achieves its purpose. Similarly, a barrier system can be defined as the basis for the barrier function, i.e., the characteristics of a system without which the barrier function could not be accomplished. It is possible from this basis to develop a systematic description of various types of barrier systems and barrier functions, which can be used as a starting point for practical methods (Hollnagel, 1999).

Despite the importance of the barrier concept, the accident literature only contains a small number of studies (Leveson, 1995; Svenson, 1991, 1997; Taylor, 1998; Trost & Nertney, 1985). Other uses have been in relation to the notion of defences, for instance by Reason (1997) and in the Japanese approach to accident prevention named *hiyari-hatto* (Kawano, 1999). The classifications proposed by these studies have been quite diverse, partly because of the lack of a common conceptual background, and partly because they have been developed for specific purposes within quite diverse fields. The most explicit attempt of developing a theory of barriers has been the work of Svenson (1991), which also was the basis for the field studies of Kecklund, Edland, Wedin, and Svenson (1996).

Types of Barrier Systems

An analytical description of barrier systems can be based on different concepts, such as their origin (e.g., whether they are created by organisations or individuals), their purpose, their location or focus (relative to e.g., the source or target), and their nature. Of these only the concept of the nature of the barrier system is rich enough to support an extensive classification. The nature of a barrier system is furthermore independent of its origin, its purpose (e.g., as preventive or protective), and its location. Although there initially may seem to be many different types of barrier systems, ranging from physical hindrances (walls, cages) to ethereal rules and laws, experience shows that the four categories shown in Table 7.2 are sufficient.

1. Material barrier systems physically prevent an action from being carried out or the consequences from spreading. Examples are buildings, walls, fences, railings, bars, cages, gates, etc. A material barrier system presents an actual physical hindrance for the action or event in question and although it may not prevent it under all circumstances, it will at least slow it down or delay it. Furthermore, a material barrier system does not have to be perceived or interpreted by the acting agent in order to serve its purpose. A wall will prevent movement of an agent (or a substance) from one location to another even if the agent cannot see the wall – provided, of course, that it is strong enough.
2. Functional (active or dynamic) barrier systems work by impeding the action to be carried out, for instance by establishing logical or temporal interlocks. A functional barrier system effectively sets up one or more pre-conditions that must be met before something can happen. These pre-conditions need not be

interpreted by a human, but may be interrogated or sensed by technological artefacts. Functional barrier systems may not always be visible or discernible, although their presence often is indicated to human users in one way or another and may require one or more actions to be overcome.

3. Symbolic barrier systems require an act of interpretation to achieve their purpose, hence an 'intelligent' agent that can react or respond to the barrier system. Whereas a functional barrier system works by establishing a pre-condition that must be met by the acting agent or user before further actions can be carried out, limitations or constraints indicated by a symbolic barrier system may be disregarded or neglected. For instance, the railing along a road constitutes a material and a symbolic barrier system at the same time, while reflective posts or markers are only a symbolic barrier system. The reflective markers indicate where the edge of the road is but are by themselves insufficient to prevent a car from going off the road. Although all kinds of signs and signals are symbolic barriers systems, visual and auditory signals play a special role in normal work environments as part of warnings (texts, symbols, sounds), interface layout, information presented on the interface, visual demarcations, etc.
4. Immaterial (or nonmaterial) barrier systems are not physically present or represented in the situation, but depend on the knowledge of the user to achieve their purpose. Immaterial barrier systems usually also have a physical representation, such as a book or a memorandum, but this is normally not present when their use is mandated. Typical immaterial barrier systems are rules, guidelines, restrictions, and laws.

The classification of barriers is not always a simple matter. A wall is, of course, an example of a physical barrier system and a law is an example of an immaterial barrier system. But what about something like a procedure? A procedure by itself is an instruction for how to do something, hence not primarily a barrier (except in the sense that performing the right actions rules out performing the incorrect ones). Procedures may, however, include warnings and cautions, as well as conditional actions (pre-conditions). Although the procedure may exist as a physical document, other formats are also possible, such as computerised procedures. The procedure therefore works by virtue of its contents or meaning rather than by virtue of its physical characteristics. The warnings, cautions, and conditions of a procedure are therefore classified as examples of a symbolic barrier system, i.e., they require an act of interpretation in order to work.

Immaterial barriers are often complemented by symbolic barriers. For instance, general speed limits as given by the traffic laws are supplemented by road signs (a symbolic barrier system) and at times enforced by traffic police (performing the immaterial monitoring function, perhaps supplemented by physical barriers such as road blocks or speed bumps). Material barriers may also be complemented by symbolic barriers that encourage their use. Seat belts are material barriers, but can only serve their purpose if they are actually used. In commercial aeroplanes, the use of the seat belt is supported by both static cautions (text, icons) and dynamic signals (seat belt sign), as well as verbal instructions, demonstrations, and visual

Table 7.2 Barrier systems and barrier functions

Barrier system	Barrier function	Example
Material, physical	*Containing* or protecting. Physical obstacle, either to prevent transporting something from the present location (e.g., release) or into the present location (penetration).	Walls, doors, buildings, restricted physical access, railings, fences, filters, containers, tanks, valves, rectifiers, etc.
	Restraining or preventing movement or transportation.	Safety belts, harnesses, fences, cages, restricted physical movements, spatial distance (gulfs, gaps), etc.
	Keeping together: Cohesion, resilience, indestructibility.	Components that do not break or fracture easily, e.g., safety glass.
	Dissipating energy, protecting, quenching, extinguishing.	Air bags, crumble zones, sprinklers, scrubbers, filters, etc.
Functional	*Preventing* movement or action (*mechanical, hard*).	Locks, equipment alignment, physical interlocking, equipment match, etc.
	Preventing movement or action (*logical, soft*).	Passwords, entry codes, action sequences, pre-conditions, physiological matching (iris, fingerprint, alcohol level), etc.
	Hindering or impeding actions (spatio-temporal).	Distance (too far for a single person to reach), persistence (dead-man-button), delays, synchronisation, etc.
Symbolic	*Countering*, preventing, or thwarting actions (visual, tactile interface design).	Coding of functions (colour, shape, spatial layout), demarcations, labels & warnings (static), etc. *Facilitating correct actions may be as effective as countering incorrect actions.*
	Regulating actions.	Instructions, procedures, precautions / conditions, dialogues, etc.
	Indicating system status or condition (signs, signals and symbols).	Signs (e.g., traffic signs), signals (visual, auditory), warnings, alarms, etc.

Table 7.2 Continued

Barrier system	Barrier function	Example
	Permission or authorisation (or the lack thereof).	Work permit, work order.
	Communication, interpersonal dependency.	Clearance, approval, (on-line or off-line), in the sense that the lack of clearance etc., is a barrier.
Immaterial	*Monitoring*, supervision.	Check (by oneself or another a.k.a. visual inspection), checklists, alarms (dynamic), etc.
	Prescribing: rules, laws, guidelines, prohibitions.	Rules, restrictions, laws (all either conditional or unconditional), ethics, etc.

inspection. In private cars the material barrier is normally only supported by the immaterial barrier, i.e., the traffic laws, although some models of cars also have a warning signal. On the whole, the result is less than satisfactory, especially since the use of a safety belt seems to be influenced by cultural norms as well.

The four types of barrier systems may, at a first glance, seem to be incomplete since common types such as organisational barriers and technical barriers are missing. This lack is, however, only an apparent one since a combination of barrier systems and barrier functions will suffice to account for every type of barrier. An organisational barrier such as a rule or guideline should therefore be considered as an example of a symbolic barrier system. The proposed definitions also mean that more than one barrier system may be present in the same physical artefact or object. For instance, a door may have on it a written warning and include a lock that requires a key to be opened. Here the door is a material barrier system, the written warning is a symbolic barrier system, and the lock requiring a key is a functional barrier system. It is probably the rule rather than the exception that more than one barrier system is used at the same time, at least for the first three categories.

Types of Barrier Functions

Whereas it was possible to make do with only four different barrier systems, Table 7.2 shows that there are several more barrier functions. As a start, barrier functions can either prevent an accident from taking place or protect against the consequences. The overall functions of prevention and protection can be further specialised, depending on the domain and on the type of barrier system. All four types of barrier system, for instance, can accomplish prevention – although not with the same degree of efficiency. Protection, on the other hand, cannot be provided by symbolic or immaterial barrier systems.

In the development of an accident, prevention refers to what may be done before the accident occurs while protection refers to what may be done after. Considering the following four stages can refine this distinction:

1. Steady-state performance. Here the main concern is to monitor how the system performs. While monitoring does not constitute a class of barrier function as such, effective monitoring can prevent accidents. Monitoring may involve functions such as observation, confirmation, managing, and recording.
2. Pre-accident build-up. At this stage the main objective is to detect variations or deviations in system performance. This is a genuine part of accident prevention, and many system design features (technological and organisational) relate to that. Detection may involve functions such as identification, verification, and questioning or probing, and typically relies heavily on functional barrier systems.
3. When the accident happens, the first stage of protection is to reduce or deflect the immediate consequences. Some characteristic functions here are attenuating, partitioning, and reducing the direct effects, as well as strengthening defences and resources.
4. The post-accident or recovery period constitutes the second stage of protection. This covers a longer time period and can be seen as a way of correcting what went wrong, involving replacement, modification and improvement of both barrier systems and barrier functions.

It is possible to develop this classification further, and thereby provide the basis for a comprehensive approach to barrier analysis and accident prevention. The principles for this are outlined in Table 7.3 but space prohibits a full treatment here. The high-level barrier functions of monitoring – detecting – deflecting – correcting, describes the characteristics stages in the development of an accident. Each of the high-level barrier functions can be implemented by one or more barrier systems; the choice of the most effective solution depends on the requirements from the stakeholders as well as the prerequisites for the individual functions. A correction that is implemented by an immaterial barrier system, such as a new rule, can be put in place very quickly and inexpensively. On the other hand, immaterial barriers are rarely effective in the long run, since they require a high degree of compliance by those involved.

Table 7.3 Barrier functions and accident prevention

	Barrier system			
Barrier function (high level)	**Material**	**Functional**	**Symbolic**	**Immaterial**
Monitoring	Monitoring involves an act of reasoning, hence a function. It cannot be provided by a material barrier system .	Monitoring can be provided by a functional barrier system, which checks (logical) conditions.	Monitoring can be provided by a symbolic barrier system, which interprets signs and indicators.	Monitoring involves an act of reasoning, hence a function. It cannot be provided by an immaterial barrier system .
Detecting	Detection cannot be provided by a material barrier system, since it involves an act of reasoning, hence a function.	Detection is usually functional, and can be implemented by technology with or without human collaboration.	Detection by humans (unaided or aided by machines) can be provided by symbols and interpretation.	Detection cannot be provided by an immaterial barrier system, since it requires an act of identification / interpretation.

Table 7.3 Continued

	Barrier system			
Barrier function (high level)	**Material**	**Functional**	**Symbolic**	**Immaterial**
Deflecting or reducing	Deflection is usually provided by a material barrier system, such as a firewall, a fire belt, etc.	Deflection can be provided by a functional barrier system, such as interlocks, airbags or sprinklers.	Deflection cannot be provided by a symbolic barrier system, since it means changes in the direction of matter and energy.	Deflection cannot be provided by an immaterial barrier system, since it means changes in the direction of matter and energy.
Correcting	Correction can be provided by a material barrier system, such as restoring.	Correction can be provided by a functional barrier system, such as developing new interlock.	Correction can be provided by a symbolic barrier system, such as developing new signs and symbols.	Correction can be provided by an immaterial barrier system, such as instituting new laws.

From 'Error' Management To Performance Variability

The argument of this chapter is that it may be more efficient to prevent accidents through the judicious use of barrier systems and barrier functions than to identify and eliminate specific causes. This corresponds to the view that causes represent an *ex post facto* attribution, based on an oversimplified understanding of the nature of accidents. Given that accidents more often are due to complex coincidences than well-defined cause-effect relations, it makes sense to approach accident prevention by managing the variability of the system. This in turn can be accomplished by considering the various characteristic stages of an accident, as described above, and by applying the barrier functions that are most effective at each stage. Many of the functions necessary for managing system performance variability are already in place in both the technological and organisational area. They have, however, usually been implemented in a piecemeal fashion and without the benefit of an overall perspective. The further application of performance variability management is therefore more a question of using existing principles in an integrated fashion than of inventing completely new ways of doing things. Efforts in this direction are currently underway in a number of domains, ranging from industrial manufacturing and production to traffic management.

References

Amalberti, R. (1996). *La Conduite des Systèmes à Risques*. Paris: PUF.

Green, A. E. (1988). Human factors in industrial risk assessment - Some early work. In L. P. Goodstein, H. B. Andersen, & S. E. Olsen (Eds.), *Task, Errors, and Mental Models*. London: Taylor & Francis.

Heinrich, H. (1931). *Industrial Accident Prevention*. New York: McGraw-Hill.

Hollnagel, E. (1993). *Human Reliability Analysis: Context and Control*. London: Academic Press.

Hollnagel, E. (1998). *Cognitive Reliability and Error Analysis Method – CREAM*. Oxford: Elsevier Science.

Hollnagel, E. (1999). Accidents and barriers. In J.-M. Hoc, P. Millot, E. Hollnagel, & P. C. Cacciabue (Eds.), Proceedings of *Lez Valenciennes: Vol. 28* (pp. 175-182). Presses Universitaires de Valenciennes.

Kawano, R. (1999, July). *Studies on human-machine interface for safer operations in nuclear power plants*. Proceedings of International Workshop on Human Factors in Space, Tokyo.

Kecklund, L. J., Edland, A., Wedin, P., & Svenson, O. (1996). Safety barrier function analysis in a process industry: A nuclear power application. *Industrial Ergonomics, 17*, 275-284.

Leveson, N. (1995). *Safeware: System Safety and Computers*. Reading, MA: Addison-Wesley Publishing Company.

Perrow, C. (1986), *Complex Organizations: A Critical Essay* (3rd ed.). New York: Random House.

Reason, J. (1990). The contribution of latent human failures to the break down of complex systems. *Philosophical Transactions of the Royal Society* (Series B. 327: 475-484). London.

Reason, J. T. (1997). *Managing the Risks of Organizational Accidents.* Aldershot, UK: Ashgate.

Senders, J. W., & Moray, N. P. (1991). *Human Error: Cause, Prediction, and Reduction.* Hillsdale, NJ: Lawrence Erlbaum.

Svenson, O. (1991). The accident evolution and barrier function (AEB) model applied to incident analysis in the processing industries. *Risk Analysis*, *11*(3), 499-507.

Svenson, O. (1997). Safety barrier function analysis for evaluation of new systems in a process industry: How can expert judgment be used? *Proceedings of Society for Risk Analysis Europe Conference*, Stockholm, June 15-18, 1997.

Taylor, R. J. (1988). *Analysemetoder til vurdering af våbensikkerhed.* Glumsø, DK: Institute for Technical Systems Analysis.

Trost, W. A., & Nertney, R. J. (1985). *Barrier Analysis* (DOE 76-45/29). Idaho Falls, Idaho: EG&G Idaho, Inc.

Woods, D. D., Johannesen, L. J., Cook, R. I., & Sarter, N. B. (1994). *Behind Human Error: Cognitive Systems, Computers, and Hindsight.* Columbus, Ohio: CSERIAC.

Chapter 8

Cognitive Work Analysis across the System Life-Cycle: Achievements, Challenges and Prospects in Aviation

Penelope M. Sanderson
Swinburne Computer Human Interaction Laboratory, Melbourne, Australia

Introduction

Although technical systems such as aircraft are becoming increasingly automated and their hardware components increasingly reliable, human operators are retained to ensure smooth operation in the face of "normal" disturbances and to handle unanticipated situations. Many Human Factors techniques indicate how to provide support for human monitoring and intervention under normal disturbances, but few techniques indicate how to support the human operator when a system such as a flight management system encounters extraordinary conditions that have not been anticipated during design.

Cognitive Work Analysis (CWA) is an approach to analysing, modelling, designing, and evaluating complex systems. Proponents of CWA claim that it leads to designs that are particularly useful when people have to adapt to unanticipated situations (Rasmussen, Pejtersen, & Goodstein, 1994; Vicente, 1999). CWA does not focus on how human-system interaction should proceed (normative modelling) or how human-system interaction currently works (descriptive modelling). Instead, it focuses on identifying properties of the work environment and of the workers themselves that define possible boundaries on the ways that human-system interaction might reasonably proceed, without explicitly identifying specific sequences of actions (formative modelling).

From 1998 onwards, researchers at Swinburne Computer Human Interaction Laboratory (SCHIL) and the Defence Science and Technology Organisation (DSTO) have used some aspects of CWA in a variety of work domains, including air defence, anaesthesia, intensive care, and continuous process control. We have applied CWA to problems at different points in the system life cycle, including tender evaluation, instrumentation engineering, definition of crewing needs in C2 environments, training needs, visual and auditory display design, and forecasting the impact of new technologies on work domains (Naikar, Lintern, & Sanderson,

2002). Others have applied it to visual display design, including the design of cockpit displays for the C130J (Dinadis & Vicente, 1999).

In this paper we provide some of CWA in use in aviation domains from our investigations. We then borrow some ideas from Lakatos (1974) and Chalmers (1982) to assess the effectiveness of CWA. From our experience, we indicate where CWA is likely to be most beneficial, where its strengths and weaknesses appear to lie, and what the prospects are for its future development and use.

Overview of Cognitive Work Analysis

CWA orients the analyst towards five different factors that need to be taken into account when analysing human work in complex socio-technical systems such as an air defence environment, the cockpit, or an ATC environment. Each factor captures a different, but important set of considerations that will affect what kind of human activity is possible and sensible. Figure 8.1 illustrates the CWA framework as conceived by Rasmussen et al. (1994) and by Vicente (1999). At the centre of the Figure at right is a grey area that represents – in the most general way – human activity in some work context. The arrows inside the grey area represent different possible activity sequences. Around the outside are five factors that interact to define the activity sequences that are possible and reasonable. In the remainder of this section, these five shaping or constraining factors are described in more detail.

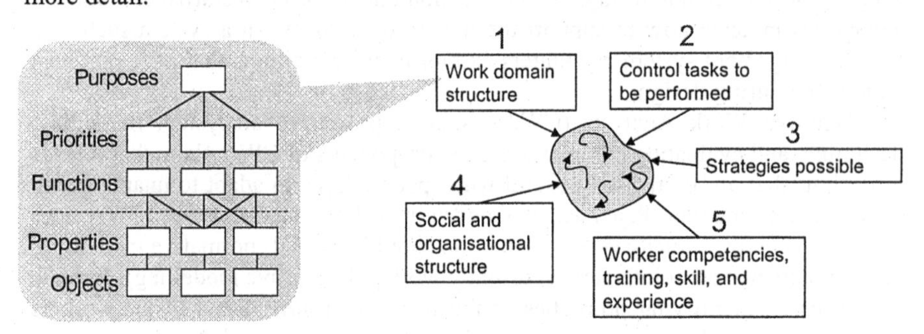

Figure 8.1 Cognitive Work Analysis framework and structure of Work Domain Analysis

First, the structure of a work domain will partly constrain what is sensible behaviour. For example, a pilot will not be able to travel from origin A to destination B in a time less than is technically possible given the capability of his aircraft. Similarly, through conventional rules and practices, a pilot will be constrained from flying into controlled airspace without a clearance, even though it might be physically possible. The "physical" limits of the aircraft and the "intentional" limits of aviation practice therefore constrain the pilot's activity. An important part of CWA involves identifying such properties of the work domain itself because they constrain the possibilities for action. This is termed Work Domain Analysis (WDA).

Second, if the system is to achieve its purpose, various control tasks need to be performed. At the coarsest level, control tasks in aviation include familiar operations such as taxi, take off, level out, navigate, communicate, descend, approach, and land. Control tasks are described in general terms, but they will shape the kinds of activities seen. Therefore, navigation is a necessary control task for aviation, regardless of how it is done and by whom, and cockpit activity will reflect that necessity. This is termed Control Task Analysis (CTA).

Third, activity sequences will be constrained and shaped by the strategies a work crew chooses for executing control tasks. As designers, we can shape strategies for carrying out control tasks by varying the kind of human-system interface and the kind of decision support tools we make available. This is termed Strategies Analysis (SA).

Fourth, activity will be constrained and shaped by how control tasks are shared between members of a team (e.g., captain and first officer) and between humans and flight systems. For example, activity that is possible and reasonable will vary significantly between a manual landing and an automated landing. This is termed Social-Organisational Analysis (SOA).

Fifth, activity will be constrained and shaped by the degree of training and experience that human operators bring to their tasks. This is termed Worker Competencies Analysis (WCA).

These five general classes of constraints form the basis of the CWA approach to analysing human-system interaction. In this paper we are principally concerned with the WDA phase, which is the phase that differs most from other modelling methods. The abstraction hierarchy framework that usually underlies WDA is indicated at right in Figure 8.1 and as the first of the five major columns in Table 8.1. The abstraction hierarchy is a way of describing the physical and intentional constraints in a domain of work. The bottom two layers (objects and properties in hierarchy at left of Figure 8.1) provide information about the physical elements and physical properties that make up a work domain. The top three layers (functions, priorities, and purposes) provide information about how the physical properties of a work domain are put to use to serve human purposes.

In the abstraction hierarchy, physical properties are coordinated to support the basic functions or operations of the work domain. To give an aviation example, control surfaces are configured so that, when interacting with the laws of aerodynamics, flight is achieved along a chosen route. Functions are supported in a way that respects the priorities and values of the work domain (flight is constrained to a routing allocated by ATC and aircraft position remains within defined boundaries). When functions are achieved in a way that is consistent with the priorities and values, the overall purpose(s) of the work domain are achieved. Links between nodes provide 'what-why-how' relations. For any given node, nodes linked to it from a lower level indicate 'how' the property, function, priority, or purpose of the node is achieved, whereas nodes linked to it from a higher level indicate 'why' the object, property, function, or priority is being included in the definition of the work domain.

In the following section there are some examples of CWA in action from some of our recently performed work. Finally we evaluate how useful the CWA approach is.

Examples of CWA in use

We have used CWA at various points across the system life-cycle (Sanderson, Naikar, Lintern, & Goss, 1999). Table 8.1 shows the phases of CWA in the columns and the steps of the system life-cycle in the rows. Rows 1-5 indicate system design and evaluation steps undertaken before a system is implemented; rows 6-7 the implementation and test steps; rows 8-10 the selection of personnel and development of training systems; rows 11-14 normal use and evaluation; and rows 15-16 the response to changing conditions during the system's lifetime. The arrow at right of Table 8.1 indicates that analyses performed at one part of the system life cycle (1-16) can usually be reused after minimal adjustment for another part of the life cycle.

In this section we provide three examples of CWA in action: one from knowledge elicitation work with search-and-rescue (SAR) crews, one from work on Australia's proposed Airborne Early Warning and Control (AEW&C) platform (both tender evaluation and team design), and one developed in work on the F/A-18 (Hornet) upgrade. These examples are mapped onto Table 8.1. Further information about the AEW&C and SAR work can be found in the conference proceedings (Naikar, Drumm, Pearce, & Sanderson, 2000; Elliott, Watson, Crawford, Sanderson, & Naikar, 2000). Yet another instance of CWA in action for design are found in the conference proceedings for patient monitoring in medicine and for approach and landing monitoring and information systems (Watson, Sanderson, & Anderson, 2000).

Table 8.1 Cognitive Work Analysis phases over the System Life-cycle

Phase		Step in System Life-Cycle	Work Domain Analysis	Control Task Analysis	Strategies Analysis	Social-Organisational Analysis	Worker Competencies Analysis
Design	1	Requirements					
	2	Specification					
	3	Design	Proc Cntrl				
	4	Modeling and simulation					
	5	Design evaluation	AEW&C	AEW&C		AEW&C	
Development	6	Implementation					
	7	Test					
Operational preparation	8	Simulator development	F/A-18				
	9	Operator selection					
	10	Operator training	F/A-18				
Use	11	Routine use	SAR	SAR			
	12	Non-routine use					
	13	Maintenance					
	14	Performance evaluation					
Re-evaluation	15	System upgrade					
	16	System retirement					

Knowledge Elicitation for SAR using CWA Concepts

Analyses that emerge from CWA are based on a variety of sources, including examination of documents (for example, concept of operations, incident reports, manuals, and operating procedures); structured interviews with subject matter experts and other stakeholders, participant observation, and so on. An example of how knowledge can be elicited within a CWA framework comes from some structured interviews we conducted with Search and Rescue (SAR) pilots (Elliott, Crawford, Watson, Sanderson, & Naikar, 2000). The goal of the interviews was to develop a framework for evaluating SAR crew performance.

The critical decision methodology (CDM) was adapted for the purpose (Klein, Calderwood, & MacGregor, 1989). Consistent with the CDM, we took a case-based approach in which pilots were asked to recount an incident or episode that was non-routine in some way. However, unlike the CDM which uses probe questions to enable participants to reflect on their through processes when making important decisions, our adaptation of the methodology probes for information that would help us build CWA-based analyses. For example, at suitable points, the interviewer asked questions that probed each of the five levels of abstraction in the WDA, as follows:

- Functional purpose: 'If you were to sum up the overall purpose of your role in one sentence, what would that be?'
- Priorities and values: 'What aspects of your environment are you trying to maximise and minimise? What are your priorities in order to achieve the mission goal?'
- Purpose-related function: 'What is the goal of doing that?'
- Physical function: 'What does that piece of equipment actually do? What function does it actually carry out?'
- Physical objects: 'What physical objects are you exploiting at this stage?'

The probe questions were designed to elicit information about each level of abstraction in the WDA abstraction hierarchy.

After the interview, analysts coded utterances for material relevant to the five levels of abstraction in the WDA. In some cases the evidence was directly in the answers to the probe questions. In other cases the evidence emerged from the SAR pilot's account. An example of how an utterance from the pilot's general account was decomposed into evidence at each of these levels is given in Figure 8.2.

On this basis, a WDA was initially constructed that distinguished the domain of risk and the domain of resources for mitigating the possible consequences of risk. A representative sample is given in Figure 8.3. The sample consisted of 94 separate nodes, with one functional purpose, five priorities and values, 11 purpose-related functions, 35 object related functions, and 50 physical objects. In Figure 8.3, some of the nodes are bold outlined. The company may be interested in improving their operations. Starting from nodes higher in the abstraction hierarchy, such as 'minimise time to care needed' the company can seek the various functions and tools that contribute to that function, and start to speculate on alternative arrangements that might minimise the time. In the example in Figure 8.3, the company might decide to upgrade beacon homers or to explore the market for new equipment to detect and locate emergency beacons.

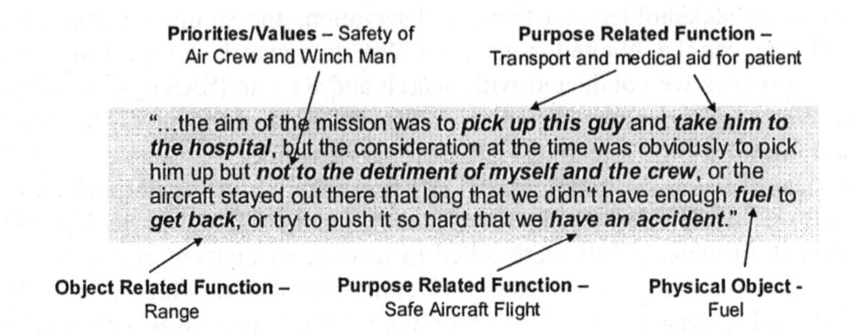

Figure 8.2 Coding of an utterance by SAR pilot using WDA levels of abstraction

Alternatively, the company many be interested in the possible impact of an upgrade in technology. A change in beacon homer performance characteristics will affect a variety of processes, functions, and priorities at the higher levels. The WDA allows the analyst to trace through the areas of possible impact, and to make

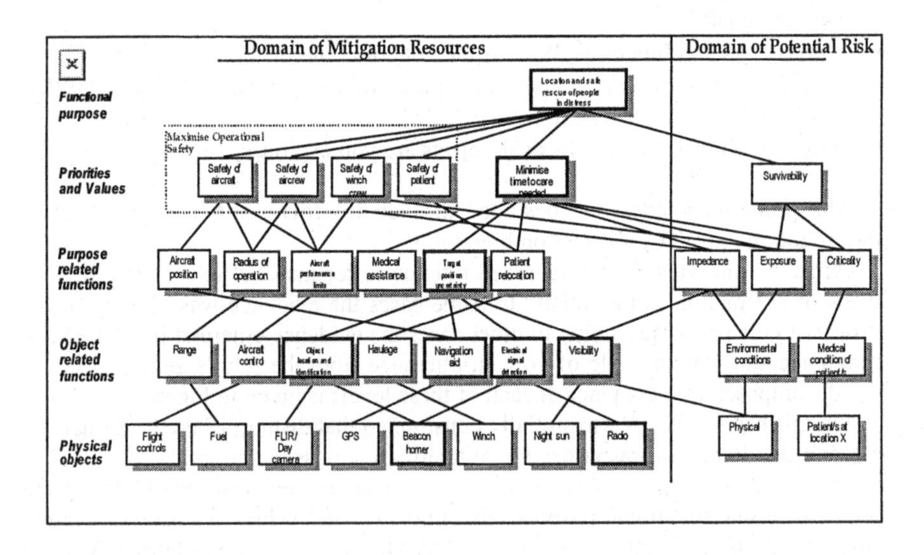

Figure 8.3 Representative WDA that emerged from the SAR interviews

a judgement about impact. When a full CWA has been done, other phases of analysis of CWA, such as an understanding of operator control tasks or strategies, can be enlisted to help with such judgements.

AEW&C

Starting with methods such as that shown for SAR, we have used CWA – and specifically WDA – at the tender evaluation stage for AEW&C. AEW&C was many years away from existence when CWA modelling started. No truly comparable system existed to the one envisaged. Normally, tender evaluation tends to emphasise physical functionality of a system – the objects and properties level of the WDA. This does not provide any evaluation of whether the physical properties of the system will work together effectively to help human operators coordinate the functions of the work domain most effectively, and to do so according to the priorities of the domain so that the overall purpose of the system is reliably achieved.

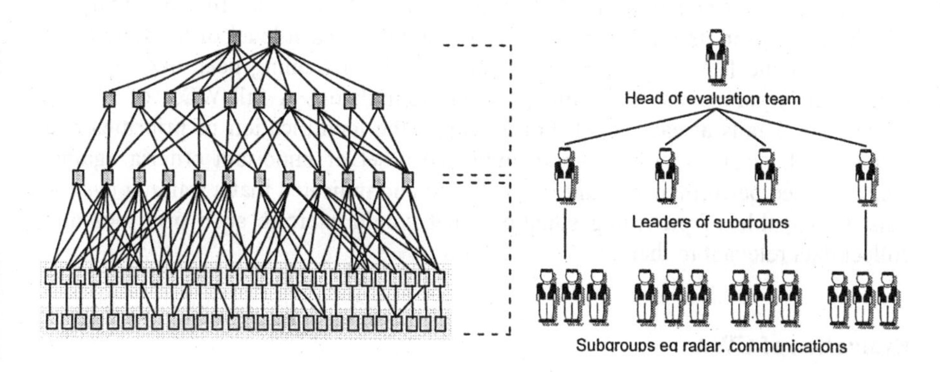

Figure 8.4 Evaluative framework used for AEW&C tender evaluation: WDA at left (schematic simplification); structure of tender evaluation working groups (TEWGs) at right

By evaluating the effectiveness of the proposed designs not only at the lower two physical levels of the abstraction hierarchy, but also at the top three purposive levels, a more comprehensive evaluation against objectives was achieved. As Figure 8.4 shows, a collection of Tender Evaluation Working Groups (TEWGs) evaluated design proposals for AEW&C using the WDA framework, alongside other methods. Each sub-sub-TEWG evaluated specified technical properties and provided advice to Sub-TEWGs. Each Sub-TEWG evaluated overall performance of a technical system, but also performed the evaluation against multiple functions of AEW&C (see left side of Figure 8.4) rather than just physical capabilities. An OPS/TECH TEWG then evaluated mission functions against priorities and overall purpose of AEW&C. This allowed a comprehensive and objective comparison of the three proposals to be considered. This procedure was the first time that CWA had been used for tender evaluation and was considered to be valuable (Naikar, & Sanderson, 2001).

More recently, Naikar and colleagues used CWA to evaluate options for crew composition, training, and crew workstation configuration for AEW&C, again in the absence of any existing equivalent of the platform (Naikar, Pearce, Drumm, & Sanderson, 2002). An extension of CWA control task analysis (CTA) was used that – for each arrangement of crew composition, training and workstation configuration – outlined which crew-member would be tasked with which work functions at which phase of mission. With this technique, Naikar and colleagues were able to identify the factors that defined the boundaries on practical crewing solutions and proposed a possible crewing solution for AEW&C that had not previously been considered. This is a form of social-organisational analysis.

F/A-18 and training

Naikar and Sanderson (1999) used the WDA framework to guide the identification of training needs and training system (e.g., simulator) needs. In their analysis, which is shown in its most general form in Table 8.2, each level of the functional structure of the F/A-18 platform was equated with a training need and with the functional requirements of a training system (simulator). A full WDA of F/A-18 guided the details at each level. For example, the identification of priorities and values in the F/A-18 work domain indicated a set of important criteria against which trainee performance could be evaluated and indicated that a training system must be capable of providing situations that would exercise such priorities and collect data relevant to them.

Evaluation of CWA

The previous section outlines three recent applications of CWA to human-system integration issues that have hitherto not explicitly been addressed with CWA. Table 8.1 shows that they represent only a small fraction of the possible applications of CWA in the analysis, modelling, design, and evaluation of human-machine systems.

How much better might CWA be doing than other approaches to such issues? What follows is a series of observations based on seeing many CWA efforts over the last six years or so, many at first hand but many reported by others.

On the positive side, CWA appears to provide a clear framework for analysing the main factors influencing human-system effectiveness (see the five factors in Figure 8.1). The framework is a helpful guide to where effort should be expended in getting further information and balancing different forms of information. Moreover, CWA analyses developed in one context (row of Table 8.1) 'roll over' to help solve problems in other contexts. In addition, CWA appears to be helpful in synthesising the results of analyses performed with other techniques.

Table 8.2 Framework for inferring training and simulator needs from a WDA

Functional Structure	Training Needs	Functional Requirements
Functional Purposes: why a work domain exists or the reasons for its design	*Training Objectives*: purpose for training workers is to fulfil the functional purposes of a work domain	*Design Objectives*: training system must be designed to satisfy the training objectives of the work domain
Priorities and Values: criteria for ensuring that purpose-related functions satisfy system objectives	*Measures of Performance*: criteria for evaluating trainee performance or the effectiveness of training programs	*Data Collection*: training system must be capable of collecting data related to measures of performance
Purpose-related Functions: functions that must be executed and coordinated	*Basic Training Functions*: functions that workers must be competent in executing and coordinating	*Scenario Generation*: training system must be capable of generating scenarios for practising basic training functions
Physical Functions: functionality afforded by physical devices in the work domain and significant environmental conditions	*Physical Functionality*: workers must be trained to exploit the functionality of physical devices and operate under various environmental conditions	*Physical Functionality*: training systems must simulate the functionality of physical devices and significant environmental conditions
Physical Form: physical devices of the work domain and significant environmental features	*Physical Context*: workers must be trained to recognise functionally-relevant properties of physical devices and significant environmental features	*Physical Attributes*: training system must recreate functionally-relevant properties of physical devices and significant features of the environment

CWA can also speak effectively to different communities. Research and development communities composed of people with different scientific backgrounds often find CWA a useful framework for integrating their concerns because of its 'systems' qualities. For Human Factors psychologists, CWA can provide a framework for scaling up an analysis from a concern solely with individual cognition to a concern with teams in a rich real-world context. For engineers, CWA can provide a simple way to introduce key factors relating to human cognition and decision making that will influence the effectiveness of human-system interaction. For systems developers and software engineers, there is a structural similarity between CWA and existing software engineering techniques, and notions of abstraction on both sides are easy to confuse. However, as Leveson (2000) has noted in her important application of CWA to the US Traffic Collision Avoidance System (TCAS), CWA can provide a framework for capturing the design intention behind a proposed system that can guide the technical evolution of the system throughout its lifetime.

On the negative side, CWA has been criticised for the apparent imprecision and time-consuming nature of its methods in the face of possible alternatives (Lind, 1999). There is not space to do much more than point to these issues here and to acknowledge that there are certainly areas where further formal definition and methodological precision would help. Overall, evaluating CWA is complex because CWA has the following properties, any of which could be subject for evaluation:

- It is based in a particular scientific model of human-environment interaction.
- It is a philosophy of engineering human-machine systems.
- It is an organisational framework for modelling techniques associated with CWA.
- It includes particular modelling techniques that impose certain syntactic requirements.

More broadly, to evaluate CWA we might invoke Lakatos' (1974) notion of a scientific research program, apply the notion to an applied research and development program (see Figure 8.5), and see how CWA fares. Lakatos focuses on a whole program, which is a series of investigations informed by a particular theoretical orientation, rather than on particular investigations within the program. The program has a *hard core* of theoretical assumptions that cannot be questioned without bringing the whole enterprise into question. Around the hard core is a *protective belt of auxiliary hypotheses* that have emerged from investigations that extend or better define the scope and applicability of the hard core. The *positive heuristic* consists of rough guidelines on profitable investigative paths to pursue to expand the protective belt, indicated in Figure 8.5 by long arrows pointing rightwards, extending the reach of the theory underlying the program. The *negative heuristic* indicates investigative paths that are unprofitable or premature under the assumptions of the program, because they bring the hard core into question (indicated by short arrows at left).

Lakatos' framework may help to describe CWA as an engineering framework, and some of the sociology of the R&D community associated with CWA. The hard core of CWA could be considered CWAs theoretical roots in an ecologically-oriented view of human-environment interaction, and its commitment to 'formative' rather than normative or descriptive modelling. The protective belt could represent current hypotheses about the logical sequelae of CWAs theoretical roots and the range of its application. The positive heuristic currently points to testing hard core predictions about the organisation of complex human-system interaction, applying CWA in new domains and extending it to new problems in the system life cycle. The negative heuristic may include issues relating to the syntactic structure of WDA models (always five levels?), the handling of control mechanisms (part of the work domain or not?), and questions relating to the primacy of ecological over cognitive considerations.

Lakatos does not give clear guidelines on how a research program is to be evaluated against rivals but it probably lies in the balance between progressive and degenerative aspects of the program. A progressive program is one in which the theoretical framework gives rise to questions and predictions that have not been prefigured by other theories or frameworks, whereas a degenerative program is one that fails to do so. Chalmers (1982) refers to this balance as the degree of fertility of a research program. It could be argued that the degree of fertility of the ideas underlying CWA is high in terms of its ability to make novel predictions and provide theoretical synthesis on issues relating to psychology and human-environment adaptation (Vicente & Wang, 1998; Yu, Lau, Vicente, & Carter, 1998). Although no formal comparison of methods has been done, it could also be argued that CWA also has a degree of fertility in the sense that questions relating to the human-system engineering of as-yet nonexistent systems appear to be attacked more easily with CWA than with other methods (see examples referred to herein).

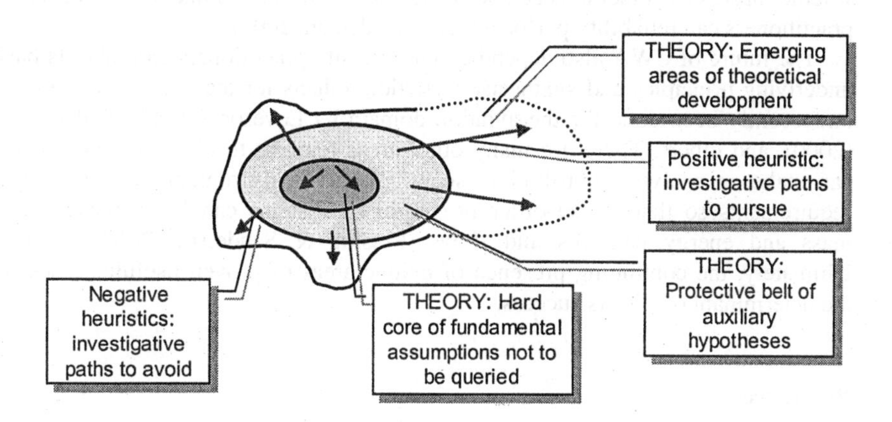

Figure 8.5 Lakatos' conceptualisation of scientific research programs

It is probably in relation to the apparent degree of fertility that some of the zeal associated with CWA arises. Reasonably enough, theorists and practitioners may not wish to abandon a program of investigation with a high perceived degree of fertility. They may tolerate some level of incomplete theoretical closure, the challenge of rival approaches, and methodological difficulties, in the interests of reaping the benefits. It is important for theorists and practitioners using CWA continually to assess whether the level of tolerance of these factors is justified. For example, for many people, learning to conduct WDA in particular is difficult. Moreover the reliability of the technique across different analysts is only just starting to be evaluated (Bisantz, Burns, & Roth, 2002). Lind (1999) has recently noted methodological and conceptual problems with using the abstraction hierarchy to perform WDA of physically engineered systems – systems hitherto believed to be the most straightforward to represent. Although the CWA community will question the details of Lind's criticisms and his suggested solutions, the fact that they have been posed is salutary for CWA as a whole.

Conclusions

CWA has proven to be useful in air defence and ATC contexts as well as in many other domains. At present, CWA is practised by a relatively small but growing 'school' of cognitive engineers. The future of CWA depends on (1) whether it continues to provide conceptual tools to handle new problems in the design of human-system integration environments, (2) whether it does so sufficiently better than other techniques that the effort of learning it is justified, (3) whether CWA analytic products prove to be useful across the whole system lifecycle, and (4) whether basic CWA methods can be sufficiently well-defined that a wide variety of practitioners can reliability perform them (Sanderson, 2003).

The future of CWA also depends on how easily practitioners can take its basic underlying principles and synthesise analytic products for the purpose at hand, as has recently been done for the aviation domain by Leveson (2000), Naikar et al. (2000), and others. There are many other areas relevant to aviation where CWA can make an analytic contribution, such as defining instrumentation and sensor requirements so that higher-order properties of systems can be displayed (e.g., mass and energy balances and flows; Reising & Sanderson 2002a, 2002b). Ultimately, the continuing presence of unique areas of proven usefulness will be the determinant of CWAs success.

References

Bisantz, A., Burns, C., & Roth, E. (2002). Validating methods in cognitive engineering: A comparison of two work domain models. *Proceedings of the Human Factors and Ergonomics Society 46th Annual Meeting* (pp. 521-525). Santa Monica, CA: HFES.

Chalmers, A. (1982). *What is This Thing Called Science?* Brisbane, Australia: University of Queensland Press.

Elliott, G., Watson, M., Crawford, J., Sanderson, P., & Naikar, N. (2000, November). Knowledge elicitation techniques for modelling intentional systems. *Proceedings of the Fifth Australian Aviation Psychology Symposium*. Manly, Australia.

Klein, G., Calderwood, R., & MacGregor, D. (1989). Critical decision method for eliciting knowledge. *IEEE Transactions on Systems, Man, and Cybernetics, 19*(3), 462-472.

Lakatos, I. (1974). Falsification and the methodology of scientific research programs. In I. Lakatos & A. Musgrave (Eds.), *Criticism and The Growth of Knowledge.* Cambridge, UK: Cambridge University Press.

Leveson, N. G. (2000). Intent specifications: An approach to building human-centred specifications. *IEEE Trans. on Software Engineering, 26*(1), 15-35.

Lind, M. (1999, September). Making sense of the abstraction hierarchy. *Proceedings of the Cognitive Science Approaches to Process Control conference (CSAPC99).* Villeneuve d'Ascq, France.

Naikar, N., & Sanderson, P. M. (1999). Work domain analysis for training-system definition and acquisition. *International Journal of Aviation Psychology, 9*(3), 271-290.

Naikar, N., & Sanderson, P. (2001). Evaluating system design proposals with work domain analysis. *Human Factors, 43*(4), 529-542.

Naikar, N., Drumm, D., Pearce, B., & Sanderson, P. (2000, November). Designing new teams with cognitive work analysis. *Proceedings of the Fifth Australian Aviation Psychology Symposium*. Manly, Australia.

Naikar, N., Pearce, B., Drumm, D., & Sanderson, P. (2002, July). A formative approach to designing teams for first-of-a-kind, complex systems. *Proceedings of the 21st European Annual Conference on Human Decision Making and Control* (pp. 162-168). University of Glasgow.

Naikar, N., Drumm, D., Pearce, B., & Sanderson, P. (in press). Designing teams for first-of-a-kind, complex systems with Cognitive Work Analysis. *Human Factors.*

Naikar, N., Lintern, G., & Sanderson, P. (2002). Cognitive Work Analysis for air defence applications in Australia. *State of the Art (SOAR) Report on Cognitive Systems Engineering*, Human Systems Information Analysis Center [HSIAC] (pp. 162-199). Wright-Patterson AFB: U.S. Department of Defense.

Rasmussen, J., Pejtersen, A. M., & Goodstein, L. P. (1994). *Cognitive Systems Engineering.* New York: John Wiley & Sons.

Reising, D. C., & Sanderson, P. (2002a). Work domain analysis and sensors I: Principles and simple example. *International Journal of Human-Computer Studies, 56*(6), 569-596.

Reising, D. C., & Sanderson, P. (2002b). Work domain analysis and sensors II: Pasteurizer II case study. *International Journal of Human-Computer Studies, 56*(6), 597-637.

Sanderson, P. M. (2003). Cognitive Work Analysis. In J. Carroll (Ed.), *HCI Models, Theories, and Frameworks: Toward an Interdisciplinary Science.* New York: Morgan-Kaufman.

Sanderson, P., Naikar, N., Lintern, G., & Goss, S., (1999). Use of Cognitive Work Analysis across the system life cycle: Requirements to decommissioning. *Proceedings of the 43rd Annual Meeting of the Human Factors and Ergonomics Society* (pp. 318-322). Santa Monica, CA: HFES.

Vicente, K., & Wang, J. (1998). An ecological theory of expertise effects in memory recall. *Psychological Review, 105*, 33-57.

Vicente, K. (1999). *Cognitive Work Analysis: Toward Safe, Productive, and Healthy Computer-based Work.* Mahweh, NJ: LEA.

Watson, M., Sanderson, P., & Anderson, J. (2000, November). Designing auditory displays for team environments. *Proceedings of the Fifth Australian Aviation Psychology Symposium*. Manly, Australia.

Yu, X., Lau, E., Vicente, K. J., & Carter, M. (1998). Advancing performance measurement in cognitive engineering: The abstraction hierarchy as a framework for dynamical systems analysis. *Proceedings of the Human Factors and Ergonomics Society 42nd Annual Meeting* (pp. 359-363). Santa Monica, CA: HFES.

Chapter 9

Human Factors Reporting and Situation Awareness

Mike O'Leary
British Airways Safety Services

Introduction

History, they say is a great teacher. So this paper will begin with a brief analysis of accident data over the last 40 years. The data is provided by IATA (2000) and a summary is provided in Figure 9.1 below. The data indicate the variation in aviation 'safety' using as a measure the number of aircraft hulls lost as a proportion of the number of sectors flown. To smooth out the large year by year variation in this measure the data has been averaged across eight five-year intervals.

As we can see, aviation is safer today than ever. The graph illustrates the hull loss rate from 1960 to 1999 and it shows an 80 percent reduction in hull loss rate over the period. There are three areas of the chart that are particularly interesting. The first is the phenomenal improvement across the first two five-year intervals in the 1960s with the second half indicating a more than 55 percent decrease in the accident rate during this period. This success was very largely due to the enormous effort invested in research and development of aircraft structures and propulsion units. The introduction of the jet engine was probably the dominant event in that respect.

The next area of the chart that is particularly interesting is the second largest accident rate reduction that occurred in the seventies. Although this part of the graph does not look particularly dramatic in comparison with the previous decade, the hull loss rate reduced by 30 percent between the early and late 1970s. This improvement also coincided with a technological development – the Ground Proximity Warning System (GPWS) - another dramatic technological success story for the industry.

The third interesting area in the graph is the remaining 20 years. Since the successful introduction of the GPWS the safety statistics shown above have remained relatively level. There are variations but they take the accident rate up as well as down. Across the four five-year periods since the late 1970s the accident rates have, on average, decreased by only eight percent per period.

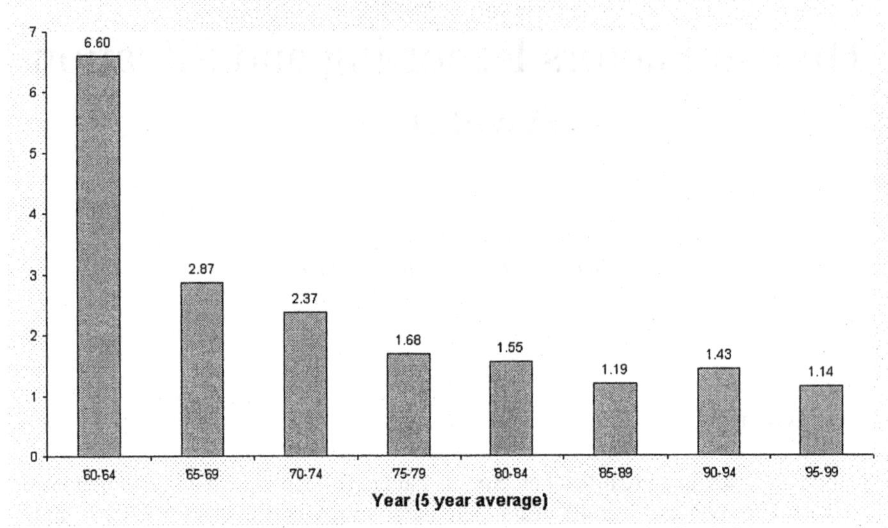

Figure 9.1 Operational total Loss Rates per 106 Sectors: 1960–1999

By comparison with the first 20 years depicted in the graph, this rather slow reduction in accident rate is frustrating. The major factors in the early years were the result of great technological improvement and although much more technological development has occurred in the final two decades of the 20th century it has had nowhere near the earlier success. What is true is that it is now rare for an aircraft failure to result in a crash. What seems to have become the typical accident is some kind of human failure that results in the crew being unaware of a dangerous situation they have been put in, or have inadvertently put them selves in. Accidents such as controlled flight into terrain (CFIT) seem to be the archetypal accident of today and rarely do we see the structural failures so commonplace 40 years ago. Consequently, industry concerns about human weaknesses have now overtaken the concerns with aircraft structural weakness. However, could it be that the industry is persevering, almost pathologically, in trying to use its engineering philosophy – so successful previously – to cure itself of the newly apparent operator failures.

It has yet to be shown whether the technological successes of manufacturers and systems designers have led the industry to the belief that they can design out – or 'automate out' – human error. Certainly, the GPWS has without doubt saved many lives over the last quarter of a century and will continue to do so in the future. But this success has been in mitigating human error not in preventing it. As an attempt to automate out human error, the GPWS has failed.

To those involved in system design however, this lesson appears to have gone unnoticed, or at least unheeded. The pace of technological development seems to have accelerated both in extent and complexity as ever more digital devices attempt to reduce human activity in the flight deck with the laudable ambition of reducing the potential for human error.

Aviation Safety and Automation

The question whether automation has been totally beneficial was posed as long as 20 years ago by Wiener and Curry (1980). They examined automation's 'Promises and Problems' and suggested that the benefits of automation related to reliability and economy whereas the problems related to the lack of integration of the abilities and competence of the human operator with those of the automation. Since then there is hardly a single area of flight that has not been touched by the onward march of automation, not just of aircraft control but of the collection, prioritisation, and display of information. Table 9.1 below gives some feel for the range of functions that are now automated.

Table 9.1 Extent of Control and Display Automation

Flightpath & systems control	Information displays & warnings
Vertical & horizontal flightpath	Terrain
Flightpath limits	Weather
Thrust	Traffic
FMS modes	Abnormal system synoptics
Navigation	Lateral and vertical navigation
Radio Nav selection	Geography

As noted above the archetypal accident is now one related to human error and often more specifically to a loss of situation awareness. The term 'Situation Awareness' is one that leads to misunderstanding because of lack of definition. To avoid this an informal definition is offered here with which the reader can decide to agree or not, but at least the author's view will be clarified. Situation awareness is a state of mind – a dynamic mental model of relevant aspects of an individual's (or team's) 'World'. It requires both physical and cognitive activity to ensure its currency and relevance.

Given that situation awareness seems to have become a common feature in aircraft accidents such as CFIT, the question has to be asked whether the onward march of automation has a positive or negative impact it. More specifically, what is the impact on situation awareness of the pilot's increasing responsibility in the role of system monitor. Automated warning and monitoring systems now abound on the flight deck and the pilot has become almost twice removed as the monitor of the automated monitoring systems. Indeed it could be argued that the pilot could be three times removed in situations where the flight management system undertakes an unannounced and unrequested mode change. Systems feed the pilot with information on navigation – both lateral and vertical; external conditions such as wind speed, terrain features, and conflict with other aircraft; and aircraft systems (fuel, hydraulic, electric, etc.). In non-normal situations the latter information is usually supported with the automatic appearance of a synoptic display of the relevant problem system. Moreover, navigation is achieved through an inertial or global positioning system monitored by reference to ground based navigation beacons that are automatically tuned. The pilot of a modern aircraft is, in reality, required to undertake very little activity (apart of course from the actions required by the company's standard operating procedures [SOPs]).

In view of the extent of the automation of the guidance, monitoring, and warning systems in modern aircraft, one could understandably ask how much the pilot has benefited from the technology. Has the pilot become a safer operator, or whether, on the contrary, is in danger of succumbing to complacency, delegating situation awareness and responsibility for the safe operation of the flight to the computers.

This question arose in British Airways a couple of years ago after a number of disturbing incidents involving apparent loss of situation awareness. On the surface it looked as if the frequency of these events, although low, was greater on the more modern aircraft. Were old-fashioned mistakes still being made on new technology aircraft? To answer this question Flight Operations turned to sources of incident information both within Flight Operations and within Safety Services.

British Airways Safety Information Systems British Airways (BA) traditionally uses two important safety data collection programs; the Air Safety Reporting program (ASR) and the Flight Data Recording program (SESMA). Both programs give an excellent view of the safety health of BAs flight operations and offer a wealth of detail on our current and past operational problems. Initially Flight Operations turned to these programs to assess the importance of their concerns over their incidents and the result was not reassuring. The big surprise was that the incident rates were no better in the modern aircraft (the 'Glass' fleets) than they were in the older technology aircraft aircraft (the 'Steam' fleets). If anything, the incident rates were lower and less severe on the steam fleets. For instance: (1) the occurrence rates of rushed approaches were roughly equal in the Glass and Steam fleets, (2) the rate at which hard GPWS warning occurred was significantly greater in the Glass fleets, and (3) whilst more navigation errors occurred in the Steam fleets they were normally picked up by the crew immediately, whereas on the Glass fleets they more frequently resulted in flight path deviations.

The interesting aspect of these three kinds of incidents is that they all involved, to some extent, a loss of situation awareness. Given the nature and relative frequency of these incidents on the Glass and Steam fleets the data seemed to suggest that situation awareness on the Glass fleets had not benefited from the new technology.

However, as the FDR program is effectively an anonymous program, feedback on the causes of incidents is very scarce. The ASR is somewhat better off in this respect but it is an 'open' program and reporters tend to offer factual descriptive accounts of incidents and very rarely describe factors such as a loss of situation awareness. Consequently, Flight Operations turned to the Safety Services confidential Human Factors Reporting (HFR) program.

Table 9.2 Examples of Crew Behaviour and the Influences

CREW BEHAVIOURS		
CRM Teamskills	**Handling Skills**	**Errors & Violations**
Briefing	Handling - Manual	Mistake
Vigilance	System Handling	Mis-recognition
Assertiveness		Violation
INFLUENCES ON CREW BEHAVIOUR		
Environment	**Organisation**	**Person**
Met Conditions	Commercial Pressure	Stress
Workspace Ergonomics	Training	Morale
Technical Failures	Maintenance	Mode Awareness

The HFR program was introduced to help overcome just this kind of difficulty by focussing the questionnaire on why problems occurred and how the crew overcame them. The filing of an ASR is mandatory after any safety incident and HFR questionnaires are sent out to all crew members on a flight from which an ASR originates. Some filtering does take place such that HFRs are not requested in the more trivial or purely technical incidents. A significant difference from the ASR is that, as completing a HF questionnaire could require some 'embarrassing' disclosures by the flight crew, the program was made both voluntary and confidential.

The HF reports are analysed by trained volunteer flight crew. The analysis requires the reports to be encoded using a set of 'Factors' which describe flight crew behaviour and the influences on that behaviour. The behaviours include Crew Resource Management (CRM) Teamskills derived from the work of Bob Helmreich, e.g., Butler, Taggart, and Wilhelm (1995). The behavioural influences include categories that define Environmental, Organisational, and Personal factors. Most of these factors can be assigned either positively, i.e., the factors indicate a safety enhancing action or influence, or negatively, in a safety degrading sense. A few examples of these factors are shown in Table 9.2.

The complete set of factors number 64 but the analysis of the reports normally goes beyond just the assignment of factors. It is complete only when the factors are all causally sequenced in an Event Sequence Diagram. However, this paper does not require explanation of this facet of the program.

Situation Awareness Study

One of the factors in the above table, Mode Awareness, is of particular interest here as, along with Environment Awareness and System Awareness, it is one of the three Personal Influences that describe Situation Awareness (SA). Personal factors are different from all the others in that they are 'internal' or 'subjective' and in all but a few exceptional cases have to be described in the report or during a call-back session with the reporter. In assigning a Personal factor we make no inferences of the 'it must have been' variety. These three factors combine to offer a complete representation of good or bad SA as reported in an incident. Table 9.3 shows these three factors with their definitions as used in the HFR analysis process.

Table 9.3 Definitions of SA Factors

Environment Awareness
Crew awareness of environment, e.g., other aircraft, communications between ATC and other aircraft, met conditions, geographical position, terrain features, and MSA.
Mode Awareness
Crew awareness of aircraft configuration, flight and powerplant parameters, flight control system modes, and the dynamic aspects of all of these.
System Awareness
Crew awareness of the state of the aircraft's technical systems, e.g., fuel, electric, hydraulic, air.

In this study we were interested only in the relative frequency of positive and negative assignment of the SA factors in the modern (Glass) and older (Steam) fleets. The Glass fleets consist of A320, B737-400, B747-400, and B757/67/77s. Steam fleets include B737-200, B747-100/200, Concorde, and DC10. This study uses data collected since January 1997 and within this period the number of reports was approximately 3,700 with over two-thirds from the Glass fleets. It is unlikely that this will be updated as, in common with most airlines, the number of 'Steam' aircraft still operating will be too small to make any useful comparisons.

Analysis

Situation Awareness Despite having this number of reports the majority did not report SA directly. A total of 396 SA factors were assigned in the analyses of which approximately one quarter of these were from the Steam fleets. In view of this, no fleet by fleet comparisons were attempted, comparisons were made only between the Glass and Steam groups.

Another issue is the comparison method. Direct comparison of assignment frequency is not possible because of the different sizes of the groups and the different reporting frequency of the three SA factors. A measure was calculated by taking, for each of the two groups and the three factors, the number of positive assignments as a percentage of the total number of assignments (positive plus negative) for that group / factor combination. Figure 9.2 shows the percentage positive assignment for the three factors, and a composite of all three, for the Glass and Steam groups.

It is clear from the chart in Figure 9.2 that this study offers no evidence that Glass cockpits enhance SA. Overall SA for Steam indicates 79 percent positive assignments compared to 60 percent for Glass. That is, 79 percent of the Steam SA factors were coded positively with 21 percent negative. For Glass the corresponding figures were 60 percent positive and 40 percent negative. At this point it is worth reminding the reader that the HF questionnaires are invariably completed after some problem has occurred. The above figures (hopefully) do not indicate the state of SA during majority of flights in which no reportable problem has occurred. It is, however, quite likely to be indicative of the kinds of issues arising in or causing our incidents.

The three sub-factors confirmed the above finding individually. Environment Awareness was only slightly superior in the Steam fleets but given the benefit of the navigational and environmental displays of the Glass cockpit, the slight difference is not necessarily good news for Glass. The figure for Systems Awareness in the Glass group is relatively high, however it is particularly impressive for Steam with no negative assignments for this factor in the group. Perhaps we are seeing here the benefit in four out of the five Steam fleets of the inclusion of a Flight Engineer as part of the crew.

Finally the Mode awareness scores were high in neither group. However, even here, the Steam fleets do substantially better than the Glass fleets, with scores of 58 percent and 42 percent positive respectively. Whilst the immediate instinct may be to blame the usual problems of hidden FMS mode changes in the modern flight deck, a look at the definition of Mode awareness above will remind the reader that this factor is not solely concerned with FMS modes.

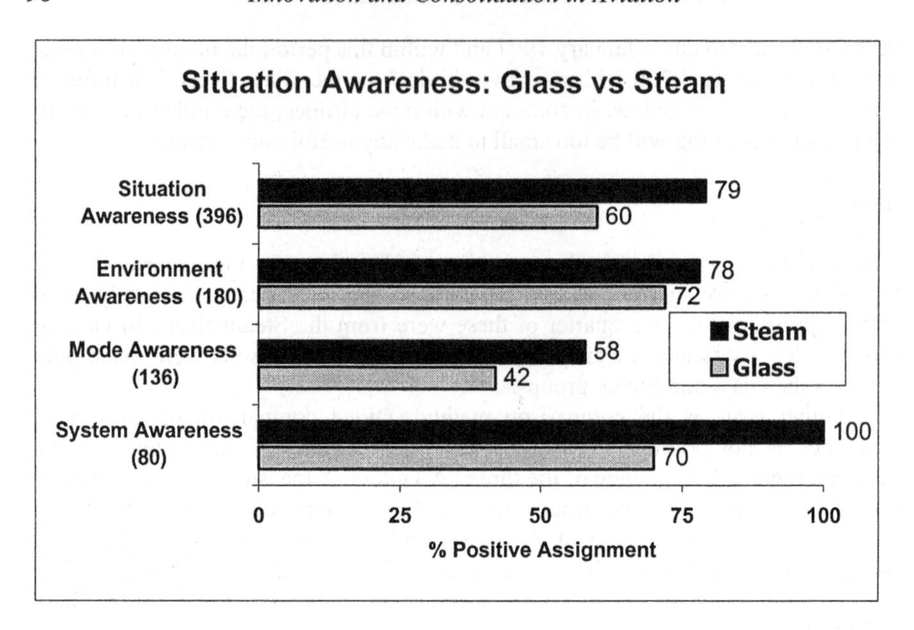

Figure 9.2 Aspects of Situation Awareness for Glass and Steam aircraft fleets

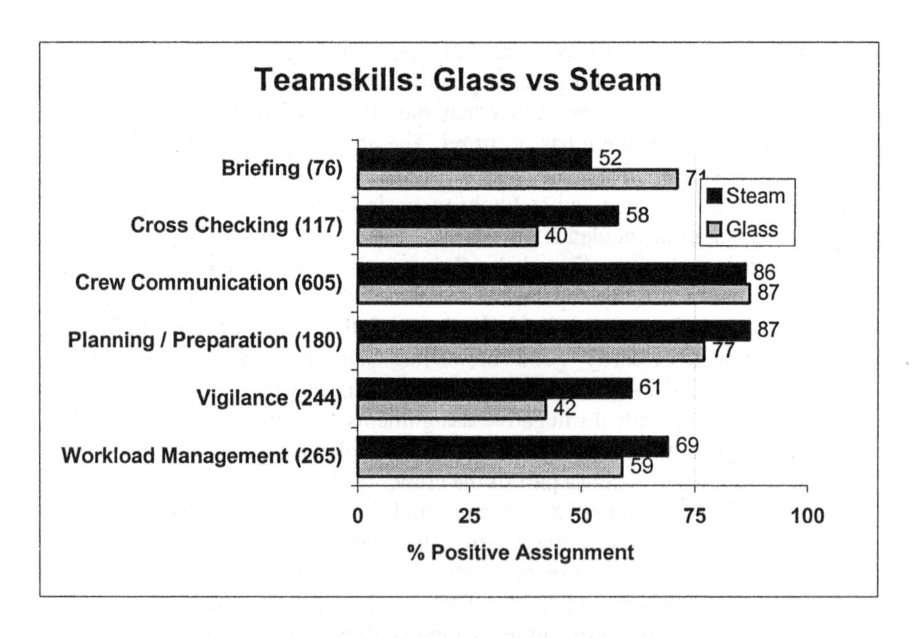

Figure 9.3 CRM Teamskills reported for Glass and Steam fleets

Teamskills The clear result above was somewhat unexpected and further evidence was sought. Above, SA was characterised as a state of mind supported by certain kinds of activities that create and maintain the desired state of awareness. The CRM Teamskills that are also encoded by our analysts are largely concerned with communications and other processes necessary for the establishment and sharing of SA. Consequently, relevant Teamskills were treated to the same analysis as for the awareness factors. Much more data is available in this analysis as activities are much easier to recognise, remember, and report than are states of awareness.

Figure 9.3 shows the same comparison for the Teamskills as Figure 9.2 does for SA. Of the six factors shown only 'Briefing' scores higher for the Glass group with 'Crew Communication' more or less a tie. General communications therefore appear to be somewhat better in the Glass group particularly with Briefing in Glass cockpits showing 71 percent positive against 52 percent for the Steam group. However, the two lowest scores in the whole data set are in the Glass group and involve 'Vigilance' and 'Cross Checking'. These Teamskills are arguably the most important for SA as Vigilance is the process whereby the raw data for SA is collected and Cross Checking is the process of ensuring that the individuals in the team share the same understanding of the world. In the Glass group Vigilance has 42 percent assigned positively and Cross Checking only 40 percent. For the Steam group the scores are 61 percent and 58 percent respectively. The remaining two factors, 'Planning and Preparation' and 'Workload Management' both indicate a 10 percent positive advantage for the Steam group.

Discussion

These studies concern the question of whether or not the 'Glass' cockpit has benefited from modern technology and delivered its promise of enhanced SA in comparison with the older generation 'Steam' cockpits. The results of this study do not support the contention that Glass is better than Steam.

All three of the SA factors (Mode, Environmental, and Systems Awareness) were assessed as better in the Steam group, with the finding of an overall rate of positive assignment nearly 20 percent greater than for Glass. Although the majority of the factors were encoded positively in both groups, it is clear that the Steam group consistently reported better SA than the Glass group.

The story is much the same with the Teamskills. Of the Teamskills relating to situational awareness, only one, Briefing, was scored more positively in the Glass group. However, apart from a tie in 'Crew Communications', the advantage for the Steam group in Vigilance, Cross Checking, Workload Management, and Planning and Preparation was clearly defined.

The interpretation of the Teamskills data is not as clear as that of the SA factors. The latter give a clear measure of the reporters' perceived level of SA in the incidents they report. However, the Teamskills do not indicate the level of situational awareness directly but are representative of the behaviours that bring about SA. From this perspective the low score for Vigilance in the Glass group is particularly significant.

The fact that SA and the supporting Teamskills are reportedly better in Steam than in Glass cockpits is not of itself a criticism of Glass cockpits. It may be that the Glass environment is sufficiently different from the Steam that the Teamskills need to be modified to be better adapted for the Glass environment. Another plausible explanation is that training programs for Glass cockpits have not been designed to take into account fully enough the particular demands of the Glass environment.

An alternative view has already been alluded to the above. Automation has not only relieved the pilot of the necessity of manually controlling the aircraft but has also largely replaced the need to undertake any monitoring or navigational activity. It was argued above that SA depended on a process of both physical and cognitive exploration of the 'world'. Modern trends towards automation have obviated much of the need for this activity. By creating an expectation that all necessary information will be automatically presented to the crew, automation may have reduced the will to undertake the activities that would turn the display of data into useful information and knowledge. Situation awareness does not reside in the retina but in the recognition of the implications of what the retina holds. This view is particularly well supported by the finding that essential Teamskills are less well performed in the more automated cockpits.

The view also has support in academic circles. For instance Billings (1997), and Endsley (1996) who states , 'Turning a human operator from a performer into an observer can, in and of itself, negatively affect situation awareness...'. However, both purchasers and producers of aircraft collude in the belief that more technology means more safety. If they are wrong then we will have to cope as best we can with the present generation of aircraft for the next couple of decades. The expense of changing our new technology to support a more human centred operation would be prohibitive. In that event we will have to resort again to a philosophy of SOPs and training programs that are designed to make the man fit the machine.

References

Billings, C. E. (1997). *Aviation Automation: The Search for a Human-centered Approach*. Mahwah, New Jersey: Erlbaum.

Endsley, M. R. (1996). Automation and situation awareness. In R. Parasuraman & M. Mouloua (Eds.), *Automation and Human Performance*. Mahwah, New Jersey: Earlbaum.

Helmreich, R. L., Butler, R. A., Taggart, W. R., & Wilhelm, J. A. (1995, March). *Behavioral Markers in Accidents and Incidents: Reference List* (Tech. Rep. No 95-1). NASA/University of Texas/FAA Aerospace Crew Research Project.

IATA. (2000). *IATA Safety Report (Jet) 1999*. Montreal, Canada: IATA.

Wiener, E. L., & Curry, R. E. (1980). *Flight-deck Automation: Promises and Problems. Ergonomics, 23,* 995-1011.

Chapter 10

Threat and Error in Aviation and Medicine: Similar and Different[1]

Robert L. Helmreich
University of Texas Human Factors Research Project
Department of Psychology
The University of Texas at Austin, USA

Introduction

Most functions in socio-technical environments require groups to work together effectively to accomplish their mission safely. The record of disasters in the air, at sea, and on the rails is evidence of the catastrophic effects of failed teamwork. NASA research in the late 1970s demonstrated that more than two-thirds of all air crashes involve human error, especially failures in communication and team coordination (Helmreich & Foushee, 1993; Helmreich & Merritt, 1998). Examination of training practices in aviation revealed that the *technical* aspects of flying were addressed effectively, but little attention was paid to developing, assessing, and reinforcing the interpersonal skills that are central in most accidents.

Team performance consists of three main elements, *inputs* that reflect the history and characteristics of the group, *processes* that reflect both the interpersonal and technical aspects of the task, and *outcomes* that include not only task fulfilment but also the attitudes and morale of the participants. A conceptual model of team performance shown in Figure 10.1 illustrates the recursive nature of these activities and applies to both the medical and aviation domains.

Human Error

Errors, both by individuals and teams, at the 'sharp end' of operations have roots in human limitations (Reason, 1990). Humans have restricted memory and information processing capacity. Limits are also posed by stressors that induce

[1] Support for the author's research reported here was provided by grants from the US Federal Aviation Administration and the US Agency For Healthcare Research and Quality.

'tunnel vision', the inability to deal simultaneously with multiple stimuli characteristic of most tasks in complex environments. Fatigue and other physiological factors also lead to error. At the team level, group dynamics and the effects of culture play a role in determining how effectively threat and error are managed.

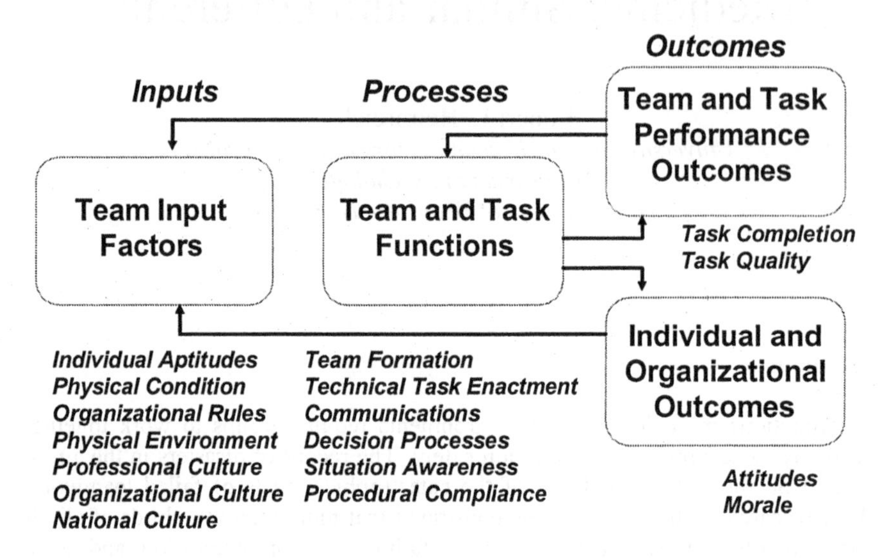

Figure 10.1 A model of team performance

Several generalisations about error can be made: it is ubiquitous, it is inevitable, and it increases the probability of incidents and accidents. The elimination of error is not a reasonable expectation unless a new breed of human can be developed. However, threat and error management strategies are effective both in reducing the incidence of error and the severity of outcomes of those that do occur (Helmreich & Wilhelm, 1991).

Aviation's Approach to Threat and Error Management

Three factors have helped aviation deal with threat and error. The first was the initiation of formal training programs known as Crew Resource Management or CRM to address the interpersonal aspects of flight operations. The second was the collection and analysis of data to provide an accurate picture of the strengths and weaknesses of organisations and the larger aviation system. The third was the development of safety cultures that cope with the sources of threat and error.

Central to understanding systemic and cultural issues is the concept of latent threats or factors defined by Reason (1990) in his seminal research on human error. Latent threats are underlying conditions such as organisational norms, cultural issues, design characteristics, management policies, training practices, etc., that may come together under special circumstances to induce error or catastrophe.

CRM as Organisational Strategy

Crew Resource Management training originated in the early 1980s as a response to data showing the high percentage of 'pilot error' in air crashes. The International Civil Aviation Organisation (ICAO) now requires CRM worldwide. The training concepts of CRM are being extended into other domains including maritime operations, nuclear power plants, and hospitals, especially in operating theatres and intensive care units (Helmreich, Wilhelm, Klinect, Merritt, 2001). Effective CRM programs are data driven, using information from surveys, observations of normal operations, and detailed analyses of errors, accidents, and incidents. Effective programs are both specific and practical. They deal with observable behaviours and eschew vague generalities and what is often called 'psychobabble'. Most important, programs that have a positive impact are ongoing and embedded in the organisational culture (Helmreich & Merritt, 1997).

Training issues in contemporary CRM programs include human limitations as sources of error, the nature of error and error management, expert decision making, conflict resolution, the use of specific strategies as threat and error countermeasures, formal review of relevant accidents and incidents, and practice in employing error countermeasures (for example, in simulation) with reinforcement for threat and error recognition and management.

Assessment of Threat and Error Management

Our research group at The University of Texas at Austin has developed a methodology to assess team performance during normal flight operations. This process, called the Line Operations Safety Audit (LOSA: Helmreich et al., 2001) involves placing expert observers in the cockpit on regular line flights. These observers record team behaviours as well as threats and errors and their management in the operational environment. Observers are trained to a high level of reliability and their observations are conducted with strict guarantees of confidentiality. The methodology has been endorsed as an optimal approach to assessing system safety in aviation by the International Civil Aviation Organisation (ICAO), the International Air Transport Association (IATA), and pilots' organisations. To date, our group, in cooperation with US and international airlines, has collected data on more than 5,000 flight segments. Having achieved a high level of trust with flight crews, LOSA provides an accurate picture of flight operations, and one that often differs from what assessment of performance would suggest either in training or during formal evaluation.

A Typology of Error

From the observation of errors in normal operations, a typology of team error was developed that classifies all of the errors that have been identified. The typology consists of four categories (see also Chapter 3). These are: (a) *procedural errors,* which are what most people think of as errors. These involve situations where the team is trying to do the right thing but makes a mistake or suffers from a slip or lapse of attention as described by Reason (1997); (b) *communications errors* where critical information is incorrectly transmitted or is misunderstood; (c) *decision errors* which are discretionary decisions made by crews that unnecessarily increase risk; and (d) intentional non-compliance errors or violations of formal procedures or regulations.[2]

The most frequent types of errors observed are procedural mistakes and intentional non-compliance (see Chapter 3). Knowing the types of errors committed gives organisations guidance as to what actions can best be taken to improve error management. For example, procedural errors may indicate poor workload management, poor CRM, and/or poor procedures, while proficiency errors may reflect pressures to qualify individuals and/or the need for higher standards. Communications errors may reflect inadequate team training (CRM) or complacency. Decision errors point to the need for more CRM training in expert decision making and risk assessment. Finally, violations may reflect poor procedures, weak leadership, or a culture of non-compliance.

Culture

One of the significant findings regarding error was the fact that there were large differences between organisations and subgroups (aircraft fleets) in the number and nature of errors. Clearly, organisational cultures and subcultures are major aspects of error and error management. By culture we mean the values, beliefs, and behaviours that we share with other members of groups to which we belong. Culture binds us together as a group and provides cues and clues as to how we should behave in a given situation. In particular, culture influences how juniors relate to their seniors and how information is shared. Culture impacts how people relate to technology, including computers. Importantly, it also influences willingness to adhere to rules (Helmreich & Merritt, 1998).

Investigations of aviation accidents have shown poor organisational cultures to be precursors to disaster as reflected in a lack of safety concerns, pressures to operate even under risky conditions, poor leadership, and an environment of conflict between pilots and management (Helmreich & Merritt, 1997). Related to organisational culture is organisational climate, which is defined by pride and a sense of family in the organisation and liking for the job (or the lack of these characteristics). When the organisational climate is positive, harmony exists

---

[2] The empirical typology of error is still emerging. Recently, a lack of technical proficiency has been re-classified as a threat rather than an error.

between the subcultures of an organisation and better teamwork, and increased safety result (Helmreich & Merritt, 1997).

Accidents Illustrating the Role of Organisational Factors

Two well-studied accidents illustrate the role of organisational culture in safety. The loss of the space shuttle Challenger has been extensively investigated by Vaughan (1996). She identified one of the most significant factors as a high level of organisational pressure to start the mission, despite adverse weather, because of prior delays and the enormous publicity surrounding the 'teacher in space' program. There was a known history of gas leaking past o-rings on the solid rocket boosters and awareness that the leakage was exacerbated by low temperatures. Because of the risks associated with o-ring leakage, a ban on launching at temperatures below 10-12C had been instituted. However, on a series of missions, waivers had been granted and shuttles launched at ever lower temperatures. Vaughan defines the reaction to this pattern of changing standards as the *normalisation of deviance.* On each occasion when a successful launch was accomplished at a lower temperature, the new, lower, launch temperature became the de facto norm. As a result, when engineers objected to the Challenger launch (at a temperature of -2.8C), they were overruled by management and the launch proceeded with a disastrous outcome.

A second accident also illustrates the power of organisational cultures. Air Ontario Flight 1363 crashed on takeoff from Dryden, Ontario on a snowy day in March 1989 (Helmreich & Merritt, 1997). The plane was a used Fokker F-28 recently bought from a Turkish airline. The aircraft had a number of defects including an inoperative auxiliary power unit that made it impossible to restart engines without ground power should they be shut down for de-icing. When the plane landed at Dryden, the weather was deteriorating and heavy snow was beginning to fall. The crew asked about the possibility of de-icing, but did not have it done (there was no ground power available at Dryden). The plane taxied out in increasing snowfall and was further delayed awaiting the landing of a small plane that had been lost in the storm. A tragic aspect of the accident was the fact that several pilots who were passengers mentioned to the flight attendants the need to de-ice the aircraft. This information was not relayed to the pilots.

The flight took off, but crashed, and burned, killing the majority of passengers and crew. Although the proximal cause of the crash was almost immediately determined to be a stall caused by ice on the wings, a Royal Commission was formed and a massive investigation of all aspects of the crew, the organisation and the Canadian aviation system was initiated. The Commission found that the crew committed four errors, one procedural and three decision, and that there were also errors by flight dispatch in the flight plan and in launching the flight with maintenance problems. However, the heart of the investigation centred on the latent threats that had been allowed to accumulate in the Air Ontario operation and the Canadian air transport system. These were identified at the regulatory, organisational, and crew (professional) levels. At the regulatory level, errors included certification of a wing design on the aircraft that induced ice accumulation, the failure to define de-icing requirements, and the failure to audit the F-28 operation that had just begun, and was the first jet operated by the airline. At the organisational level, Air Ontario had no company manuals for the aircraft using instead those of two different airlines that had trained their pilots. The culture of the company tolerated a habit by pilots of taking off in the company's propeller aircraft with ice on the wings, assuming that it would blow off during the take off roll (swept wing jets like the F-28 are more sensitive to icing than the turbo-props the pilots were used to); and instead of entering defects (snags) in the logbook, pilots were encouraged to write them on cocktail napkins and pass them on to relieving crews (thus avoiding grounding for maintenance problems).

Both pilots had just qualified on the aircraft and had received inconsistent training at different airlines. In addition, the company trained its flight attendants 'not to bother' the pilots with passenger concerns, a fact that may account for the failure to pass on the critical information about ice on the wings. Finally, at the individual level, the captain was scheduled to get married on arrival of the flight at its final destination, a fact that may have clouded his judgement about the necessity to delay further to de-ice. In summary, a series of errors by the crew had roots in latent factors that left them without adequate defences against the threat and error in their operating environment.

The University of Texas Threat and Error Management Model (TEMM)

Data from observations of normal flights, as well as analyses of causal factors in accidents and incidents, have aided the development of a conceptual model of threat and error management grounded in empirical data. The model, which is shown in Figure 10.2, includes both latent and overt threats along with threat and error management strategies.

Organisations use the model as a guide for the analysis of accidents and incidents with the goals of detecting latent threats and assessing the effectiveness of countermeasures that are designed to avoid error and manage threat, error, and undesired aircraft states.

Threat and Error Countermeasures

A number of factors have been identified as threat and error countermeasures (or negatively as contributors to adverse outcomes of threat and error management). These consist of practicing active leadership including being open to questions and suggestions, briefing threats anticipated in the environment, junior crewmembers asking questions without hesitation, clearly communicating operational plans and perceptions of the situation, making decisions that are shared and revisited, managing workload effectively, monitoring other team members for signs of stress and fatigue, maintaining vigilance, and debriefing flights on their completion to facilitate learning. Overall, it can be concluded that the combination of effective

CRM, adequate data, and a willingness to address systemic problems has enhanced safety. Error has not been eliminated, but effective error management has been created.

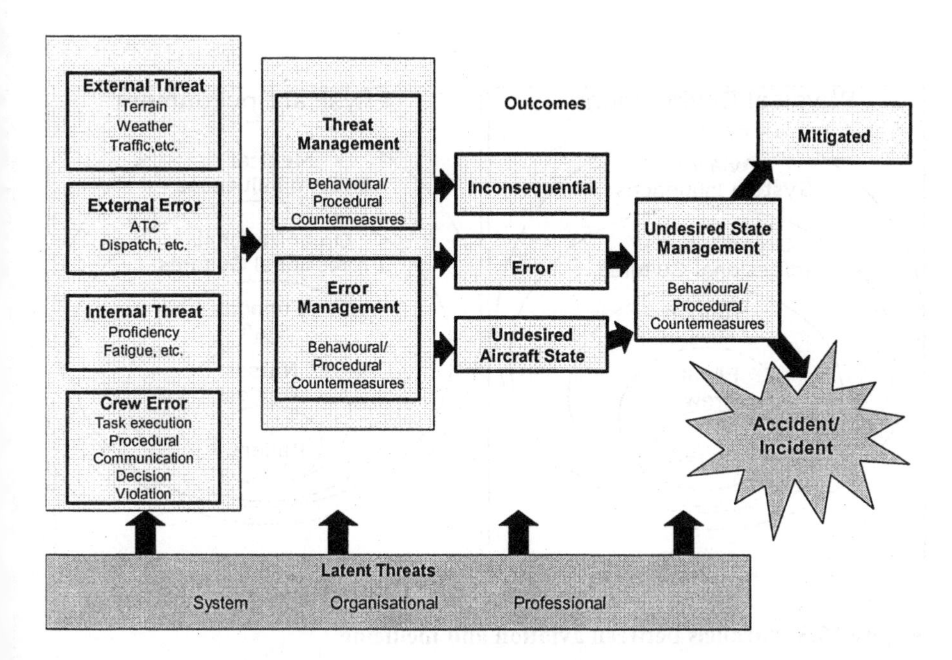

Figure 10.2 The University of Texas threat and error management model

Transferring Aviation's Experience to Medicine

The examination of Human Factors problems in medicine reveals that they have much in common with aviation. To explore these issues, we modified the data collection methods of aviation to assess the special environments of the operating theatre and intensive care unit. To do this we revised a survey, the *Flight Management Attitudes Questionnaire* (FMAQ: Merritt, Helmreich, Wilhelm & Sherman, 1996) to measure medical staff perceptions of safety, Human Factors issues, and personal capabilities. We also utilised a systematic observational methodology to record team interactions during surgery. Results support the view that similar approaches may be effective for enhancing performance and error management in medicine. In both domains, safety is a superordinate goal, but cost factors limit safety-related activities. Risk varies from low to high and threats come from multiple sources. Human issues of teamwork, communication, leadership, situation awareness, and decision making are highly similar. Data from observations reveal failures in communication including a lack of briefings, hostility and frustration due to poor team coordination, and a loss of situation awareness from poor planning and vigilance. We have also found that a comparable model of threat and error management can be developed for the medical environment (Helmreich & Musson, 2000a, 2000b).

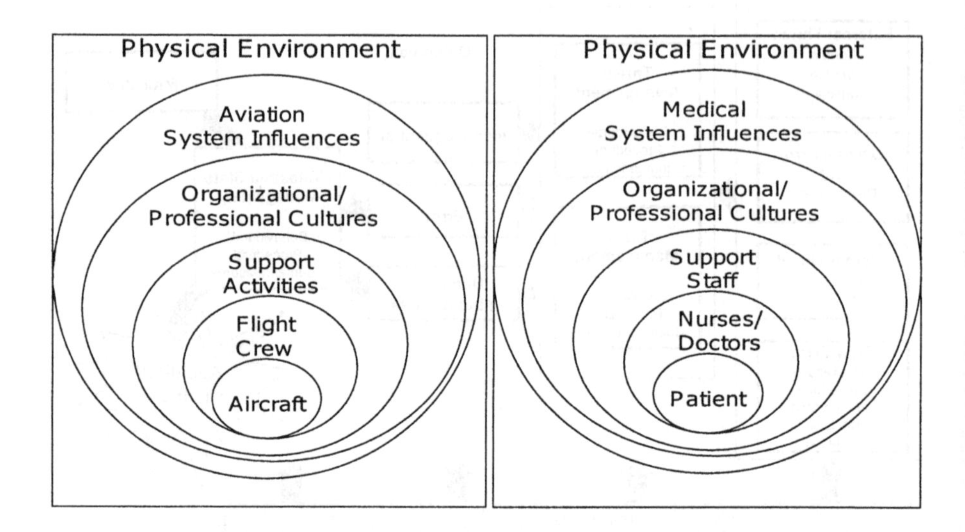

Figure 10.3 Parallels between aviation and medicine

Recognising the commonality of problems, a number of medical organisations are developing coordinated programs that build on aviation's experience and utilise its approaches including training that has much in common with CRM. The more thoughtful programs recognise that, despite the obvious similarities, there are cultural and contextual factors unique to medicine, including aspects of the professional culture, patient variability, and the complex interface between disciplines and specialties. The parallels between aviation and medicine showing the context of team performance are shown in Figure 10.3. As programs are initiated in leading hospitals, a number of threats and barriers to their effectiveness can be seen. One is the sudden emergence of consultants who are ready to sell aviation programs promising quick fixes. Typically, these efforts fail to diagnose the context and the culture, and many are mindless applications of aviation training. In addition, one rapidly encounters resistance on the part of the medical community to outsiders who claim that they have ready solutions to longstanding problems. In aviation, there was a similar proliferation of mediocre programs developed by consultants primarily concerned with personal gain.

Making the Transfer of Human Factors Work

Despite these risks, medicine should be able to use the aviation experience to reach the same goals more rapidly and with fewer missteps. The common elements of teams dealing with technology and known human limitations imply that training that enhances the awareness of the sources of error and provides countermeasures against both threat and error has the capacity to impact safety positively. Aviation's approach to gathering data and interpreting reliable information in a model of threat and error should be equally effective in medicine. Finally, demonstrated organisational commitment to address latent threats identified in data should prove to be as critical in building support from front line personnel in medicine as it has been in aviation.

Table 10.1 Developing and maintaining a safety culture: A six step program

1.	Take history of the organisation
2	Diagnose organisation including nature of threats and errors in the environment
3.	Change the organisational culture including procedures and policies
4.	Provide training in threat and error countermeasures for frontline staff
5.	Provide staff with feedback and reinforcement on technical and interpersonal performance
6.	Demonstrate ongoing commitment to a safety culture

From experience in both aviation and medicine, a six-step program can be outlined that should apply equally in other domains. Efforts should begin with (a) a complete *history* of issues in the organisation. Guided by the history, the second step; (b) *diagnosis* can be undertaken. Diagnosis should be based on current, reliable data indicating the nature and prevalence of problems, including threats, errors, and their management. Non-punitive incident reporting systems are a particularly important component of diagnosis. Diagnosis should be followed by (c) *organisational culture change*. In this step, the organisation should define clear standards for performance, including its interpersonal components, and a credible policy that reflects acceptance of the inevitability of error and a non-punitive stance toward error (this does not imply, however, that any organisation should tolerate intentional non-compliance with its rules and procedures). Also, as part of organisational change, procedures should be reviewed and made more consistent and user friendly as needed. The next step, (d) *training*, consists of providing formal, practical instruction in the human/interpersonal aspects of the job. CRM and its countermeasures that reflect the operational environment should prove effective if the organisational culture is supportive of such efforts. The fifth element; (e) *feedback and reinforcement*, provides front-line personnel with accurate assessment of the interpersonal, as well as technical aspects of their performance and rewards effective teamwork. Accomplishing this requires sophisticated performance appraisal methods that depict the social psychological aspects of teamwork. Finally, (f) the organisation must recognise the fact that these efforts will not work if they are one-shot interventions. Success requires *ongoing commitment.* One-time efforts will not work. Recurrent training, regular assessment of the organisation and its personnel, and continuous monitoring of the system through active data collection must be part of the organisational culture. Table 10.1 lists these six steps that can help an organisation create and maintain a safety culture.

Conclusion

Will the approach taken by aviation be successful in other domains such as medicine? All evidence suggests that it will be if the conditions outlined are met and sufficient resources are made available. Given the nature of litigation, it is also likely that, at least in the United States, organisations that do not have programs to address threat and error will face increased risk of lawsuits and liability.

References

Helmreich, R. L., & Foushee, H. C. (1993). Why Crew Resource Management? Empirical and theoretical bases of human factors training in aviation. In E. Wiener, B. Kanki, & R. Helmreich (Eds.), *Cockpit Resource Management* (pp. 3-45). San Diego, CA: Academic Press.

Helmreich, R. L., & Merritt, A. C. (1998). *Culture at Work in Aviation and Medicine: National, Organisational, and Professional Influences.* Aldershot, UK: Ashgate.

Helmreich, R. L., & Musson, D. M. (2000a). Surgery as a team endeavor [Editorial]. *Canadian Journal of Anesthesia, 47*(5), 391-392.

Helmreich, R. L., & Musson, D. M. (2000b). Threat and error management model: Components and examples. *British Medical Journal* [On-line], bmj.com. Available: http://www.bmj.com/misc/bmj.320.7237.781/sld001.htm

Helmreich, R. L., & Wilhelm, J. A. (1991). Outcomes of Crew Resource Management training. *International Journal of Aviation Psychology, 1*(4), 287-300.

Helmreich, R. L., Wilhelm, J. A., Klinect, J. R., & Merritt, A. C. (2001). Culture, error, and Crew Resource Management. In E. Salas, C.A. Bowers, & E. Edens (Eds.), *Improving Teamwork in Organisations: Applications of Resource Management Training* (pp. 305-331). Hillsdale, NJ: Erlbaum.

Merritt, A. C., Helmreich, R. L., Wilhelm, J. A., & Sherman, P. J. (1996). Flight Management Attitudes Questionnaire 2.0 (International) and 2.1 (USA/Anglo). The University of Texas Aerospace Crew Research Project (Tech. Rep. No. 96-4). Austin, TX: The University of Texas at Austin.

Reason, J. (1990). *Human Error.* New York: Cambridge University Press.

Reason, J. (1997). *Managing the Risks of Organisational Accidents.* Aldershot, UK: Ashgate.

Vaughan, D. (1996). *The Challenger Launch Decision: Risky Technology, Culture, and Deviance at NASA.* Chicago: University of Chicago Press.

Chapter 11

Managing Human Factors at Qantas: An Investment in the Future

Ian Lucas and Graham Edkins
Qantas Airways, Australia

Introduction

According to the International Civil Aviation Organisation, the management of human error within the aviation industry is one of the greatest challenges facing airline managers in the new millennium (ICAO, 1993). Human error represents a real threat to airline safety and efficiency and in building a motivated workforce that understand the challenges of today's and tomorrow's business environment.

This paper will reflect on the past and present Human Factors challenges faced by Qantas. The goals of Qantas in meeting these Human Factors challenges will also be outlined, including the need to better integrate Human Factors principles into existing business and safety systems. In addition, a blueprint to further enhance our approach to Human Factors, specifically a new corporate behaviour based training and assessment program, will be presented.

Finally, this paper will look beyond the needs of Qantas and critically focus on some of the key Human Factors challenges that need to be addressed if further advancements in safety are to be made in the Australian aviation industry.

Past Human Factors Challenges at Qantas

In response to the evolution of early cockpit resource management (CRM) programs in the late 1970s by airlines like United and KLM, cockpit training programs were implemented and well established for Qantas flight crew by the mid 1980s. The merger of Australian Airlines and Qantas in 1992 brought together two organisations with a developing but successful application of CRM concepts. Integrated annual emergency procedures training for all cabin and flight crew was introduced in 1994, which ensured training shifted from the cockpit to an emphasis on the whole crew and team performance.

While initial training efforts within Qantas were met with initial scepticism from some line crew, the development of practical CRM skills based on established psychological knowledge has ensured a greater acceptance in more recent times. For example, 'managing upwards' or assertiveness training for flight crew and the integration of these principles into standard operating procedures has helped to reinforce and maintain an ideal cockpit authority gradient within line operations. In addition, the introduction of Line Orientated Flight Training (LOFT) has provided flight crew with the opportunity to practice in the simulator their communication, team decision-making, and leadership skills in specially developed scenarios. This training has demonstrated dividends in the professional way crew continue to manage emergencies on the line.

Mixed Reaction to CRM

The successful application of CRM within Qantas has been offset by the mixed reaction of crews to this type of training. Cabin crew appear to have readily accepted many of the core principles and perhaps this is attributable to their wide and diverse backgrounds in areas other than aviation. While Human Factors concepts are understood and endorsed by the majority of crew, the fact remains that it does not reach everyone. The non-acceptance by some individuals of CRM is a problem not confined to Qantas but a continued challenge that plagues the industry.

The non-acceptance of CRM by some has partly been reinforced by the 'special' subject status that has traditionally been attached to this training. The perception that CRM is an 'add on' to the normal aircrew training curriculum, is thankfully beginning to disappear with the realisation that Human Factors skills are part of the normal repertoire of professional aircrew.

Narrow Focus

Qantas has recognised for some time that its existing CRM focus is too narrow (mainly aircrew) and needs to be broadened. Safety incidents continue to demonstrate that errors are not confined to the aircraft but their origin has a much wider area of involvement.

In the past, Qantas has had elements of Human Factors orientated programs, policies, procedures, documents, and training scattered throughout the organisation. It is apparent that a new generation Human Factors program should be extended well beyond the cockpit to include error management across the company. Future efforts need to bring consistency to the human performance programs that currently exist within the company and to develop programs in other areas that are lacking.

A Good Safety Record can be a Burden

An additional challenge facing Qantas in advancing future Human Factors efforts, is the enviable safety record the Australian aviation industry enjoys. Australian aviation, Qantas included, remains the safest operating environment on the world stage. While these safety achievements are something to be proud of, there is further room for improvement. A lack of serious safety incidents has the potential to reinforce a culture that we are immune to airline accidents. It is difficult to introduce continuous improvement strategies when everything appears to be working. The concept of Human Factors represents one important area where further safety gains can be achieved.

Present Human Factors Challenges at Qantas

International Developments

Recent international developments in Europe and the United States on CRM skill assessment has led Qantas to reconsider its approach to Human Factors training. There is now a general acceptance that expert performance of tasks involves both technical and non-technical skills. Technical skills are typically the procedural and factual knowledge required to complete a task, such as the highly practiced manipulation skills of flight crew. In contrast non-technical skills may involve reasoning, judgement, and communication skills.

As a result of the work conducted by the Joint Aviation Authority (JAA) on NOTECHS (Non-technical skill evaluation in JAR-FCL; van Avermaete, 1998) and research by Helmreich (2000) on the Line Orientated Safety Audit (LOSA); individual and attitude based training and evaluation has been replaced by team orientated behavioural assessment. This has been the motivation behind a Qantas Flight Training development program called Advanced Proficiency Training (APT) which was initiated in early 1999.

Regulatory Drivers

Apart from international developments, there have also been some local regulatory changes, which are indicative of the increasing importance of Human Factors for Part 121 operators. For example, the Australian Civil Aviation Safety Authority (CASA) has released a discussion paper, Civil Aviation Safety Regulation (CASR) Part 121A, which proposes mandating Crew Resource Management (CRM)/Human Factors training for flight crew based on nine modules. These nine modules specify the basic training curriculum required during initial, recurrent, command, and aircraft type conversion stages and follow that recommended by the JAA.

CASR Part 121A also proposes that flight crew should be regularly assessed on their CRM competencies during line flying, consistent with that recommended by the European Union countries. This means that the formal evaluation of CRM behaviour will become the norm for flight crew. However, it is the intention of Qantas to go much further than flight crew and make behavioural assessment a regular practice for all its operational divisions.

Financial Drivers

A worldwide trend within the aviation industry is the rising cost of safety incidents in both flight and ground operations. Can a Human Factors program help the bottom line?

Lost time injury data shows some potential. For example, 1999/2000 figures from the Corporate Safety Department, indicate that there were around 1,510 lost time injury occurrences recorded across the Qantas Group, resulting in 18,324 lost days. This is conservatively estimated to have incurred a direct cost of AUD$40 million dollars in workers compensation. Research from Dupont Safety Resources suggests that 96 percent of lost time injuries are the result of worker Human Factors/behaviour problems (Hainsworth, 2000). If a behaviour based Human Factors program within Qantas could reduce workers compensation costs by 10-20 percent, this would represent a substantial annual cost saving (Edkins, 2000).

Organisational Goals Driving Change

The above discussion has implied that some staff are hard pressed to acknowledge the value by which Human Factors training can add to the organisation. Therefore, the challenge for operators is to provide Human Factors training that is not only meaningful for operational staff but also useful for improving core safety and business objectives. The following goals outline the direction the company is headed in regard to Human Factors, which are reflective of recent international developments in this field.

Goal 1. Identify the key competencies of domain experts

We need to identify and clearly define what behavioural competencies are representative of safe and efficient performance within the key operational areas of the company.

Goal 2. Develop a better understanding of human performance issues

The focus of most airlines has been to collect and trend data on human performance problems from reactive, event based information, such as reported safety incidents and accidents. Given the unique human capability to detect and recover from error, there is a lot of information we are not collecting because an incident may never have occurred.

Goal 3. Integrate Human Factors principles into existing business and safety systems

Human Factors must not be isolated from core business and safety objectives. There is a need to establish a more systematic method of ensuring that the operational risks identified can be used as learning tools by various operational areas.

Goal 4. Enhance operational and business outcomes through Human Factors initiatives

While financial imperatives should not be the sole reason for embarking on Human Factors initiatives, we need to get smarter at identifying what has been the return on investment.

A Blueprint for Change

The question facing many airline managers is not if, but rather how to integrate Human Factors principles with current resource management programs. This section details how Qantas proposes to achieve the above mentioned goals.

A Corporate Approach to Human Factors

In recognition of the potential impact of Human Factors on Qantas business objectives, a Human Factors Steering Committee was formed in April 1998 to guide the extension of a new generation corporate Human Factors program well beyond the cockpit to include error management across various operational divisions. The structure of this group is outlined below:

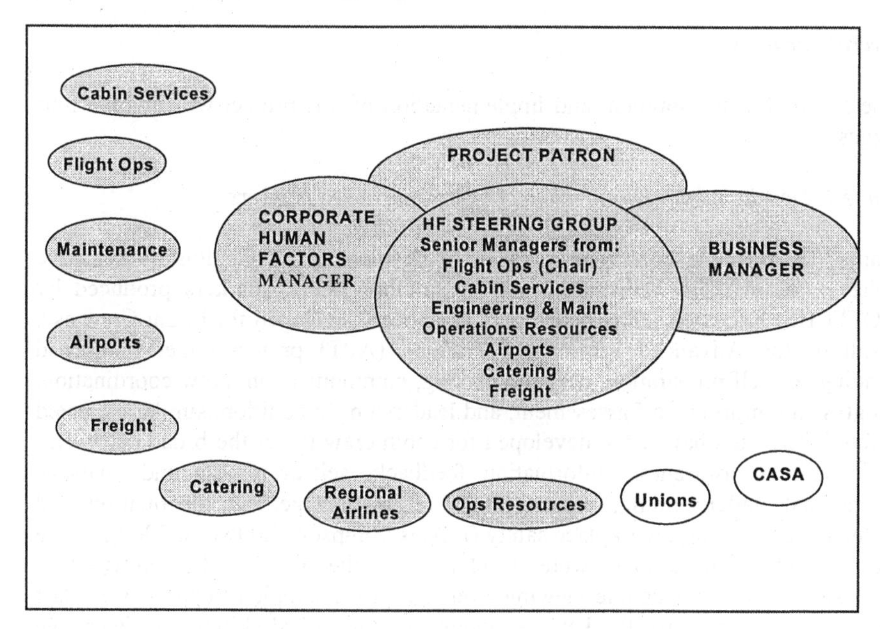

Figure 11.1 Corporate structure for Human Factors

The role of the Human Factors Steering Committee is to:

1. approve and direct fiscal resources at specific Human Factors programs; and
2. maintain a corporate level focal point to ensure Human Factors program consistency, while identifying and reducing costly duplication of effort.

The committee is comprised of senior management from various operational divisions to ensure company wide commitment at a high level.

Development and Assessment of Non-technical Skills

To provide the foundation for corporate Human Factors excellence, a behavioural training and assessment project across various operational divisions is proposed. This project will be managed by the Corporate Human Factors Manager, Dr Graham Edkins, and include Regional Airline, Union, and CASA involvement to help ensure commitment to this program. This new approach to Human Factors has the dual objectives of:

1. providing staff with enhanced skills to avoid, trap, and mitigate human error; and
2. providing management with a structured process to assess specific behavioural performance indicators based on core safety and business objectives.

Overall Strategy

The successful development and implementation of this project will involve four stages.

Stage 1. Development of non-technical behaviour based markers

Qantas has made a conscious decision to develop its own culturally sensitive behavioural markers, rather than relying on the generic markers produced by NOTECHS. Expected safety behaviours have been developed for Flight Crew as a result of the Advanced Proficiency Training (APT) project under the broad headings of self preparation, decision making, communication, crew coordination, progress monitoring, self assessment, and leadership. In addition, similar expected safety behaviours have been developed for cabin crew under the broad categories of situational awareness, information feedback, self-awareness and analysis, operational understanding, passenger and crew management, negotiation and influencing skills, and workplace safety (Edkins, Simpson, & Owens, 2002). These expected safety behaviours were developed via the process of cognitive task analysis, which involves interviewing a representative sample of experienced staff in each area to identify the key behaviours (non-technical skills) that determine effective and safe job performance. Other operational divisions progressively over the next two years will adopt the same process.

Stage 2. Development of Human Factors training material to support behavioural markers

Human Factors training material based on the behavioural markers identified in Stage 1 will be developed, to ensure that new hire staff are provided with a framework to develop expert behaviour during initial training stages. In addition, this material will help to reinforce the behavioural skill repertoire of existing staff during annual refresher and command/upgrade stages of training. While the focus of the training will be on reinforcing a sound understanding of the behavioural markers, it is expected that much of the psychological theory underpinning the material will be based on that specified in the CASR Part 121A requirements.

Stage 3. Development and trial of behaviour based performance appraisal system

Behaviour based performance appraisal systems will be developed so that management can assess that staff are demonstrating the behavioural markers trained in Stage 2. Qantas has for many years assessed many of its operational staff, such as cabin crew and airports personnel, on customer service skills. Assessment of behavioural skills underpinning safe performance is seen as a natural extension of our existing staff appraisal systems.

One of the more promising platforms to collect data on safety behaviour has been the recent development of the Line Operations Safety Audit (LOSA) program. LOSA utilises trained observers to collect data about flight crew behaviour on normal flights under non-jeopardy conditions. Observers record potential threats to safety and how the flight crew manage errors. In addition, behavioural markers are used repeatedly for every flight phase and rated on a four-point scale (poor – outstanding). While, the validation of the LOSA program and the use of behavioural markers is ongoing (Helmreich, Klinect, & Wilhelm, in press), it is expected that LOSA may provide a useful tool to assess the safety behaviour of Qantas operational staff on a non-jeopardy basis.

Stage 4. Implementation of the behaviour based performance appraisal system

Behaviour based performance appraisal systems will be tailored to the requirements of each Division. For example, flight crew are currently assessed on non-technical skills during route checks. The behavioural markers will essentially form a framework to analyse performance at the individual, divisional, and organisational level, which can then be compared to behavioural data reported via safety incidents and accidents, and through scheduled audits. The integration of these two historically isolated areas of trend information will provide Qantas with an enhanced ability to identify the major Human Factors risks to safety and efficiency across its various operational areas. The organisation of this system is presented below in Figure 11.2.

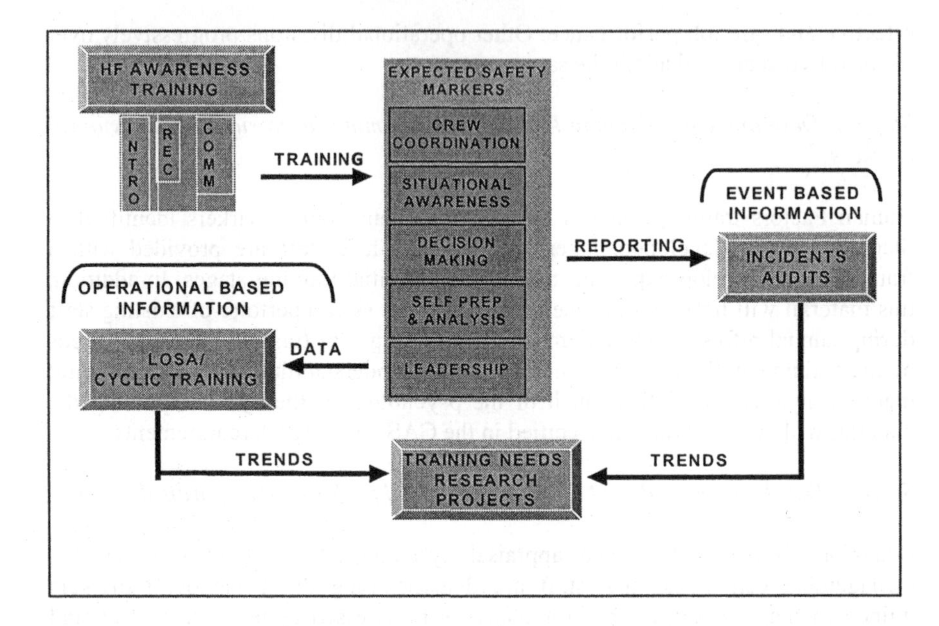

Figure 11.2 Integration of Human Factors training and safety

Evaluation of Change

Stage 5. Longitudinal evaluation of the program

Any change program of this magnitude needs to be measured against expected improvements in operating safety and efficiency. This is where academic expertise has much to offer. To this end a joint Qantas/University of Newcastle PhD scholarship, funded by the Civil Aviation Safety Authority, is being conducted, to determine the impact of the Human Factors program on safety and business objectives within Qantas ramp operations (Piotrowicz, Edkins, & Pfister, 2002). To date the aviation industry has yet to develop a robust evaluation methodology for assessing the effectiveness of Human Factors training (Edkins, 2002; Flin & Martin, 2001).

Future Challenges

The corporate Human Factors program outlined above demonstrates that Qantas is committed to investing in Human Factors. However, Qantas does not operate in a vacuum and there are a number of challenges that need to be collectively addressed by the industry if we are to further improve safety in the region.

Regulations

The recent proposal by CASA to mandate Human Factors training is welcomed. However, we must avoid being overly prescriptive and legislating an exact regimen for Human Factors training across Australia. Compliance should ensure that any training is conducted in accordance with organisational culture. For operators with limited resources and an unrefined approach to Human Factors, prescribing specific topics to train will be of value. However, organisations like Qantas who have well developed and mature training approaches require the flexibility to cover the material based on years of cultural introspection. The Human Factors Advisory Group (HFAG) established by CASA, of which Qantas is an active participant, has an important role in ensuring that industry input into this important issue continues.

Return on Investment

The wider implementation of Human Factors programs within our industry is still plagued by a lack of studies demonstrating a return on investment. The continued development of Human Factors 'technology' depends on the establishment of a stronger link between Human Factors programs and operational benefits. A number of promising results have been achieved in airline maintenance. For example, Stelly and Poehlmann (2000) report on a two-day Human Factors training, costing US$25,000, developed for maintenance management and technicians for a major US airline. The authors claim a reduction in ground damage incidents by 68 percent, on the job injuries by 12 percent and overtime by 10 percent resulting in a cost saving of US$300K over 5 years. Similarly, Taylor (2000) claimed an 80 percent reduction in lost time injury, following the implementation of a two-day maintenance resource management program (US$251,660) resulting in a US$1.3M saving over 2 years.

More studies like this, particularly in other operational areas such as flight operations and airport services are needed. The open exchange of information relating to cost effectiveness must be promoted throughout the industry.

Conclusion

This paper has traced the historical journey of CRM development at Qantas and discussed some of the challenges that have impeded past and current progress.

The maturity of our aircrew body in relation to Human Factors principles means that gains in this area will continue to be small. It is other operational areas within Qantas, with less refined Human Factors development, that hold the greatest promise.

A strategy for the implementation of an ambitious company wide Human Factors program has been outlined. This program is expected to take 3 years but the pay off will be a more strategic method of identifying major human risk factors that may threaten core business objectives.

The main objective of this paper has been to demonstrate that we take Human Factors seriously at Qantas. In most respects, Human Factors is the last frontier in the battle to work safer and smarter. This recognition at the highest corporate level guarantees that Human Factors will become an integrated way of life at Qantas.

References

Edkins, G. (2000). *Business Case: Development, Implementation, and Evaluation of a Human Factors Training and Assessment Program for Operational Staff.* Qantas Airways, Sydney.

Edkins, G. (2002). A review of the benefits of aviation human factors training. *Human Factors and Aerospace Safety, 2*(3), 201-216.

Edkins, G., Simpson, P., & Owens, C. (2002, June). *Development and evaluation of cabin crew expected safety behaviours.* Paper presented at the Human Error, Safety System Development Workshop, Newcastle, Australia.

Flin, R., & Martin, L. (2001). Behavioural markers for Crew Resource Management: A survey of current practice. *International Journal of Aviation Psychology, 11*, 95-118.

Hainsworth, D. (2000, October). *Dupont Safety Resources.* Briefing to Qantas on Safety Management.

Helmreich, B. (2000, August). *Culture and error: The bridge from research to safety.* Paper presented at the IATA/ICAO Flight Safety and Human Factors Regional Seminar, Rio de Janeiro.

Helmreich, R. L., Klinect, J. R., & Wilhelm, J. A. (in press). System safety and threat and error management: The Line Operations Safety Audit (LOSA). In *Proceedings of the Eleventh International Symposium in Aviation Psychology.* Columbus, OH: The Ohio State University.

ICAO. (1993). *Human Factors, Management, and Organisation* (Human Factors Digest No. 10). International Civil Aviation Organisation: Montreal, Canada.

Piotrowicz, M., Edkins, G., & Pfister, P. (2002, June). *The monitoring of safe behaviour in ramp operations.* Paper presented at the Human Error, Safety System Development Workshop, Newcastle, Australia.

Stelly, J., & Poehlmann, K. L. (2000, April). *Investing in human factors training: Assessing the bottom line.* Paper presented at the 14th Annual Human Factors in Aviation Maintenance Symposium, Vancouver BC.

Taylor, J. C. (2000, April). *A new model of return on investment for MRM programs.* Paper presented at the 14th Annual Human Factors in Aviation Maintenance Symposium, Vancouver BC.

van Amermaete, J. A. G. (1998). *NOTECHS: Non-technical Skill Evaluation in JAR-FCL.* National Aerospace Laboratory NLR.

Chapter 12

CRM Behaviour and Team Performance under High Workload: Outline and Implications of a Simulator Study

Barbara Klampfer, *Swiss Federal Institute of Technology*
Ruth Haeusler, *University of Bern*
and Werner Naef, *gemako*

Introduction

This paper introduces a currently running project ('GIHRE-aviation') aimed at investigating the use of CRM (Crew Resource Management) behavioural markers in the simulator environment. The project is part of an interdisciplinary project group entitled 'Group Interaction in High Risk Environments' (GIHRE), which was launched by the Daimler-Benz Foundation.

The database for analysis consists of videotaped simulator sessions from the recurrent training of an airlines A320 fleet. Three predefined scenarios with different demands were recorded. The video data collected were then analysed with two behavioural maker sets NOTECHS and LOSA to assess CRM performance of 46 crews.

Many organisations – from aviation to medicine, from nuclear power plants to shipping, rely on teams in order to accomplish important tasks in an environment where safety is concerned. Successful team functioning in demanding situations is known to be a crucial factor. Therefore training and assessment of CRM (Crew Resource Management) skills and associated behaviours need to be promoted and proven to be relevant in different settings. In the Joint Aviation Requirements, CRM is defined as '... the effective utilisation of all available resources (e.g., crew members, aeroplane systems, and supporting facilities) to achieve safe and efficient operation. The objective of CRM is to enhance the communication and management skills of the flight crew member concerned. The emphasis is placed on the non-technical aspects of flight performance'.

In order to enhance flight safety, the airline industry is convinced that Human Factors skills need to be promoted and cultivated just as much as technical skills. Only the combination of profound technical knowledge and skills with adequate

Crew Resource Management (CRM) skills empowers crew members to operate safely and efficiently under heavy workload. With this in mind, regulatory background for a multi-modular Human Factors training system was developed and integrated into the new European aviation regulatory system under Joint Aviation Authorities (JAA). This scheme mandates the development and implementation of process oriented, soft skills training throughout the career of flight crew members. In consequence, the assessment and measurement of team performance is central. Despite the reliance on teams, there is still little known about the processes that occur within a team that help account for real differences in outcomes. Components (behavioural markers) and their measure focus on what team members do to attain various levels of performance. Therefore, measures of teamwork must be behaviourally oriented in content, which leads to the development of behavioural markers. The term 'behavioural markers' refers to 'observable, non-technical behaviours that contribute to superior or substandard performance within a work environment (for example, as contributing factors enhancing safety or in accidents and incidents in aviation)' (Klampfer, Flin, Helmreich et al., 2001, p. 10).

LOSA and NOTECHS – Two Instruments for the Measurement of Team Performance

First research efforts on behavioural markers generated the Line/Los Checklist (LLC) (Helmreich, Wilhelm, Kello, Taggart, & Butler, 1990), which is frequently cited and well known in the area of aviation psychology. The LLC is the basis of the Line Operations Safety Audit (LOSA) Human Factors Checklist, which is the first of three parts of LOSA (the second and the third part refer to threat and error management). The LOSA program is an organisational strategy for identifying threats to safety on the basis of routine flight observations. LOSA has become the central focus of the ICAO flight safety and Human Factors program for the period 2000-04, and it is anticipated that the implementation of LOSA will be recommended by 2004 (ICAO, 2002). The purpose of the LOSA Human Factors Checklist is the evaluation of CRM skills at the crew level in Human Factors line audits.

In Europe, Joint Aviation Requirements Flight Operations (JAR-OPS) and Joint Aviation Requirements Flight Crew Licensing (JAR-FCL) require CRM training and evaluation of CRM skills in multi-crew operations.

In 1996 the European Project NOTECHS was initiated by the JAA to provide background information for the JAR-FCL in relation to the evaluation of a pilot's non-technical skills. 'The goal of the project was to develop a methodology for assessing pilots' non-technical skills during flight and simulator checks' (Flin & Martin, 2001, p.114). The product has been a new draft standard for the assessment of individual pilot's non-technical skill on the basis of observable behaviours. The evaluation of NOTECHS, in regards to applicability, reliability, and the influence of cultural differences, was accomplished in the JAR TEL Project (e.g., Delsart & Andlauer, 1998; Flin & Martin, 1998).

One of the foremost issues in team performance assessment is the purpose of measurement (e.g., certification, feedback for training, training evaluation, etc.). It cannot be assumed that measures collected for one purpose will also directly serve another. The above mentioned behavioural marker systems (NOTECHS and LOSA Human Factors Checklist) were designed for different purposes, which has to be taken into account, but, content-wise, they both measure an essentially similar set of component behaviours (cf. Klampfer et al., 2001).

GIHRE-Aviation

Over the last years much research has been devoted to team training and performance. Based on practical interests regarding the assessment of CRM performance the main objective of GIHRE-aviation is the investigation of existing behavioural marker sets in the simulator environment under conditions of high workload. The significance of behavioural markers for team work in high risk environments is investigated, where time pressure and unexpected events play an important role. The project sheds light on the interplay of individual pilots working as a crew (NOTECHS markers for individual pilots) and the team work processes on the level of the crew (LOSA markers for crew).

The advantage of using two different behavioural marker sets for the assessment of each crew is seen in the possibility of a cross-validation of general results because of similar core concepts. Additionally, since the two systems used differ in respects to the unit of analysis, a 'comparison' between individual ratings and crew ratings can be made.

Because co-ordinated teamwork is seen as an important determinant of team performance, a component of teamwork should discriminate between teams that perform well and those that perform poorly. Thus one matter of interest besides the observability of markers (frequency of ratings), is the variance of their ratings and their predictive value for the overall performance of a crew.

The main objectives of GIHRE-aviation are:

1. Investigation (validation) of the usefulness of CRM behavioural markers for assessment of CRM performance under high workload conditions in the simulator.
2. Examination of the stability of CRM performance across different scenarios of high and moderate workload.
3. Identification of strengths of excellent crews and deficiencies of poor crews.

Training methods and qualification procedures are based on assumptions about the skills to be exercised and assessed. Despite the great relevance that accident and incident reports attribute to deficient CRM, there is still not much known about the nature of CRM skills until now. An important question, in regards to the significance of CRM behaviours for coping with high workload under certain conditions, is which CRM behaviours are highly related to crew effectiveness under high workload conditions. These analyses are complemented with general examinations of the observability and variability of behavioural markers (i.e., which behavioural markers are easy to observe and differentiate best between crews). Up to now important questions like the stability of CRM behaviours across different situations or the combination of individual behaviours to crew performance are not sufficiently clarified. The question of stability is of high practical relevance, because if CRM performance was situation dependent, the use of diverse training scenarios and multiple measurements for qualification would be necessary to make judgements valid. If CRM skills were rather stable, it would not be necessary to exercise it under several different conditions and qualification could be based on single measurements. Other questions are directed towards the ways in which CRM performance is affected by workload. How do teams react to high workload, i.e., how do they strategically manage their tasks or their workload. For instance situation awareness is one of the generic tasks that crews neglect under exceeding workload (Wickens, 2001). Whether or not demanding situations lead to changes in CRM performance regarding the emphasis on specific behaviours, leads to another question, namely, whether certain situations call for specific CRM behaviours. Furthermore, characteristics of crews that have generally high performance ratings and characteristics of those crews that have a general poor performance across the three scenarios should be identified.

Method

A simulator study with a quasi-experimental field design was conducted that allowed a systematic variation in the amount of workload. Data were collected by agreement of the crews concerned during simulator training days.

Data Collection

The data collection was comprised of the following data sources and instruments:

1. video taped simulator sessions including basic flight parameters;
2. NASA Task Load Index (TLX) for subjective workload measurement;
3. questionnaires for the crews (self-assessment and socio-demographic information);
4. questionnaires for the simulator instructors (expert ratings on workload and performance aspects);
5. NOTECHS (non-technical skills) for behavioural observation; and
6. LOSA (line operations safety audit) ‘Human Factors Checklist’ markers for behavioural observation.

The data collected are video taped simulator sessions from a training day of Swissair A320 crews. The video recording additionally includes main flight parameters like speed, altitude, rate of climb or descent, heading, and power setting, all of which are visible at the bottom of the screen.

The simulator scenarios consist of two high workload situations and one moderate workload situation. All crews were flying the same scenarios. Immediately following the end of the simulator session each crew member filled out the TLX (as a measure for subjective workload) for each of the three scenarios, as well as a short questionnaire regarding self-assessment of team and individual performance and socio-demographic information. In a separate questionnaire, the instructor gives an expert rating regarding the CRM and technical performance of the crew for each scenario, as well as estimates for the workload of the three scenarios.

The TLX-ratings are used to certify the variation in workload between the different scenarios. The flight parameters and the expert ratings are supposed to serve as additional performance indicators.

Sample

The number of crews that participated in the study is N = 80 A320 crews. After the exclusion of video tapes that had low quality (missing tone or flight data) or deviations from schedule (missing scenarios, variation in scenario), the sample of the simulator study consists of N = 46 A320 crews.

Scenarios

Scenario 1 'handling problem' (high workload) An ILS approach in manual back up law has to be performed after the loss of flight control computers, which means the crew is controlling the aircraft with rudder, pitch trim and power only. This scenario is characterised by a technical failure, leading to a non-certified emergency procedure; the problem itself is well defined and action requirements are obvious for the crew. The workload is especially high in regards to psychomotor (flying) skills. The Pilot flying is the Captain.

Scenario 2 'diagnosis problem' (high workload) An ILS approach in direct law has to be performed after several failures in the Air Data Unit of the ADIRU. In this scenario it is the diagnosis which is of primary interest. This scenario is also characterised by a technical failure, but, in this case, the diagnosis is the most challenging part of the problem. Good system knowledge and a systematic approach to problem solving are crucial. The Pilot flying is the First Officer; the analysis of the problem is primarily carried out by the Captain.

Scenario 3 'difficult approach' (moderate workload) The task is a non-precision (NDB) approach with subsequent circling to the opposite runway. In this scenario no technical failure occurs. However, this kind of approach is rarely flown and can be especially demanding for inexperienced first officers. The Pilot flying is the First Officer.

CRM Assessment Tools used for Video Analysis

As mentioned above, NOTECHS and LOSA Human Factors Checklist were used in parallel for the analysis of the video material. Both instruments measure CRM aspects with behavioural markers.

To ensure the quality of the analyses, the observers[1] have been formally trained in LOSA and NOTECHS. A systematic variation of observers and scenarios was applied so that each of the three scenarios per crew was analysed by a different rater. In addition, the independence of LOSA and NOTECHS observations was secured i.e., each scenario was rated with LOSA and NOTECHS by two different raters. To maintain high interrater reliability, several sessions were run to test interrater agreement. Agreement between the three raters was good (average rwg = 0.80). In addition to that, regular expert consultations took place to enable discussion in ambiguous cases. In the case of an overall poor crew performance rating, the video containing the scenario in question was reviewed by an expert.

NOTECHS (Non-Technical Skills; edited by Avermate and Kruijsen, 1998) NOTECHS has four main categories 'co-operation', 'leadership and managerial skills', 'situation awareness' and 'decision making'. They are each linked with 3–4 behavioural elements. In the validation phase the rating is done on a five-point scale: 'very poor', 'poor', 'acceptable', 'good', and 'very good'. Categories and elements of NOTECHS are listed in Table 12.1.

Table 12.1 NOTECHS categories and behavioural elments

Co-operation	*Leadership, Managerial Skills*
Team building and maintaining	Use of authority and assertiveness
Consideration of others	Providing and maintaining standards
Support of others	Planning and co-ordination
Conflict solving	Workload management
Situation Awareness	*Decision Making*
Awareness of aircraft systems	Problem definition and diagnosis
Awareness of external environment	Option generation
Awareness of time	Risk assessment and option selection
	Outcome review

LOSA (Line Oriented Safety Audit; Helmreich et al., in press) This instrument consists of three complementary parts: the Human Factors Checklist, the External Threat Management Worksheet, and the Flight crew Error Management Worksheet. The first component includes the behavioural markers, which are used for the video analysis in this project. The second part deals with the description of

[1] Video analysis with the two behavioural marker systems was accomplished by altogether three raters.

external threats (e.g., weather, ATC, etc.), which influence the performance of the crew. The third part is used for the description of errors made by the pilots or other people involved as well as for the description of the management of these errors.

The behavioural markers are assigned to one of three action sequences (planning, execution, review/modify) and are repeatedly observed for the flight phases pre-departure, take off/climb, and approach/land. LOSA is rated on a four point rating scale: 'poor', 'marginal', 'good', and 'outstanding'.

Table 12.2 LOSA categories and behavioural markers

Planning	Execute	Review/Modify Plans
Briefing	Monitor/cross-check	Evaluation of plans
Contingency planning	Workload management	Inquiry
Workload assignment	Vigilance	Assertiveness
Plans stated	Automation management	

To a large extent LOSA and NOTECHS comprise similar behavioural markers. Despite these similarities in content, they differ in a number of aspects. Firstly, they were designed for different purposes. LOSA was mainly designed for organisational diagnosis (audits) and research and is used in normal flight operations. NOTECHS was designed to assess the CRM performance of individual pilots for qualification purposes. Secondly, LOSA and NOTECHS measure CRM performance on different levels. For LOSA, the crew is the main unit of analysis, complemented by an additional overall rating for each of the individual crew members. NOTECHS measures CRM behaviours on the individual level as a consequence of qualification intention.

The third difference is that LOSA takes repeated measures of the crew with behavioural markers attributed to flight phases, whereas NOTECHS addresses the assessment of CRM behaviours on the category level only, unless a crew member is judged to have failed. In the case of failure, the elements of the respective category were observed to be insufficient and need to be explained (Avermate & Kruijsen, 1998). A fourth difference is that NOTECHS and LOSA use a different number of scale points for the assessment of behavioural markers. CRM assessment with LOSA is done on a four point rating scale, ranging from 'poor', 'marginal', 'good' to 'outstanding'. For NOTECHS a two point rating scale ('acceptable/non acceptable') is proposed when used by practitioners and additionally for validation purposes a five point rating scale is presently used ('very poor', 'poor', 'acceptable', 'good', and 'very good').

Results

Overall Performance of the Sample

The majority of the crews have a good overall CRM performance in all three scenarios (according to LOSA as well as to NOTECHS), which corresponds to the expected distribution for this population. However, performance varies slightly across the three scenarios. A few poorly performing crews were found in each of the three scenarios. In the sample of 46 crews (each flying three scenarios) this was the case for 9 crews and 2 captains overall. Each of the 11 crews concerned performed poorly in only one of the three scenarios. Captains were most often judged as performing poorly in scenario 1 in which the manual workload was very high and technical consequences were most likely.

When looking at the four NOTECHS categories on the individual level, the general picture, that the majority of the crews show good performance, remains the same. A more detailed look at the categories shows that the performance regarding co-operation varies least for captains and first officers (F/Os). Their performance in leadership and situation awareness shows much more variation. Decision making was the category that was the most infrequently observed (Figure 12.1).

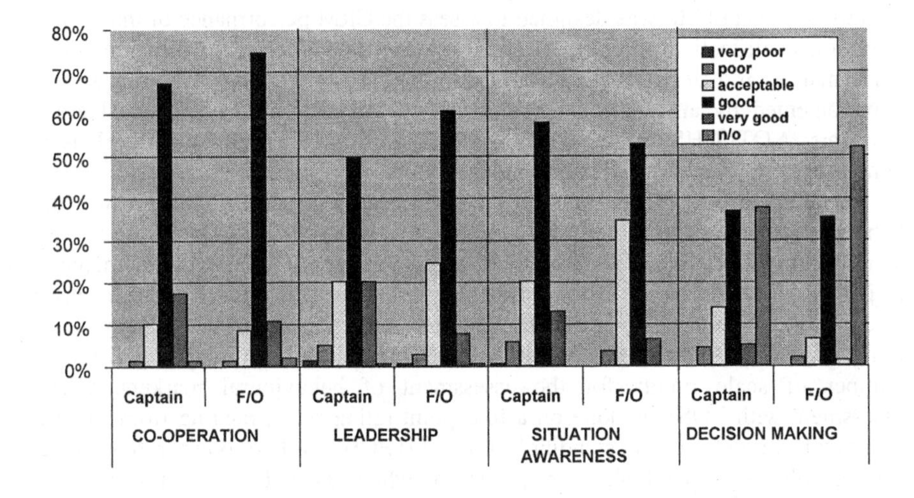

Figure 12.1 Ratings of NOTECHS categories (N=46)

Frequencies of Observations of the Single Behavioural Markers

The five NOTECHS elements most often observed for the Captain were *authority*, *maintaining standards*, *planning/co-ordination*, *awareness of A/C systems,* and *awareness of environment*. Generally, more observations across all elements could be made for the captains than for the F/Os. For the F/Os, the five most frequent observed NOTECHS elements were *maintaining standards*, *awareness of A/C systems*, *planning/co-ordination*, *awareness of environment*, and *support of others*.

The five LOSA behavioural markers that were most often observed were *briefing*, *plans stated*, *monitoring/cross check*, *workload management*, and *vigilance*.

The frequency of observation points towards how easily a specific behaviour can be observed. It does not necessarily tell something about its importance, since there are markers which are more seldom observed, but if they are observed, they are of great relevance. Only one behavioural marker, *conflict solving* (NOTECHS), could never been observed in the simulator.

Variability of Behavioural Markers in Different Flight Phases

The variability of ratings for all behavioural markers (LOSA and NOTECHS) is much higher in the phases of descent and landing than in the other flight phases (pre-departure and take-off) which clearly indicates that descent and landing are the phases of flight where most differences in performance occur during the simulator scenarios.

Instructor Ratings as Performance Indicators

Generally, the study used different kinds of performance indicators. Flight data (integrated into the videos) was used to support the rater by providing objective information (e.g., information regarding the rating of markers such as situation awareness, standards, vigilance), but they did not provide a performance reference independent of the ratings.

Instructor judgements of the performance of every crew were collected in each situation. Contrary to expectations, results clearly showed that overall judgement of the CRM performance from instructors (little variance regarding situations and crews) does not correspond to the NOTECHS or LOSA ratings. This could be due to a number of possible reasons:

1. The global judgement not referring to specific behaviours is too general and, therefore, cannot adequately differ between better and poorer performing crews.
2. The chosen scenarios are difficult for the instructors to rate because scenario 1 is a non-certified emergency procedure and in scenario 2 some instructors seemed to have problems themselves understanding the interconnections of the systems concerned.
3. Instructors (from the same airline and fleet) tend to rate their trainees as being better than they are (i.e., 'too good').

However, it becomes obvious that adequate and sufficient rater training for instructors is essential for a differentiated and valid CRM assessment of crews.

Risk Index

For phase specific analysis of the data a risk index was built which represents the total number of instances encountered across all behavioural markers during an entire flight, when safety of the flight was jeopardised, resulting in below standard assessment (poor and marginal ratings for LOSA/very poor and poor ratings for NOTECHS). If a crew observed with LOSA was rated poor for briefing and marginal for plans stated in the descent phase and was rated marginal for vigilance in the landing phase, the risk index of the crew was 3. This does not necessarily mean that the overall rating of the crew's performance in the scenario concerned was poor. The risk index is mainly a measure for the frequency of occurring deficiencies during the process.

For NOTECHS, a separate risk index was computed for each crew member, as well as, a risk index for the crew. This kind of index provides information regarding the overall process performance of the crew and additionally enables a direct comparison of NOTECHS and LOSA results on a general level. Correlation of the Risk Index with the judgement of the overall effectiveness of the single scenarios proved to be highly significant. This confirms the usefulness of the risk index. Table 12.3 shows the descriptive statistics for the NOTECHS and LOSA risk indices for each of the three scenarios.

Table 12.3 Descriptive statistics of risk indices for NOTECHS and LOSA

	NOTECHS Risk Index									**N**
	Captain			First Officer			Crew			
	Freq. RI	Mean *(SD)*	*Min/ Max*	*Freq. RI*	Mean *(SD)*	*Min/ Max*	*Freq. RI*	*Mean (SD)*	*Min/ Max*	
S1	0: n=24 1: n=12 ≥2: n=9	1.04 *(2.00)*	0/11	0: n=29 1: n=9 ≥2: n=7	0.67 *(1.11)*	0/4	0: n=19 1-3: n=16 ≥4: n=10	2.24 *(3.30)*	0/14	45
S2	0: n=34 1: n=7 ≥2: n=4	0.73 *(2.31)*	0/12	0: n=36 1: n=5 ≥2: n=4	0.38 *(0.94)*	0/4	0: n=14 1-3: n=17 ≥4: n=14	2.84 *(3.63)*	0/18	45
S3	0: n=42 1: n=0 ≥2: n=3	0.38 *(1.56)*	0/9	0: n=40 1: n=2 ≥2; n=3	0.38 *(1.27)*	0/5	0: n=23 1-3: n=15 ≥4: n=7	1.84 *(3.28)*	0/14	45
S1-S3 total	0: n=19 1-4:n=19 5-8:n=4 ≥9: n=4	2.109 *(3.143)*	0/12	0: n=24 1-4:n=17 5-8:n=5 ≥9: n=0	1.391 *(1.972)*	0/7	0: n=6 1-4: n=15 5-11 n=15 ≥12: n=10	6.78 *(6.24)*	0/21	46

It is important to notice that the absolute values of the NOTECHS and LOSA risk indices are not directly comparable because of the different scales. Since point three ('acceptable') of the five-point NOTECHS scale is not included in the risk index, the risk indices of NOTECHS represent a more rigorous judgement than the

LOSA risk indices (including 'marginal' ratings, point two on a four-point scale). The NOTECHS individual risk indices show that captains have a higher overall value than first officers and that the maximal for the captains are higher in each situation. Here one must take the fact that the number of observations was higher overall for the captains into account. Table 12.4 also shows that the risk indices for LOSA and NOTECHS were lower for scenario 3 in which no technical problem occurred.

The NOTECHS risk index shows a high correlation between the Captain and the First Officer, which could be confirmed on the level of the four categories (Table 12.4).

Table 12.4 NOTECHS: Correlation between Captain's and First Officer's ratings

	S1	S2	S3
Risk Index [a]	.746*** p= .000 n= 45	.920*** p= .000 n= 45	.906*** p= .000 n= 45
NOTECHS co-operation	.576*** p= .000 n= 43	.628*** p= .000 n= 43	.384** p= .005 n= 44
NOTECHS leadership	.508*** p= .000 n= 43	.293* p= .025 n= 45	.549*** p= .000 n= 45
NOTECHS situation awareness	.683*** p= .000 n= 45	.447*** p= .001 n= 45	.494*** p= .000 n= 45
NOTECHS decision making	.659*** p= .000 n= 23	.535** p= .002 n= 28	- - n= 4

Correlation: Spearman's rho

*** Correlation is significant at the .001 level (1-tailed).

** Correlation is significant at the .01 level (1-tailed).

* Correlation is significant at the .05 level (1-tailed).

[a] Correlation: Pearson (1-tailed)

Stability of CRM Performance

CRM performance was rated separately for each scenario to compare the ratings for each behavioural marker measured in three different scenarios. Stability of CRM skills is represented by a high correlation between the three ratings of a specific behavioural marker across the scenarios.

In general, for both LOSA behavioural markers and NOTECHS categories (cf. Table 12.5), the results show correlations that are not statistically significant.

After finding little consistency of CRM performance across different scenarios on the level of behavioural markers, the relationship between risk indexes for each scenario was analysed. The question was whether crews show a similar level of performance (i.e., a constant number of poor or marginal ratings across the three scenarios) independent of the single CRM behaviours concerned. The results of LOSA, as well as of NOTECHS, showed statistically insignificant relationships between the risk indices of the three scenarios (Table 12.6).

Table 12.5 Stability of CRM performance across different scenarios NOTECHS, Captain, and First Officer

	Captain			First Officer		
	S1-S2	**S1-S3**	**S2-S3**	**S1-S2**	**S1-S3**	**S2-S3**
Co-operation	.088 p= .290 n= 42	.342* p= .013 n= 42	-.022 p= .444 n= 44	-.102 p= .261 n= 42	.032 p= .420 n= 43	-.014 p= .464 n= 41
Leadership	.282* p= .034 n= 43	.127 p= .208 n= 43	.245 p= .054 n= 44	.106 p= .250 n= 43	.148 p= .172 n= 43	.060 p= .350 n= 44
Situation awareness	.110 p= .238 n= 44	.039 p= .400 n= 44	.170 p= .135 n= 44	-.118 p= .223 n= 44	.035 p= .411 n= 44	.003 p= .491 n= 44
Decision making	-.059 p= .376 n= 31	.219 p= .338 n= 6	-.335 p= .258 n= 6	.207 p= .198 n= 19	-.745 p= .074 n= 5	n= 3

Correlation: Spearman's rho
** Correlation is significant at the .01 level (1-tailed).
* Correlation is significant at the .05 level (1-tailed).

Both levels of measurement – behavioural markers and risk index – showed inconsistency in CRM performance across the three scenarios for the whole sample. This led to the analysis of subgroups with specific performance patterns across the scenarios. As criterion for identifying patterns of CRM performance across the three scenarios, the risk indices were used and the following three subgroups with different patterns of CRM performance were identified:

1. Excellent crews (n = 8): more than 95 percent of the ratings across all three scenarios in the 'good/very good' area (NOTECHS), 'good/outstanding' area (LOSA), respectively:
2. Poor crews (n = 9): Constantly a risk index above the average (NOTECHS and LOSA) in all three scenarios; and
3. Other Crews (n = 26): all other crews not meeting the criteria defined for the extreme performing crews.

Table 12.6 Stability of Risk Index across different scenarios

	S1-S2	S1-S3	S2-S3
LOSA Risk Index	.056 p= .359 n= 44	.011 p= .471 n= 44	.105 p= .248 n= 44
NOTECHS Risk Index (CAPTAIN)	-.105 p= .249 n= 44	-.071 p= .323 n= 44	-.053 p= .367 n= 44
NOTECHS Risk Index (FIRST OFFICER)	.031 p= .421 n= 44	-.010 p= .475 n= 44	.066 p= .335 n= 44

Correlations: Pearson, N = 46
** Correlation is significant at the .01 level (1-tailed).
* Correlation is significant at the .05 level (1-tailed).

The majority of the crews (n = 26) across the three scenarios showed no specific pattern concerning their CRM performance that varied between marginal and outstanding (LOSA), acceptable and very good (NOTECHS). The two extreme groups show more stable CRM performance, one in the good to very good area, the other at the lower end of the performance scale. The following section will further elaborate on the characteristics of these two groups.

Characteristics of Excellent and Poor Crews

In this section particular strengths were found in this study for excellent crews and deficiencies of poor crews are reported. Strengths correspond to areas in which the majority of the excellent crews had outstanding ratings. Deficiencies of poor crews are indicated when the majority of them had marginal or poor ratings on specific behavioural markers (cf. Table 12.7).

Excellent crews show particular strengths in aspects of planning and anticipation. In respect to NOTECHS, captains of those crews generally have very good ratings on leadership and situation awareness elements. In the scenario with the diagnosis problem they additionally distinguish themselves through highly adequate decision making behaviours (problem definition, option generation) while the respective first officers provide good support for the Captain. In regards to LOSA it was observed that most of the excellent crews had an outstanding contingency management. In scenario 2 ('diagnosis problem'), these crews are characterised by strengths in that planning (plans stated, workload assignment, contingency management) and execution (monitoring/x-check), as well as review/modify (inquiry) behavioural markers are concerned.

In general, poor crews show specific deficiencies in both their briefing/planning behaviour and in aspects of situation awareness/vigilance. There are instances in which they fail to fulfil and maintain standards.

Table 12.7 Strengths and deficiencies of poor vs. excellent crews

		Strengths of excellent crews	Deficiencies of poor crews
S1	**NOTECHS**: Captain	authority planning & co-ordination awareness of time	maintaining standards workload management awareness of A/C systems awareness of environment
	NOTECHS: First Officer		support of others
	LOSA	contingency management	briefing
S2	**NOTECHS**: Captain	authority awareness of A/C systems problem definition option generation	authority maintaining standards planning & co-ordination awareness of A/C systems problem definition option generation risk assessment
	NOTECHS: First Officer	support of others	awareness of A/C systems
	LOSA	plans stated workload assignment contingency management monitoring, x-check inquiry	briefing plans stated vigilance automation management evaluation of plans
S3	**NOTECHS**: Captain	authority	support of others authority maintaining standards awareness of A/C systems awareness of environment
	NOTECHS: First Officer	planning & co-ordination	maintaining standards planning & co-ordination awareness of A/C systems awareness of environment awareness of time
	LOSA		briefing vigilance evaluation of plans

When taking a closer look at the NOTECHS ratings, it is obvious that captains of poor crews show insufficient leadership behaviours (authority, maintaining standards, planning/co-ordination). In scenario 2 ('diagnosis problem'), they also fail to make decisions in a well structured and appropriate way (problem definition, option generation, risk assessment). The captains and first officers lack adequate situation awareness in respects to the A/C systems and the environment in several instances across all three scenarios. A further problem is the insufficient support provided by the Pilot non-flying (PNF).

The striking deficiencies in respect to the LOSA behavioural markers are planning elements (briefing, plans stated), vigilance, and the evaluation of plans. Furthermore, in scenario 2, automation management showed to be inadequate within the sub-sample of poor crews.

Correspondence Between NOTECHS and LOSA

NOTECHS and LOSA give a very similar picture of the sample regarding the overall performance of the crews that were rated. Since the NOTECHS and LOSA ratings were done independent of each other, the correspondence between similar aspects of NOTECHS and LOSA can be considered to be a proof of the validity of the measurements.

There is good correspondence between the risk indices measured with NOTECHS and LOSA. They come to an overall similar result considering the number of times the performance of a crew drops below an acceptable level as shown by the statistically significant correlation of the risk indices in Table 12.8.

Table 12.8 Correlation between NOTECHS and LOSA risk indices

	LOSA risk index S1	**LOSA risk index S2**	**LOSA risk index S3**
NOTECHS risk index crew S1	.867** p= .000 n= 45		
NOTECHS risk index crew S2		.502** p= .000 n= 45	
NOTECHS risk index crew S3			.880** p= .000 n= 45

Pearson correlations
** Correlation is significant at the .01 level (1-tailed).
* Correlation is significant at the .05 level (1-tailed).

Also the finding of instability of performance across different situations could be identified in both behavioural marker systems. Overall pictures regarding strengths and deficiencies of poor and excellent crews correspond well but can be complemented where the concepts are not covered by the other marker system (e.g., decision making).

Summary and Implications for Practice

The Use of Behavioural Markers (LOSA/NOTECHS) in the Simulator Environment

In summary the study shows that both behavioural marker systems are fully applicable for CRM assessment in the simulator under conditions of high task load. Generally it can be said the most important observations that discriminate best between crews were made during descent and approach, where CRM markers had highest variability. Elements regarding leadership and situation awareness (NOTECHS), respectively markers concerning planning and execution (LOSA) were some of the most frequently observed behaviours. Moreover it was recognised that each scenario had special demands concerning certain behavioural markers which clearly shows that the variety of markers used is necessary to cover different situations. So possible implications to support instructors in assessing CRM performance are the reduction respectively of the emphasis on specific markers that are relevant in specific simulator scenarios and the selection of relevant flight phases for observations.

Overall the CRM assessments with LOSA markers and NOTECHS elements correspond well. The finding of instability of performance across different situations was found with both behavioural marker systems. Also overall pictures regarding strengths and deficiencies of excellent and poor crews correspond well but can be complemented if concepts are not covered by both marker system (e.g., decision making).

The question regarding the unit of analysis – individual or crew rating – cannot be clearly answered on the base of the results of the study presented. Foushee (1984) noted that when an individual team member commits an error and the team does not catch that error and correct it, the error becomes a team error. In general, the concept of CRM supports this view and puts the focus on the team. But when it comes to observations and measurement things get more complicated. Teams do not actually do anything, it is the individuals that do. On one hand the constant high correlation between captains and first officers ratings indicate that these individuals do not act independently of each other. On the other hand if crew performance is low, there are often observable differences in the behavioural markers that are rated poor for the individual pilots (e.g., poor leadership by the Captain with no intervention form the Co-pilot can result in 'poor' rating for assertiveness of the First Officer). This differentiation can play an important role, e.g., in the case of qualification or in the context of training associated with individual feedback. Taking this into account a mixture of individual and crew behaviours and associated ratings is conceivable.

Instability of CRM Performance Across Different Scenarios

Both LOSA and NOTECHS ratings of crews showed little consistency across the three scenarios neither on the level of behavioural markers nor for the risk index. The generally low correlations between CRM aspects across the three different scenarios strongly point towards situation dependency of CRM skills. The scenarios were chosen to be different so they would impose different levels and kinds of demands on the crew. It can be argued that this variation in scenario explains the inconsistency of CRM performance. Also other studies (e.g., Prince, Brannick, Prince, & Salas, 1997; Orasanu, Fisher, & Tarrel, 1993) found different performance of teams across situations even if they intended to create similar scenarios (Prince et al., 1997). This suggests that properties of the situation may have a stronger effect on team behaviour than stable team or individual differences. This points towards the conclusion that teams may not be consistent or reliable across scenarios. Another possible explanation for the instability of CRM performance takes into account that there could be a lack of rating opportunity to observe sufficient instances of behaviours. The numbers of team process behaviours called for in each scenario may not be sufficient for a reliable assessment of that dimension. The high number of missings regarding the single elements for observation of the categories would support this assumption.

The inconsistency of the CRM performance across scenarios has practical implications. It indicates that CRM performance in one situation is not necessarily predictive for CRM performance in other situations. This has implications for training and assessment practices. It is important to choose several different scenarios for training and qualification purpose in order to obtain a representative (valid and as reliable as possible) picture of a crew. Crews need to be made aware of the differences in the demands that situations provide. Briefings and debriefings should consider the situational demands.

Furthermore, in the future it seems that it will be important to distinguish those team skills that possess temporal stability from those that do not.

Deficiencies of Poor Crews and Training Needs

Despite the general inconsistency of CRM performance across different scenarios, there are two subgroups with a more stable CRM performance, whereby the group that is performing below average in all scenarios clearly points towards training needs. Poor crews have a planning deficit in all three scenarios: briefing and partly contingency management, and often fail to maintain standards. They have a lack of vigilance and of situation awareness in aircraft systems and external environment. In two of the three scenarios they perform a poor evaluation of plans. In the scenario with the diagnosis problem, they lack critical knowledge of the systems that have failed. Hence, their problem definition and option generation is poor. In summary, poor crews seem to put little effort into planning, do not have good situation awareness, and do not understand the problem at hand well. Therefore, they do not evaluate their plans as necessary. With this constellation and with captains characterised by deficits in leadership, they are not able to enhance safety in the face of difficult situations.

Rater Training

The study also showed the necessity for rater training and the use of sophisticated rating tools. Flight instructors who rated the overall performance of the crew on a general level, showed difficulties in sensitivity to performance differences. Their ratings do not correspond well with the ratings with LOSA and NOTECHS (especially not for critical and poor performance).

Pilots that were used as reviewers and who were familiar with the behavioural marker systems used in the study showed good correspondence with LOSA and NOTECHS judgements.

Outlook

This paper concludes with an outview of the second project phase. The core interests are the study of the relationships between CRM behaviour, team performance, errors, and workload. The following questions are of concern: Do good CRM performers have a better technical and overall performance? Do they commit less or different errors and do they deal differently with errors so that they get another error outcome with different consequences? Another question is the influence of crew specific factors such as crew composition (level of experience of the two pilots; e.g., experienced captain, new co-pilot) and familiarity of the crew (first leg ever flown together vs. flown together before).

References

Avermate van, J. A. G., & Kruijsen, E. A. C. (Eds). (1998). *NOTECHS - The evaluation of non-technical skills of multi-pilot aircrew in relation to the JAR-FCL requirements* (Final Report). Research Report for the European Commission (DG VII, NLR-CR-98443).

Delsart, M. C., & Andlauer, E. (1998). *Guidelines for the experimental plan of JAR TEL* (ref. JARTEL/SOF/WP2/01).

Endsley, M. R. (1995). Towards a theory of situation awareness. *Human Factors, 37*(1), 32-64.

Flin, R., & Martin, L. (2001). Behavioral markers for resource management: A review of current practice. *The International Journal of Aviation Psychology, 11*(1), 95-118.

Flin, R., & Martin, L. (1998). *Behavioural markers for Crew Resource Management* (CAA paper 98005). Civil Aviation Authority, London.

Foushee, H. C. (1984). Dyads and triads at 35,000 feet. *American Psychologist, 39*, 885-893.

Hart, S. G., & Staveland, L. E. (1988). Development of NASA-TLX (Task Load Index): Results of empirical and theoretical research. In P. A. Hancock & N. Meshkati (Eds.), *Human Mental Workload*. Amsterdam: Elsevier.

Helmreich, R. L., Klinect, J. R., & Wilhelm, J. A. (1999). Models of threat, error, and CRM in flight operations. *Proceedings of the Tenth International Symposium on Aviation Psychology* (pp. 677-682). Columbus, OH: The Ohio State University.

Helmreich, R. L., Wilhelm, J. A., Klinect, J. R., & Merritt, A. C. (in press). Culture, error, and Crew Resource Management. In E. Salas, C. A. Bowers, & E. Edens (Eds.),

Applying Resource Management in Organizations: A Guide for Professionals. Hillsdale, NJ: Erlbaum.

ICAO. (2002). Line Operations Safety Audit (LOSA) has become the central focus of ICAO's human factors programme. *ICAO Journal, 57*(4).

Klampfer, B., Flin, R., Helmreich, R. L., Häusler, R., Sexton, B., Fletcher, G., et al. (2001). *Enhancing Performance in High Risk Environments: Recommendations for the use of Behavioural Markers* [Brochure]. Berlin, Germany: Gottlieb Daimler und Karl Benz Group Interaction in High Risk Environments, Stiftung, Collegium.

Orasanu, J., Fischer, U., & Tarrel, R. (1993). A taxonomy of decision problems on the flight deck. In R. Jensen (Ed.), *Proceedings of the Seventh International Symposium on Aviation Psychology* (pp. 226-232). Columbus, OH: Ohio State University Press.

Prince, A., Brannick, M. T., Prince, C., & Salas, E. (1997). The measurement of team process behaviors in the cockpit: Lessons learned. In M. T. Brannick, E. Salas, & C. Prince (Eds.), *Team Performance Assessment and Measurement* (pp. 289-310). Mahwah, NJ: Lawrence Erlbaum Associates.

Urban, J. M., Bowers, C. A., Morgan, B. B. J., Braun, C. C., & Kline, P. B. (1992). The effects of hierarchical structure and workload on the performance of team and individual tasks. *Proceedings of the Human Factors Society 36th Meeting* (pp. 954-958). Santa Monica, CA: Human Factors Society.

Wickens, C. D. (2001). Workload and situation awareness. In P. A. Hancock & P. A. Desmond (Eds.), *Stress, Workload, and Fatigue: Human Factors in Transportation* (pp. 443-450). Mahwah, NJ: Lawrence Erlbaum Associates.

Chapter 13

Stretching the Search for the 'Why' of Error: The System Approach

Marilyn Sue Bogner
Institute for the Study of Medical Error, Bethesda, USA

Introduction

In health care as in aviation, accidents typically are attributed to human error. Not only does this occur in the media, but also people tend to blame themselves when involved in an error while working in a professional capacity. This is particularly true for health care providers who are taught they are responsible for whatever happens to the patient. The literature in a number of domains including aviation indicates that error is the result of an alignment of conditions and occurrences each of which is necessary, but none alone sufficient to cause the error. That alignment of factors creates error-provoking conditions affecting the context in which an incident occurs. The prevailing presumption in health care and its literature, however, is that the source of medical error is the care provider. This presumption is self-validating; once the source of error has been identified, there is no further search for the cause to disconfirm that finding, hence the presumption is validated. The ensuing discussion (a) illustrates how the presumption that the person is the source of error influences not only the development of research methodology, but also the interpretation of research findings, and (b) presents the systems approach to medical error (SAME) as a technique for stretching the search for the why of error beyond the care provider to factors affecting the context of care.

Medical Error

The magnitude of the problem presented in the report *To Err is Human* that 44,000 to 98,000 hospitalised patients die annually in the US as a result of error (Institute of Medicine, 1999) caused concern and consternation in policy makers and the public alike. Those figures based on findings of studies conducted in Colorado and Utah (Thomas et al., 2000) and the Harvard Medical Practice Study conducted in New York State (Leape et al., 1991) were extrapolated to hospital admissions throughout the US. The report indicated that over half of those deaths were the result of preventable medical errors.

The magnitude of the figures in the Institute of Medicine (IOM) report has been questioned by McDonald, Weiner and Hui (2000). They contend the figures are inaccurate and misleading because people who enter the hospital are ill and some are as likely to die from their illness as other reasons. There is a countervailing belief, however, that errors are under-reported perhaps by as much as a factor of 10; that is, for every error reported, 10 are not. Such under-reporting reflects care providers' fear of the consequences of committing an error or having an error attributed to them; that is, fear of blame and retribution through litigation as well as fear of loosing professional respect. Assuming that extent of under-reporting and acknowledging the concern about the projected number of deaths due to error by reducing the numbers by half, the estimated annual number of hospitalised patients deaths due to medical error becomes 220,000 to 490,000. These numbers are not to be interpreted as a condemnation of health care in America, rather the identification of a problem.

Adverse events (AEs) in health care are not unique to the US. The UK report on medical error indicates that in National Health Service hospitals alone, harm (injury as well as death), caused by human error occurs in approximately 10 percent of admissions or a rate in excess of 850,000 annually (Department of Health, 2000). Thus, error is a pervasive problem that must be addressed; it must be determined why AEs occur so that changes can be made to reduce the likelihood of recurrence. To be effective, efforts for that determination should be free of presumptions – free from the pre-conceived attitude that humans are responsible for error. That attitude, however, has deep roots.

The presumption that humans inherently are error-prone is presented as the title of the IOM report, *To err is human*, and through the focus of that report on errors *qua* errors by care providers. Error reporting as advocated in the report will provide numbers of incidents in standardised nomenclature; however, if the standardised reports focus only on care providers, the information obtained would indicate care-provider centred remediation possibly ignoring the actual cause. The potency of the presumption of human fallibility is manifest in the findings of two major studies.

The Presumption of Human Fallibility

The Harvard Medical Practice study (HMPS) addressed error by the retrospective review of a sample of charts of patients hospitalised in New York State in 1984 (Leape et al., 1991). Errors recorded on the charts were grouped in categories. The findings were presented under the heading 'Types of Errors' with the weighted proportion of the total errors reported for each category. This included the categories of total operating room (OR) errors, OR wound infections, non-OR drug related errors, and diagnostic mishaps; weighted proportions of errors for those categories were 47.7, 13.6, 19.4, and 8.1 respectively. The manner by which the data were gathered and arrayed connotes the care provider as the source of the error, hence the cause of the adverse outcome; however, that is not necessarily valid. The weighted proportion of 13.6 of errors reported for OR infections might be attributable to factors other than error. The cool temperature in the OR has been found to be conducive to wound infections (Kurz, Sessler, & Lenhardt, 1996) as have patient factors and several surgical considerations such as prophylactic antibiotics (Buggy, 2000).

The presumption of human fallibility also is evident in a study that used the HMPS methodology in a retrospective review of a sample of charts of patients hospitalised in Utah and Colorado in 1992 (Thomas et al., 2000). The percent of permanent disability attributed to error was arrayed with respect to the percent of errors in a category. Among those findings were 44.9 percent of errors occurred in the OR, 19.3 percent were drug related, and 6.9 were incorrect or delayed diagnosis; the percent of permanent disability associated with those errors are 16.6, 9.7, and 20.1 respectively. That study also presented the percent of total errors by location with respect to the percent of errors deemed to be the result of negligence. Those findings include 39.5 percent of the errors occurred in the OR, 3.4 percent in a procedure room, and 3 percent in the emergency room; the percent of those errors that involved negligence were 18.1, 5.4 and 52.6 respectively.

The findings of Thomas et al. (2000) could lead to the conclusion that incorrect or delayed diagnosis was a much worse error than drug related errors because the percent of resulting permanent disability was so much higher; however, these findings do not indicate a means for reducing the likelihood of such errors or disabilities. It also may be concluded from the findings that although OR personnel commit many errors, they are only moderately negligent; personnel in procedure rooms commit few errors which reflect a small amount of negligence. In the emergency room, however, negligent personnel appear to abound. By focussing on the care providers as the presumed cause of error, the only information these conclusions provide about how to reduce the likelihood of error is to increase staff competency. That solution may be far from fact. A comparison of the findings of this study with those of the HMPS (Leape et al., 1991) provides additional information.

The HMPS review of charts of patients hospitalised in New York State in 1984 reported that a weighted proportion of 47.7 of the total errors occurred in the OR, 19.3 were non-OR medication related, and 8.1 were diagnostic mishaps (Leape et al., 1991). The comparable review of charts of patients hospitalised in Utah and Colorado in 1992 found 44.9 percent of total errors occurred in the OR, 19.8 were medication related, and 6.9 were diagnosis related (Thomas et al., 2000). The comparability of errors in those categories over time and geographical location is striking. It may be argued that the figures were computed differently and the 1984 data were only non-OR medication errors whereas the 1992 data apparently were all medication errors; however, the methodology for collecting the data was comparable for the two studies. One interpretation of the comparability is consistency of the competency (or lack of competency) of care providers. This is not feasible given the diversity of care providers across time and location. Another interpretation stretches the search for why AEs occur beyond the care provider to factors inherent in the provision of health care as causes of error. This questions the presumption that care providers are the source of error.

Basis for the Presumption: To Err is Human

There is no evidence in the empirical literature to support the premise that to err is an inherent human trait. Rather, that statement is literary. 'To err is human, to forgive divine' appears in *An Essay on Criticism* (1711) by Alexander Pope (1688-1744). Despite the lack of empirical evidence, the cause of health care errors typically is attributed to the humans associated with the AEs often with an accompanying comment referring to the error-prone nature of people. This presumption supported only by a literary statement is neither shared nor supported by research or practice in industry (Weick, 1987). That attitude by industry may be an indication of changes in the Zeitgeist, the spirit of the times, in health care; the presumption that humans are the source of error appears to be undergoing a degree of revision.

Thomas et al., (2000) suggest that many AEs in medicine are likely the result of multiple factors rather than an individual practitioner committing an error. Similarly, the IOM report acknowledges that multiple factors can contribute to error and that blaming an individual does not change those factors (Institute of Medicine, 1999). The UK report on medical error emphasises the necessity of transcending the presumption of the human as the source of error to consider systemic factors that are deeper and more pervasive than the individual – factors which should be addressed to prevent errors (Department of Health, 2000).

Lessons learned in aviation support the emerging acknowledgement that factors other than the care provider have a role in medical error. Experience in that domain found that information about the context of accidents is necessary to understand how they occurred so that steps can be taken to prevent future accidents (Billings, 1997). The UK report proposes that to learn from and prevent AEs, the wider causes of error should be addressed (Department of Health, 2000). Those wider causes based on lessons learned from aviation are in the context of the incident. Research by industry has identified error-provoking factors in the context of accidents. Such factors when incorporated into a systems approach expand possible error sources beyond the human.

The Systems Approach

The IOM report suggests that a systems approach is required to identify and modify the conditions that contribute to error (Institute of Medicine, 1999); however, what constitutes the system is unclear. The term system is used in many ways such as to refer to a variety of entities such as the collection of facilities that provide health care – the health care system, or 'an interdependent group of items, people, or processes with a common purpose' (Leape et al., 1995). The term as used in this systems approach to medical error (SAME) refers to the complex of interacting elements (von Bertalanffy, 1968). The system may be simple consisting only of one complex of interacting elements or it may be a complicated supra-system comprised of interacting sub-systems each with its complex of interacting elements. The term approach is used because the underlying concept of systems is applicable to all targets of analysis (von Bertalanffy, 1968).

This systems approach to medical error, SAME, builds on the findings from error research in industry as identified and presented as taxonomies (Rassmussen, 1994) and as figural representations (Senders & Moray, 1991). Also incorporated are factors found by research in various disciplines such as psychology and Human Factors to impact behaviour, hence have error-provoking potential. The factors comprise categories as systems. The systems are conceptualised as concentric circles that range from distal, having an indirect impact, to proximal which have an immediate impact on the provider (Moray, 1994). Those systems starting with the most removed from the care provider, with examples of elements in parentheses, are: legal-regulatory-national culture (litigation, reimbursement, accreditation, end of life issues), organisation (workload, shifts, reports, policies for caring for uninsured, organisational culture), social (communication, interactions with patients' family members and other personnel, professional culture), physical setting (placement of equipment, room size, clutter), ambient conditions (altitude, illumination, temperature, noise), patient characteristics (presenting problem, medication history, age, body weight, co-morbidity, anxiety), means of providing care (technological sophistication, cognitive workload, time for use), and conditions of the care provider (knowledge, skills, experience, health, stamina, fatigue).

The influence of factors in each circle-as-system affects those systems within its circumstance and ultimately impacts the care provider (Bogner, 1999). This is the reverse ripple effect; rather than rippling out from the point of impact as when the surface of a pond is disturbed by a stone, the impact of a change in any system ripples inward. For example, a change in the distal legal-regulatory-national culture system that constrains reimbursement reduces the funds available to the organisation which necessitates reducing the size and educational level of the staff thus impacting workload and communication patterns, the purchasing of equipment, and patients who present at an advanced stage of illness – all affecting the care providers in the context of care. A value of the systems approach is that the analysis of the context of an AE in terms of each system interferes with manifestations of the presumption of human fallibility and as such allows identification of error-provoking factors.

The importance of the conceptualisation of the system is apparent when considering the findings of a systems approach to adverse drug events (Leape et al., 1995) in terms of the SAME technique. In that study, health care professionals trained for the purpose of the study solicited voluntary reports of adverse drug events (ADEs) or potential ADEs, The purpose was to understand why ADEs occurred in terms of system failures rather than assigning responsibility. Case investigators used an interview schedule to obtain information about the circumstances surrounding the ADE, the proximal cause. It was reported that most errors occurred in the categories of physician ordering and in nurse administration; the remaining ADEs were split between transcription and pharmacy dispensing.

The definition of the categories of AEs in terms of care providers assigns responsibility which is counter to the goal of the study (Leape et al., 1995) –

testimony to the potency of the presumption of human fallibility. The presumption is further apparent in the definition of errors associated with problems with infusion pumps (computer chip driven devices that are programmed to regulate the flow of fluid into the body) and parenteral delivery devices (the programmed delivery of nutritional fluids into a person via an infusion pump like device). The errors identified were in setting pumps, accidental tubing disconnections, and confusion between central and peripheral lines all of which imply human causality. The implication of human causality also was manifest in that study's example of the search for third-order 'whys'. The situation addressed was an elderly patient who became comatose from a drug; the third order why technique asks why the incident occurred. The response in the example that the patient received the wrong dose of the drug was questioned 'why'. The example's response was the physician did not know the proper dose of the drug for an elderly patient leading to the conclusion that the drug knowledge dissemination system was inadequate. The SAME technique provides another interpretation of the findings.

The categories of error in the Leape et al. (1995) study are transformed by SAME into drug ordering, drug administration, transcription, and drug dispensing. That change is subtly powerful; the replacement of the care provider by a care providing factor, the drug, reorients the focus and stretches the search for why the situation occurred beyond the human to system factors. That focus together with the analysis of the situation with respect to each system in SAME allows the identification of error-provoking factors affecting the context of care such as distracting sound, and identifies a target for change to reduce the likelihood of recurrence of the error. Such analysis by focussing on systems rather than the care providers prevents the presumption of human fallibility from truncating the search for factors that contribute to the error. The problems with infusion pumps defined in the study as due to errors setting up the devices would be determined by SAME as problems in the means of providing care which would lead to identifying error-provoking aspects of the design of the devices (Brown, Bogner, Parmentier, & Taylor, 1997).

The application of the SAME to the example of the elderly patient analysed by the third order why technique (Leape et al., 1995) considers the situation with respect to each system beginning with characteristics of the patient such as co-morbidity, fragility; and the means of providing care such as the name of the drug on the order, how and when was the order was placed, and how the drug was administered such as via an infusion pump. The analysis would proceed in a similar manner through the various systems. 'Why' is used in SAME to determine how an error-provoking factor occurred. If it was found that the order for the drug was difficult to read, the question why is pursued through system factors. The analysis might discover that the order was difficult to read because coffee spilled on it; why coffee had been spilled would be pursued in terms of systems, not people. The response to that question could be that the drug order had been on the desk at the nurse's station when a full cup of coffee was bumped spilling some. Querying why the cup was bumped could discover that the physical setting of the nurse's station was so small that items had to be put in perilous positions. This identification of an inappropriately small work space points to an error-provoking factor that when rectified could not only reduce the likelihood of distorted drug orders due to spilled coffee, but also the likelihood of confusion in chart entries.

The comparison illustrates that findings from systems approaches vary as to the definition of the systems and the sensitivity to the presumption of human fallibility. The SAME technique by analysing an incident by system to identify error-provoking factors not only expands the focus beyond the care provider as the source of error, it promotes the identification of a number of factors or precursor events (Bogner, 1998). To be practically viable, it is necessary the application of the systems approach to identify error-provoking factors must be easy.

The systems approach of the study of ADEs (Leape, et al., 1995) involves labour intensive individual chart review and personal interviews, so is not amenable to widespread adoption. The root cause analysis to determine the source of error in sentinel events promoted by the US Joint Commission for the Accreditation of Healthcare Organisations also is labour intensive. Root cause analysis teams focus on systems and processes, progressing from special clinical causes to common organisational causes, probing by asking why, and identifying changes to improve level of performance. Based on the social and behavioural science research literature, the findings of such an analysis probably do not reflect the factors that actually contributed to the event. Two major reasons are: memory decays over time – the closer to the event in time information is obtained, the more accurate it is; and the perspectives of different people provide different interpretations of circumstances – the most accurate description of how an incident occurred is provided by the person involved. The method discussed in the next section offers a way of avoiding those pitfalls.

Assessment Methodology

The assessment methodology for SAME is grounded in the findings that to understand a person's behaviour, it is necessary to determine how the person experienced the environment, the context in which that behaviour occurred (Gold, 1999). The aviation industry acknowledges this by providing self-reported free text to obtain comprehensive information about an event in the Aviation Safety Reporting System (ASRS) (Reynard, Billings, Cheaney, & Hardy, 1986). Free text has the advantage over a questionnaire of not channelling the respondent's attention to factors the questionnaire developer believed relevant, but may be extraneous to the person experiencing the incident. On the other hand, a totally open, free response methodology allows people to act on their tendency to blame themselves or other people for errors involving human activity. This especially is the case for health care professionals who assume responsibility for error as their professional obligation; they blame themselves even though the error involved a device or procedure that was nearly impossible to use. Another issue with free text is that the reporter may consider only the immediate factors and not factors in more distal systems that determine those proximal factors. Again benefiting from lessons learned by aviation, the design of the SAME assessment methodology acknowledges that health care providers through their experience in the context in which an AE occurs have optimal knowledge of error-provoking factors and as such are the best sources of that information.

The assessment tool for SAME is a variation of the Critical Incident Technique (Flanagan, 1954). That technique, the CIT, was effective in identifying error-provoking problems with anaesthesia machines (Cooper, Newbower, Long, & McPeek, 1978). The CIT has been modified and used to consider anaesthesia incidents in a number of studies. Short, O'Regan, Lew, and Oh (1992) developed a questionnaire based on the CIT that requested a description of the incident including associated and contributory factors. The SAME assessment tool is a further modification of the CIT that retains the CIT spirit, yet reduces the technique to a checklist. This is done to address care providers' comments that they did not have time to write details of what occurred in an incident as is necessary in a narrative yet want to record their experiences near to the time of the incident because they tend to forget aspects of the incidents after a period of time.

The checklist is comprised of the systems categories with examples of factors as memory aides for each system (Bogner, 2000). The structure of the checklist decreases the possibility of attributing error solely to an individual while at the same time stimulating recall and expanding consideration of potential contributing factors. The care provider is to note error-provoking factors on the checklist as near to the time they are experienced as possible. Those factors can be a word or two initially and later elaborated.

Identification of Error-provoking Factors

Care providers understandably are reluctant to report AEs because they typically are interpreted as an error and used in litigation. Providers also tend to be reluctant to report near misses or almost adverse events (AAEs) because of the attitude that an AAE essentially is an AE saved by good fortune. Another lesson from aviation suggests an alternative: a flight pattern approach to Dulles Airport from the west was known to be dangerous, to be an accident waiting to happen; however, there was no mechanism for bringing it to the attention of those who could change it. At 11:09 a.m. Dec. 1, 1974, TWA Flight #514 crashed into the Blue Ridge Mountains killing all 300 passengers. Had that potential adverse event (PAE) been reported, the crash may not have occurred. Thus, SAME advocates reporting PAEs as does the ASRS.

Providers are to describe PAEs in terms of categories on the checklist as they would AEs and AAEs. Promoting identification of PAEs emphasises prevention and thus avoids attribution of blame. This does not absolve care providers of responsibility for AEs and AAEs. As in the ASRS, if it becomes known that a care provider was aware of and did not identify a PAE that later became an AE or an AAE, that provider would be considered responsible and held accountable for any adverse outcome.

Implications

Health care is a frontier for change (Bogner, 1994). The magnitude and pervasiveness of adverse events demand immediate breaching of that frontier. Valuable assistance is provided through lessons learned by other domains, particularly aviation. To effectively breach that frontier, the subtle pervasiveness of the presumption that humans are responsible for error must be overcome so other factors that contribute to error can be identified and changed. The systems approach represented by SAME is a viable technique to analyse adverse events and identify those factors.

That systems approach considers the context of care analogous to a dramatic script; the individual care provider is an actor whose performance is elicited by cues from the staging, the props, and from others. Considering only the actor, even removing an actor from the performance and substituting another does not change the script; another actor will react similarly according to the script. To alter the performance of the actors, those aspects of the script that evoke undesirable or inappropriate actions must be identified and changed. Altering the performance of care provider to reduce the likelihood of error in health care requires changing or removing error-provoking factors from the script of the context of care. The SAME technique by stretching the search for why medical errors occur beyond the care provider to system factors affecting the context of care, identifies error-provoking factors in the script of care to be changed to effectively reduce error and enhance patient safety.

References

Billings, C. E. (1997). *Aviation Automation: The Search for a Human-centred Approach.* Mahwah, NJ: Lawrence Erlbaum Associates.

Bogner, M. S. (1994). Human error in medicine: A frontier for change. In M. S. Bogner (Ed.), *Human Error in Medicine* (pp. 373-383). Mahwah, NJ: Lawrence Erlbaum Associates.

Bogner, M. S. (1998). Error: It's what, not who. *TraumaCare, 8(2)*, 82-84.

Bogner, M. S. (1999, March-April). Designing medical devices to reduce the likelihood of error. *Biomedical Instrumentation & Technology*, 108-113.

Bogner, M. S. (2000). A systems approach to medical error. In C. Vincent & B. De Mol (Eds.), *Safety in Medicine* (pp. 83-101). Amsterdam: Pergamon.

Brown, S. L., Bogner, M. S., Parmentier, C. M., & Taylor, J. B. (1997). Human error and patient-controlled analgesia pumps. *Journal of Intravenous Nursing, 20*, 311-317.

Buggy, D. (2000). Can anaesthetic management influence surgical-wound healing? *The Lancet, 356,* 355-356.

Cooper, J. B., Newbower, R. C., Long, C. D., & McPeek, B. (1978). Preventable anaesthesia mishaps: A study of human factors. *Anesthesiology, 49,* 399-406.

Department of Health. (2000). *An Organization with a Memory: Report of an Expert Group on Learning from Adverse Events in the National Health Service.* Norwich, UK: The Stationery Office.

Flanagan, J. C. (1954). The critical incident technique. *Psychological Bulletin, 51*, 327-358.

Gold, M. (1999). Cassirer's philosophy of science and the social sciences. In M. Gold (Ed.). *The Complete Social Scientist* (pp. 23-37). Washington, DC: American Psychological Association.

Institute of Medicine. (1999). *To Err is Human: Building a Safer Health Care System.* Washington, DC, National Academy Press.

Kurz, A. K., Sessler, D. I., & Lenhardt, R. (1996). Perioperative normothermia to reduce the incidence of surgical-wound infection and shorten hospitalization. *New England Journal of Medicine, 334,* 1209-1215.

Leape, L. L., Bates, D. W., Cullen, D. J., Cooper, J., Demonaco, H. J., Gallivan, T., Hallisay, R., Ives, J., Laird, N., Laffel, G., Nemeski, R., Peterson, L. A., Porter, J., Serv, D., Shea, B. F., Small, S. D., Sweitzer, B. J., Thompson, T., & Vander Viet, M. (1995). Systems analysis of adverse drug events. *Journal of the American Medical Association, 274,* 35-43.

Leape, L. L., Brennan, T. A., Laird, N., Lawthers, A. G., Localio, A. R., Barnes, B. A., Hebert, L., Newhouse, J. P., Weiler, P. C., & Hiatt, H. (1991). The nature of adverse events in hospitalized patients. *New England Journal of Medicine, 324,* 377-384.

McDonald, C. J., Weiner, M., & Hui, S. L. (2000). Deaths due to medical errors are exaggerated in Institute of Medicine report. *Journal of the American Medical Association, 284,* 93-95.

Moray, N. (1994). Error reduction as a systems problem. In M. S. Bogner (Ed.), *Human Error in Medicine* (pp. 67-91). Mahwah, NJ: Lawrence Erlbaum Associates, Inc.

Rasmussen, J. (1994). Afterword. In M. S. Bogner (Ed.), *Human Error in Medicine* (pp. 385-394). Mahwah, NJ: Lawrence Erlbaum Associates, Inc.

Reynard, W. D., Billings, C. E., Cheaney, E. S., & Hardy, R. (1986). *The Development of the NASA Aviation Safety Reporting System* (NASA Reference Publication 1114). Washington, DC: National Aeronautics and Space Administration Scientific and Technical Information Branch.

Senders, J. W., & Moray, N. P. (1991). *Human Error: Cause, Prediction, and Reduction.* Mahwah, NJ: Lawrence Erlbaum Associates, Inc.

Short, T. G., O'Regan, A., Lew, J., & Oh, T. E. (1992). Critical incident reporting in an anaesthetic department quality assurance programme. *Anesthesia, 47,* 3-7.

Thomas, E. J., Stoddert, D. M., Burstin, H. R., Orav, E. J., Zeena, T., Williams, E. J., et al. (2000). Incidence and types of adverse events and negligent care in Utah and Colorado. *Medical Care, 38,* 261-271.

Von Bertalanffy, L. (1968). *General System Theory.* New York: George Braziller.

Weick, K. E. (1987). Organizational culture as a source of high reliability. *California Management Review XXXIX (2),* 112-117.

Chapter 14

Betty and the General

Jan M. Davies
University of Calgary, Alberta, Canada

Introduction

This is the story of how a chance meeting on a commercial aircraft led to the inclusion of Human Factors analysis in Canada's longest running inquest, the Pediatric Cardiac Surgery Inquiry (PCSI). In addition, a summary of some of the results of that analysis is provided.

Winnipeg and the Pediatric Cardiac Surgery Program

The story starts in Winnipeg, which is located in the Province of Manitoba at North 49.50.12 and West 97.30.43. Winnipeg lies at the junction of two murky waterways, the Red and Assiniboine Rivers, with the name Winnipeg coming from the Cree words, 'win' and 'nipee', which mean 'muddy waters'. Despite the colour of the water, the site of the present city has been a meeting place for at least 6,000 years. In 1994, the year when this particular story starts, Winnipeg had a population of 640,000.

In 1994, on Valentine's Day, a newly qualified pediatric cardiac surgeon arrived at the Winnipeg Children's Hospital. He took up the position as the solo surgeon in charge of a program that had been quiescent for several months. Over the next nine months of 1994, twelve infants and children died during or after open-heart surgery. The program was suspended in February 1995, after an external review. One of the findings of the review was that the mortality rate for infants undergoing open-heart procedures was almost three times the rate for the same age group of patients at the Hospital for Sick Children in Toronto (29 percent vs. 11 percent).

The Paediatric Cardiac Surgery Inquiry

Legal Background

A Fatality Inquiry or inquest was called in March 1995, under the terms of the Fatality Inquiries Act of Manitoba, with opening statements made in October and the first witness testifying in December. The mandate of the Inquiry was 'to examine the circumstances surrounding the deaths of those children, to file a report which makes findings where, when and by what means each of these twelve children died, the cause of their death, and the material circumstances surrounding their deaths'. Under the Fatality Inquiries Act, an Inquiry is presided over by a Provincial Court Judge and not by a Coroner. The purpose of such an Inquiry is not to lay legal blame, but to determine what happened. In addition, an Inquiry aims to determine if a death could have been prevented.

In his opening statements, the presiding Judge said that he recognised that it was 'necessary to look at individual performance'. However, under the terms of the Act, judges cannot make findings of culpability or lack of culpability. Rather, Judges can make findings of fact or law to reach conclusions, for example, that a particular act (of omission or commission) occurred. Previously when Manitoba had a Coroner's system, the Coroners had the right to accuse one or more individuals of responsibility for a death and to issue warrants for those named responsible. The Coroner's powers were removed by statute and a Medical Examiners system instituted, under the terms of the Fatality Inquiries Act.

In fact, the Judge went on to state that his interest in individual performance related to knowing that if there were 'individual problems' within a program, then the 'program should still be able to deal with it in a very effective way before too many things (went) wrong'. He added that 'one of the key issues' that he would have to address was 'what to do with a situation where over reliance on key individuals can be avoided'.

Personnel

The Judge appointed was Murray Sinclair, Associate Chief Judge of the Province. In 1988 he had been Co-commissioner of Manitoba's Aboriginal Justice Inquiry (http://www.ajic.mb.ca/). Two Crown counsel, Don Slough and Christina Kopynsky, assisted Judge Sinclair. An experienced nurse, Betty Owen-Nordrum, was seconded to join the inquest team.

Betty meets the General

After leaving health care, Betty Owen-Nordrum had taken a position as a policy analyst on the Status of Women, for the Province of Manitoba. Shortly after joining the inquest team, Betty took an unrelated commercial flight, during which she was 'bumped' up to business class. She was seated next to a Canadian Air

Force Brigadier General. During the course of the flight, she told him about the PCSI and he told her about the field of Human Factors. In particular, he described the role of Human Factors in human error and accidents.

Human Factors and System Approaches

Betty immediately recognised the importance of this concept. She related it to Judge Sinclair's opening statements in which he said '...in the course of what occurred in 1994 ... one of the most important things to consider is the whole question of system approaches and system responses'.

Judge Sinclair had stated that the inquiry also intended to examine the 'way the unit operated, the way it was managed and administered, the way the hospital managed and supervised those who were involved, and as well the relationship between the hospital and the government of Manitoba'. In describing the mandate of the Inquiry, Judge Sinclair stated that the Judge might 'also recommend changes to programs, policies or practices of the government and the relevant public agencies or institutions involved in the care of the children'. Furthermore, the Judge might 'also recommend changes to the laws of the province where and if the Judge (was) of the opinion that such changes would serve to reduce the likelihood of other deaths occurring in such circumstances'. Judge Sinclair added, 'having said all that, I am sure that will spark an interest in various people as to whether to or not participate'.

The Defence and Civil Institute of Environmental Medicine, Toronto[1]

The Brigadier General also told Betty about the Defence and Civil Institute of Environmental Medicine (DCIEM) in Toronto, where a considerable amount of research in Human Factors was being carried out. At the Crown's invitation, two Human Factors specialists from DCIEM, Keith Hendy and David Beevis, visited the Crown team in Winnipeg. Hendy and Beevis provided the Crown team with a three-day intensive 'course' in Human Factors and crew resource management. Unfortunately, neither Hendy nor Beevis were able to commit to joining the Crown team and helping with the Inquiry on a daily basis for the length of time envisaged. When asked if they could suggest anyone else, Hendy provided the name of an anaesthetist in Calgary, who had an interest in Human Factors and some experience in aviation.

The Introduction of Human Factors

I was invited to join the team in May 1996. My role was to provide expert medical evidence in the areas of quality assurance, human error, and systems analysis. In particular I was asked to evaluate the role of quality assurance and the effect of

[1] Defence R & D Canada, as of April 1, 2002.

human error and team dysfunction on the results seen in 1994 in the Pediatric Cardiac Surgery Program. I took as the definition of quality assurance the evaluation of the level of the quality of care, with the establishment of mechanisms for change and improvement. However, in addition to ensuring excellence in patient care, my definition included providing an optimal working environment for the health care providers.

I started by reviewing the written information available. This included the patients' hospital records, all transcripts of testimony, and selected exhibits. In reviewing the information I gained an overview of the cases under inquiry. In addition, themes were developed and explored as witnesses testified, for example, the attitude of witnesses to quality assurance (specifically in the form of incident reports). Five major themes were developed, one for each of the five contributory factors. These five major themes related to each contributory factor are shown in Table 14.1.

Table 14.1 Five major contributory factors and their associated themes

Contributing Factors	Theme
Patients	Differences
Personnel	Behaviour
Equipment/environment	Limitations
Organisation	Culture
Regulatory agencies	Rationale

In addition to reviewing transcripts, I was also required to develop questions about Human Factors, as well as quality assurance, for Crown counsel to ask of witnesses. These questions were first posed during what were termed 'can says', or preliminary statements, and then in Court. Developing these questions required collaboration with the PCSI team and understanding of the special demands and limitations of Inquiry process. I needed to work as part of the PCSI team to guide Crown Counsel in their questioning and to be guided by their expertise in developing questions.

I was helped in this process by reading the report of a previous inquiry, the Commission of Inquiry into the Air Ontario Crash at Dryden, Ontario (Moshansky, 1992). I was also aided by the work of two individuals, Professors Helmreich and Reason. Helmreich's Operating Room Management Attitudes Questionnaire, adapted from the Flight Management Questionnaire (Helmreich & Shaeffer, 1993), provided what I considered to be a number of important questions. The Helmreich model was developed for the Dryden Inquiry and delineated the inter-linking of Human Factors from four different environments (Helmreich, 1992). This model, showing how the behaviour of flight crews in any given situation is determined by a number of simultaneously operating factors, provided one of three components of the Winnipeg Model (Davies, 2000). This latter model was used to organise what I considered to be the important contributory factors to the twelve cardiac deaths. The work of Professor Reason also provided one-third of the components of the Winnipeg model, as well as the basis for a unique understanding of lawyers' interpretations of the term 'active failures' (Reason, 1990a, 1990b).

Two Human Factors: Noise and Fatigue

Two areas of Human Factors that were explored to some extent were noise and fatigue.

Noise Operating rooms can be very noisy. Both acute and chronic exposure to noise can lead to health problems, including irritability, hypertension, and hearing loss. In the workplace, noise can mask voices, sounds, and alarms and may contribute to difficulties in concentration (Gloag, 1980). Studies in operating rooms (ORs) in England and Canada have shown that levels of noise found in ORs are high enough to impair speech discrimination, in the range of 67-70 db[A] (Davies et al., 1989; Lewis et al., 1990). In Winnipeg, noise may have contributed to the death of one patient, Vinay Goyal, while undergoing a redo repair of patch leaks at the aortic valve and ventricular crest. Thirty days previously, Vinay had undergone complete surgical repair of a double outlet right ventricle and Tetralogy of Fallot (patch closure of a ventricular septal defect, pulmonary valvectomy, resection of some of the right ventricle to relieve the outflow tract obstruction, closure of a patent foramen ovale, and removal of previously inserted Blalock-Taussig shunts). During the redo operation, the anaesthetist did not hear the surgeon state that he had removed the aortic cannula. The surgeon testified that he did not communicate enough or loudly enough that he took the cannula out. Once the cannula was removed, the anaesthetist could no longer transfuse Vinay fast enough to keep up with the surgical bleeding. The child bled to death on the table.

The operating room normally used in Winnipeg was small and cramped, particularly when all the equipment and personnel were present. The surgeon had made previous complaints about noise in the OR. He stated that he wanted quiet 'particularly during critical point(s)' of the operation. It is possible that the surgeon's lack of experience in task performance under noisy conditions raised his levels of arousal and anxiety, making him less tolerant of noise.

Fatigue Another area of interest was that of fatigue. This is not a newly recognised problem, as Friedman and colleagues (1971) demonstrated 30 years ago. They showed that interns who had an average of 1.8 hours of sleep had a five-fold increase in physiological symptoms (weakness, nausea) in comparison to when they had seven hours of sleep. When fatigued they had a more than three-fold increase in difficulty concentrating and made almost twice as many errors in the interpretation of electrocardiograms.

Wilson and Weston (1989) applied the British Civil Aviation Authority rules for work hours to a group of six medical trainees who worked an average of 78 hours each week. A further 20 doctors would have been required to cover the work of the six, if the hours for pilots were followed.

More recently, Dawson and Reid (1995) compared sleep deprivation to alcohol intoxication. Being awake for 17 hours had similar effects to a blood alcohol concentration of 0.05 percent, whereas being awake for 24 hours was similar to a blood alcohol concentration of 0.1 percent.

In Winnipeg, the pediatric cardiologist was questioned about fatigue. He stated that from early November 1993 onward, he was the sole pediatric cardiologist at the Winnipeg Children's Hospital. He was also in charge of the Variety Heart Centre, to which almost all children from Manitoba and parts of Ontario with congenital and acquired heart defects were referred. The cardiologist was asked, 'How did your job change when you became acting head?'. He answered, 'It became unlike any job I had ever had. All of a sudden, 24 hours a day, I am on call'. He remained in the position through 1994, although another pediatric cardiologist joined him in the summer of 1994.

Difficulties Encountered

Not all the questions prepared for Crown counsel were asked of witnesses. The reasons for this are almost as complex as the Inquiry itself. There were legal constraints in that Counsel could not simply ask a witness 'Were you tired?'. Questions had to be linked to specific facts or previous testimony – not all of which were apparent at the time of questioning.

Some witnesses had already testified and one had then even left the country (for reasons unrelated to the PCSI) by the time the Human Factors questions were developed. Other witnesses were available for only a limited amount of time, and much of the questioning was related to the medical and surgical care of the patients, often in considerable detail. This detail contributed to the enormity of the task. Some three years after the start of the Inquiry, some 50,000 pages of testimony and hundreds of exhibits had been generated. The Inquiry sat for 275 days, hearing testimony from 85 witnesses, as well as sitting for an additional 25 days. Finally, this was a 'novel' situation for both the Crown Counsel and for me. Despite having and using systematic outlines for review (Armstrong & Davies, 1991; Davies, 1997; Davies & Campbell, 1998), participating in this Inquiry still proved to be a formidable task (although one which was extremely rewarding).

One example of where not all questions were asked was that of fatigue and the pediatric cardiac surgeon's ability to operate. A recent study in the literature demonstrated the effect of sleep deprivation on surgeons' dexterity. Taffinder and colleagues (1998) used a virtual-reality laparoscopic surgery simulator and measured the dexterity of six surgeons, by tracking movements of surgical instruments. The surgeons were compared after being awake all night versus having a full night's sleep. When fatigued, the surgeons made 20 percent more errors and took 14 percent more time to complete tasks. The researchers were able to determine that the increased stress of fatigue was more important than the decreased arousal state.

In Winnipeg, the surgeon was virtually always on call for pediatric cardiac surgery – a hazard of being the sole surgeon for a very small program. For some time he was also on call for the adult cardiovascular service. The night before he operated on Jesse Maguire (on November 27, 1994), the surgeon was on-call for the adult service. Jesse Maguire underwent emergency surgery for repair of an interrupted aortic arch, patch closure of a ventricular septal defect, suture closure of an atrial septal defect, and closure of a patent ductus arteriosus. During the operation the aortic cannula was dislodged and came out of the vessel. Both the surgeon and the assistant struggled for about seven minutes to replace the cannula. One of the repairs was torn and had to be re-sutured, which meant re-instituting cardiopulmonary bypass and total circulatory arrest. The total cardiopulmonary bypass time was 6 hours 12 minutes. The total circulatory arrest time was 2 hours 49 minutes. The child could not be weaned from bypass and died on the OR table, 11 hours 7 minutes after the start of the operation. The surgeon was not asked if fatigue from being on call the night before had contributed to some of the apparent surgical difficulties.

Addendum

In November 2000, Judge Sinclair's Report of the Manitoba Pediatric Cardiac Surgery Inquest was released (Sinclair, 2000). Of its 516 pages, more than 300 provided precise detailing of the medical conditions and surgical procedures of each of the 12 children who died, as well as some who did not. The last of the 10 chapters described Judge Sinclair's findings and the 36 related recommendations. These included: processes for staff recruitment; development of protocols for orientation and support for new staff; establishment of clear lines of authority and responsibility within programs; involvement of the whole team in the development of protocols for the care of patients; establishment of clear policies for the reporting by staff of concerns about risks for patients; and development of quality assurance and risk management policies.

Release of the Report was followed by an apology from the Winnipeg Regional Health Authority (WRHA) to the families of the 12 children who died. The Chief Medical Officer stated that positive things would come from the Report, that improvements had already been made to the system and that the WRHA was open to other improvements. Six months later a second report was released – by a Province of Manitoba Review and Implementation Committee, which evaluated the recommendations made in Sinclair's report and their possible implementation (Report, 2001).

To date, the pediatric cardiac surgery program has not re-opened in Winnipeg. Patients are referred either to Edmonton or Toronto. Sinclair's Report has also had effects further afield. A pediatric cardiac surgery program in London, Ontario has closed because of low volumes of cases and a second unit (in Ottawa, Ontario) may also close.

Apart from the lesson that the development and maintenance of medical and surgical expertise requires a certain volume of cases, what else can be learned from the PCSI? One major area concerns the investigation and analysis of medical catastrophes such as occurred in Winnipeg. Because these are the medical equivalent of an aircraft accident, medical, and judicial authorities with responsibility for such investigations would do well to follow the aviation model. As described in ICAOs Annex 13, there should be a systematic approach to any investigation (Annex 13, 1994). This investigation should take a systems approach, recognising the importance of both Human Factors and latent conditions in the evolution of an adverse outcome. Experts with appropriate training and experience should be involved.

Adoption of such a process will take time, in part because of the low frequency of such investigations and also because of widespread perceptions that the current faultfinding system is appropriate. For example, despite the fact that the PSCI took a systems approach, many Canadians remain convinced that the 'problem' lay solely with the incompetence of one individual – the surgeon (Davies, 2001).

This was in contrast to the reaction to the July 2001 release of a report into pediatric cardiac surgery deaths that occurred between 1989 and 1995 in Bristol, England (Bristol, 2001). Although the two surgeons were described as having made serious mistakes, concern was expressed that they could not be held solely to blame for the failings at Bristol. A consultant cardiac surgeon at another hospital stated that 'Like the captains of the ship, the surgeons got the blame. But no doubt the referring paediatricians, the paediatric cardiologists, the anaesthetists, and the intensive care staff looking after the patients following surgery were all part of the team and it is everybody's responsibility. Everybody in the team counts. Right from the very top down to the coalface, there are problems all the way. If the child dies it tends to be considered a surgical death, yet there may have been other factors'. Another surgeon stated there were 'a number of people who believe they were made scapegoats for an under-funded department' (O'Neil, 2001).

The difference in reaction to the two reports is somewhat puzzling, considering the great similarities. Both inquiries examined a series of pediatric cardiac surgery deaths, in which one of the problems was an insufficient number of cases undertaken to ensure expertise. In both centres, deaths were related to the 'learning curve'. This was defined as: 'an expected and acceptable excess of patients who will die or be harmed in the early experience of a learner but who would have fared better if they were operated upon by a surgeon who is on the plateau of experience' (Bristol, 2001). The learning curve and risk of death also related to parents not being fully informed. Risk factors for procedures were not based on the situations the children actually faced, nor did risks of mortality accurately reflect the surgeon's experience (Sinclair, 2000; Bristol, 2001).

Although the mandate of neither inquiry was to find fault or to lay blame, there were major differences that might explain the apparent variation in the reception of the reports. First, the framework of the inquiries was different. The Public Inquiry in Bristol was chaired by Professor Ian Kennedy and ran from October 1998 to July 2001, with transcripts and discussion papers available on a web site (Bristol, 2001). The Winnipeg Inquiry was an inquest under the Fatality Inquiries Act, presided over by a Provincial Court judge. Second, the Bristol Inquiry was preceded by an inquiry by the Professional Conduct Committee (PCC) of the

General Medical Council (GMC), the licensing body for all doctors in the UK (http://www.gmc-uk.org). The PCC Inquiry finished in 1998 in a flurry of media attention, when two of the doctors involved were removed from the GMC Register. No similar proceedings were undertaken in Winnipeg, although the College of Physicians and Surgeons of Manitoba (the provincial licensing body for doctors) is reviewing the role and performance of all 17 doctors named in the Report. The absence in Winnipeg of a disciplinary procedure similar to the GMCs, combined with Sinclair's evaluation that five of the 12 deaths were 'at least possibly preventable or preventable' (Sinclair, 2000), may have unwittingly led many to conclude that the deaths in Winnipeg were largely the 'fault' of the surgeon.

Despite this reaction, the Winnipeg Inquiry stands as the first formal introduction of Human Factors and systems analysis into an inquest. Nor is it the only inquest to apply the model of aviation investigation. In 2001, the Chief Coroner of Ontario presided over an inquest into the death of a 17-year-old girl after an elective gallbladder operation. Expertise in systems analysis and Human Factors was sought and used, with the jury's first recommendation being the adoption of a similar process for future inquiries. The way is now set for those in the medical community to enlist the aid of Human Factors researchers and work together in the investigation of adverse outcomes and the improvement in patient safety.

References

Annex 13. (2001). *International Standards and Recommended Practices: Aircraft Accident and Incident Investigation* (Annex 13 to the Convention on International Civil Aviation). Ninth Edition, Montreal, Quebec.

Armstrong, J. N., & Davies, J. M. (1991). A systematic method for the investigation of anaesthetic incidents. *Canadian Journal of Anaesthesia, 38,* 1033-1035.

Davies, J. M. (1996). Risk assessment and risk management in anaesthesia. In A. R. Aitkenhead (Ed.), *Quality Assurance and Risk Management in Anaesthesia: Bailliere's Clinical Anaesthesiology International Practice and Research, 10,* 357-372.

Davies, J. M. (1997). La Qualite en Anesthesie [Different axes to quality assurance in anaesthesia]. Le Conseil Scientifique des Journees d'Enseignement (pp. 39-52). Post-Universitaire d'Anesthesie et de Reanimation: Arnette, Paris.

Davies, J. M. (2001). Painful inquiries: Lessons from Winnipeg. *Canadian Medical Association Journal, 165,* 1503-1504.

Davies, J. M., & Campbell, L. A. (1998). An investigation into serial deaths during oral surgery. In H. Selby (Ed.), *The Inquest Handbook* (pp. 150-169). Leichardt, NSW, Australia.

Davies, J. M. (2000). Application of the Winnipeg model to obstetric and neonatal audit. *Topics in Health Information Management, 20,* 12-22.

Davies, J. M., Ewen, A., Cuppage, A., Gibert, D., & Winkelaar, R. (1989). Noise levels in the operating room: A comparison of Canada and England. *Anaesthesia and Intensive Care, 17,*98-99.

Dawson, D., & Reid, K. (1995). Fatigue, alcohol, and performance impairment. *Nature, 388,* 235.

Friedman, R. C., Bigger, J. T., & Kornfield, D. S. (1971). The intern and sleep loss. *New England Journal of Medicine, 285*, 201-203.

Gloag, D. (1980). Noise: Hearing loss and psychological effects. *British Medical Journal, 281*, 1325-1327.

Helmreich, R. L. (1992). Human factors aspects of the Air Ontario crash at Dryden, Ontario: Analysis and recommendations to the Commission of Inquiry into the Air Ontario crash at Dryden, Ontario. In The Hon. V. P. Commissioner Moshansky *Commission of Inquiry into the Air Ontario Crash at Dryden, Ontario* (Final Report: Technical Appendices). Ottawa, Canada: Ministry of Supply and Services.

Helmreich, R. L., & Schaefer, H. G. (1993). *The Operating Room Management Attitudes Questionnaire (ORMAQ)* (Tech. Rep. No. 93-8). Austin, Texas: NASA/University of Texas/FAA.

Lewis, P., Staniland, J., Cuppage, A., & Davies, J. M. (1990). Operating room noise. *Canadian Journal of Anaesthesia, 37*, S79.

Moshansky, The Hon. V. P. Commissioner. (1992). *Commission of Inquiry into the Air Ontario Crash at Dryden, Ontario*. Ottawa, Canada: Ministry of Supply and Services.

O'Neill, S. (2001, July 18). Bristol surgeons 'carried the blame'. *Daily Telegraph*. (www.news.telegraph.co.uk)

Reason, J. (1990a). The contribution of latent failures to the breakdown of complex systems. *Philosophical Transactions of the Royal Society of London, B327*, 475-84.

Reason, J. (1990b). *Human Error*. Cambridge: Cambridge University Press.

Report of the Review and Implementation Committee for the Report of the Manitoba Pediatric Cardiac Surgery Inquest. (2001). Winnipeg: Government of Manitoba. (http://www.gov.mb.ca/health/cardiac.html)

Sinclair, Assoc. Chief Judge Murray. (2000). *The Report of the Manitoba Pediatric Cardiac Surgery Inquest: An Inquiry into Twelve Deaths at the Winnipeg Health Sciences Centre in 1994*. Winnipeg: Provincial Court of Manitoba. (www.pediatriccardiacinquest.mb.ca)

Taffinder, N. J., McManus, I. C., Gul, Y., Russell, R. C. G., & Darzi A. (1998). Effect of sleep deprivation on surgeons' dexterity on laparoscopy simulator. *Lancet, 352*, 1191.

The Bristol Royal Infirmary Inquiry. (2001). *The Inquiry into the Management of Care of Children Receiving Complex Heart Surgery at the Bristol Royal Infirmary*. London: The Bristol Royal Infirmary Inquiry. (http://www.bristol-inquiry.org.uk)

Wilson, A. M., & Weston, G. (1989). Application of airline pilots' hours to junior doctors. *British Medical Journal, 299*, 799-81.

Chapter 15

Managing Human Performance in the Modern World: Developments in the US Nuclear Industry

John Wreathall and Ashleigh Merritt
John Wreathall & Company Inc., Dublin, OH, USA

Introduction

Work in the aviation, nuclear, pharmaceutical and other safety-critical, profit-driven industries has indicated that those who seek to be world leaders in the Modern World face common problems in meeting the challenges of accomplishing human and organisational excellence in performance. In this paper we define the common characteristics of these industries with regard to managing organisational performance, highlight areas currently overlooked, and suggest some new proactive approaches based on work currently being performed in the American nuclear industry.

The Modern World

Industries in the Modern World of safety and efficiency share the following characteristics:

1. Many industries have become increasingly well defended: Technological advancements and increasingly sophisticated thinking about safety has produced a significant decrease in accidents and those accidents that do occur are no longer the result of single system failures. Despite these improvements, many have reached a safety 'plateau' and do not seem to be able to move beyond it.
2. Figures 15.1 and 15.2 show the dramatic declines in event rates in the 1960s and 1970s for aviation, and in the 1980s for the nuclear industry.
3. The standards of performance set by regulators and the standards imposed by the industries themselves no longer allow even 'small accidents' or comparatively minor events without extended periods of intrusive

investigations, with its impact on operations, distraction of management, and a general adverse impact on morale.

4. Deregulation and strong competition have produced an environment in which economic losses can be almost as catastrophic as major accidents. For example, human errors in maintenance operations reportedly cost US commercial carriers between $US1 and $US2 billion per year because of delays and operational disruptions (Marx, 1998). Similarly in the nuclear industry, a single inadvertent reactor shutdown can cost a utility upwards of $US60 million or more in replacement electricity costs if that energy has to be bought at 'spot rates' during a summer peak demand.
5. As human error continues to dominate those failures that do occur (perhaps because many of the hardware problems have been fixed), many industries have turned their focus to the study and management of Human Performance. This work has focussed almost exclusively on individual and team training in understanding the limits of human performance and providing self-management tips and tools for performance improvement (e.g., CRM in aviation). Very little work has been done directly with managers about their role in managing Human Performance.

In the 'old world', the traditional management process was to suffer accidents and significant events and then perform root-cause investigations as a way to develop corrective actions. This *reactive* approach to managing risks and flaws in the system is no longer tolerable for several reasons. From a pragmatic standpoint, there are fewer significant events from which to draw lessons. More importantly, events are now 'unacceptable' – standards of safety set by society demand it, and the costs associated with even 'non-events' (such as near misses and event precursors) can bankrupt a company or make it vulnerable to takeover.

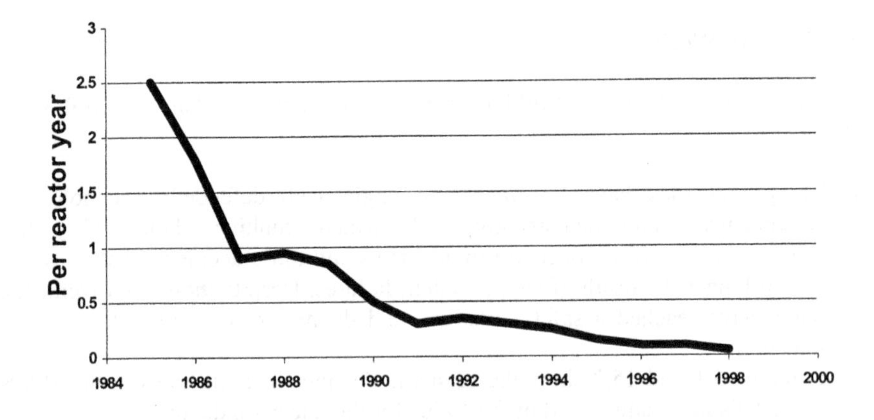

Figure 15.1 Significant event rate (US Nuclear Industry; NRC, 1999)

In the Modern World, excellence in performance must be accomplished without relying on events to provide the signs of 'what to fix'. But is improvement beyond existing plateaux possible? Have we overlooked anything in our conscientious zeal to drive down error? What *is* the role of managers with regard to human performance, and are there *proactive* tools that could help them better manage and control human performance within their organisations?

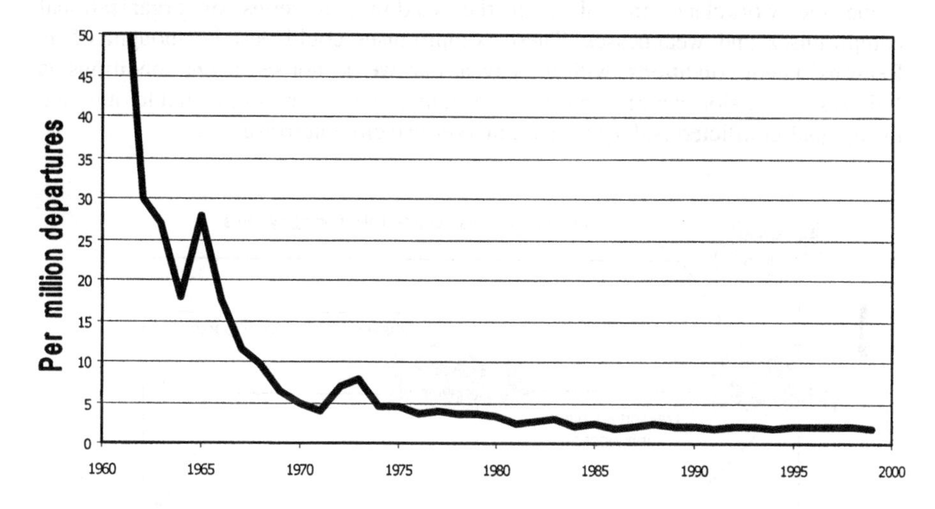

Figure 15.2 Commercial aviation worldwide accident rate (Boeing, 2000)

Framing the Modern World

In the technologies in both the aviation and nuclear power industries, the accomplishment of success (both economic and safety-related) in the end lies in the hands of individual workers – the pilots or operators, and maintenance technicians: the people who work hands-on with the equipment. While they are often considered the prime controllers of safety, we agree with Reason (Reason, 1998) that they are principally the *last* defenders of the integrity of the system, who have to cope with all the accrued 'crud' – shrinking margins of operability, poor interfaces, and so on. These people work within the framework of a workplace setting and an organisational envelope.

Figure 15.3 shows our representation of this framework of influences. A worker's performance is the result of his or her natural human abilities plus individual aptitudes and skills, *in combination with* the demands of the task in a particular work environment. Workers' attitudes and behaviour are shaped by organisational processes (work policies, training, cultural influences), which in turn are directed by Senior Managers' priorities. These priorities are a function of organisational performance (past, present and future) and the economic and regulatory environment.

Organisational processes also shape the workplace and task factors (tools, task design, procedures, physical work environment, etc.). Managers have to allocate resources, set standards, establish performance goals, and so on. All these decisions are made under conditions of uncertainty (about the marketplace, the competition, the regulatory environment, etc.) and involve tradeoffs between competing demands. Consequently, there is a series of 'best bet' decisions that shape the workplace and also, in the shadows, a series of organisational compromises and weaknesses. These compromises could best be thought of as Reasons' latent conditions, with the caveat that we do not see latent conditions as 'failures in decision-making' by the management but rather an inevitable outcome in any goal-conflicted (safety-critical, market-driven) enterprise.

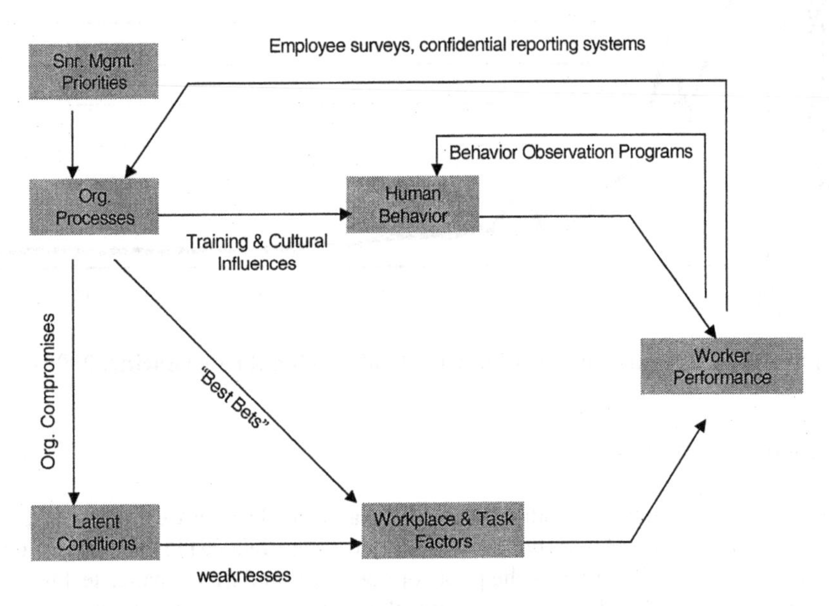

Figure 15.3 Framework of worker, workplace, and organisation

When viewed in this context, it becomes clear that the system could benefit from improved input into the organisational processes that ultimately shape workers' performance. This is the realm of the manager, yet there is often an assumption by non-managers (workers, researchers) that managers should just know how to manage. Hence, a great deal of attention has been given to individual and team management of human performance, with almost no attention given to specifying the role of managers in this process. Is it sufficient for managers to admonish everyone from the sideline to 'do their best' or is there more that they could be doing? As stated earlier, managing in the Modern World requires proactive tools for the 'dynamic non-event' (Weick, 1994) that is safety; the days of managers as reactive controllers of a closed-loop, homeostatic work environment are over.

Compounding this oversight is the standard selection process for managers in these high-technology industries. Typically, one is promoted based on technical skill and some seniority (sometimes, perhaps, some perceived people skills). When these people become managers, their natural inclination is to continue to see the organisational processes from a technology worker's vantage point – it is very difficult to shift to an organisational perspective without some guidance.

In sum, a systemic perspective on managing Human Performance quickly identifies decisional input into organisational processes – a key role for managers – as an underdeveloped area in the global endeavour of managing human performance. The next section describes some efforts that are underway in the US nuclear industry to redress this weakness.

Tools for the Modern World Manager

We are currently working with the Electric Power Research Institute (EPRI, a centralised research and development company co-sponsored by almost all the electricity producing and distributing companies in the US) to develop and implement a set of tools to assist and improve managers' focus on Human Performance issues. While the details of this work are proprietary to EPRI and its members, the following summarises some general points and lessons that may be considered for application in other industries.

Rationale

Human performance improvements are achievable in two primary modes:

- By working within the sphere of influence of individuals and teams, at the level of behaviours that can be improved, and
- By working within the sphere of influence of leaders and organisations, at the level of organisational and workplace factors that can be improved.

As noted previously, most of the work in this field has been done in the former category, and not just in the nuclear industry – Crew Resource Management is the classic example of Human Performance improvement addressed at the individual and team level. There are however, upper limits on the extent to which individual performance can be improved (i.e., attentional and physical capabilities will always be limited). This is one reason we believe that energy and resources could be spent elsewhere in an effort to advance the thresholds of Human Performance.

Our approach is more situational – it considers the context in which work is performed. As such, it looks for the source of errors not in the human so much as in *the interaction of workers with the workplace environment or situation*, hence terms such as error-inducing or error-forcing situations. This is not to discount the training efforts addressed to the individual; rather it presents a complimentary enhancement. Additionally, our approach is proactive in that it focuses on leading rather than lagging measures, or indicators, of human performance. In the next section we describe the logistics and details of this approach as currently practiced at some US nuclear power plants. Due to space constraints, each tool is discussed very briefly, followed by possible applications in the aviation environment.

Leading Indicators of Organisational Health (LIOIH)

How do we know *today* if we are making mistakes that will impact the plant's performance tomorrow? Which processes and decisions should we monitor to alert us that standards may be 'slipping'? These are some of the questions that led to the development of the leading indicators of organisational health.

The indicators are organisational self-assessment measures intended to provide information to departmental and senior management about those organisational characteristics that are strong influences on human performance. Based on a process that translates broad themes into specific plant issues, the indicators are developed and in a way that can identify potential concerns in the plant's performance.

Based on a review conducted by Reason (adviser to the project), seven themes were identified as present in all high-performing, high-reliability organisations. These themes are:

1. *Top-level commitment:* Top management recognises the human performance concerns and tries to address them, infusing the organisation with a sense of significance of human performance, providing continuous and extensive follow-through to actions related to human performance, and is seen to value human performance, both in word and deed.
2. *Just culture:* Supports the reporting of issues up through the organisation, yet not tolerating culpable behaviours. Without a just culture, the willingness of the workers to report problems will be much diminished, thereby limiting the ability of the organisation to learn about weaknesses in its current defences.
3. *Learning culture:* A shorthand version of this theme is 'How much does the organisation respond to events with denial versus repair or true reform?'
4. *Awareness:* Data gathering that provides management with insights about what is going on regarding the quality of human performance at the plant, the extent to which it is a problem, and the current state of the defences.
5. *Preparedness:* 'Being ahead' of the problems in human performance. The organisation actively anticipates problems and prepares for them.
6. *Flexibility:* It is the ability of the organisation to adapt to new or complex problems in a way that maximises its ability to solve the problem without disrupting overall functionality. It requires that people at the working level (particularly first-level supervisors) are able to make important decisions without having to wait unnecessarily for management instructions.
7. *Opacity:* The organisation is aware of the boundaries and knows how close it is to 'the edge' in terms of degraded defenses and barriers.

In a sense the easy work in the project was identifying these themes; the true challenge has been to translate the themes into observable actions – leading indicators – that can be monitored. Several plants have engaged in a workshop process to translate the themes into indicators that are relevant for their sites, and have begun to track these indicators on a monthly basis. In other words, having identified the principles of proactive management, these plants are now actively trending their plant-specific indicators to recognise when their standards are slipping, so that they can take remedial action before an event occurs. The predictive validity of these indicators is currently being tested in a parallel study.

Applications in Aviation

A Senior VP at one of the utilities commented that the 'true worth' of the Leading Indicators Project is that it has allowed a language and a dialogue to emerge in which managers can address previously unformed ideas. The framework enables connections to be made across departments, across events, and across time. It forces questions such as, 'How do we know how healthy we are (how safe and economically viable)? How do we know we have a just culture? Are we a learning organisation?'

We are not sure if what we are proposing is radical or simply management made more proactive and more explicit. We believe that many managers might think the seven themes are so obvious as to be almost 'no-brainers'. But translating the seven themes into observable processes that can be trended with data – that is where the challenge lies. Of course it is much easier to be a reactive manager; unfortunately it is also a lot riskier.

The process of translating themes into indicators is an exercise recommended for senior managers, department heads, Fleet Managers, and Quality Control. Interestingly, it can also be applied to the management of a flight. A good crew tries to stay ahead of the aircraft as much as possible – this is just another form of proactive management. A good crew asks themselves, 'how prepared are we for this trip (briefings, etc.), and how flexible are we should the unexpected arise? Are we aware of and satisfied with the risks we are taking (opacity) and are we learning from other flights, other crews?'

The leading indicators can also be incorporated into observational programs such as LOSA (Line Oriented Safety Audits). In fact the two approaches are similar in that neither wait for a negative event to occur; both are looking at today's performance for clues about tomorrow. For this reason, fleet managers or departmental heads may want to see measures of certain indicators included in a LOSA as a way to validate and track the themes at the front-line operator level.

Proactive Assessment of Organisational & Workplace Factors (PAOWF)

PAOWF is a software tool used to evaluate workplace factors by obtaining ratings of potential problem areas by the frontline workers and supervisors. With quick and simple modifications, supervisors and managers can also use the tool to assess the organisational factors that shape the work environment.

In typical applications of PAOWF, personnel enter ratings frequently, such as following completion of specific tasks, e.g., maintenance work orders. The accumulated ratings are periodically analysed and compared across factors to determine areas of strength and weakness. In addition, ratings for each factor can be trended over time to help identify growing problem areas, or to assess the success of leadership action taken to reduce or eliminate problems.

PAOWF does not require any particular set of factors to be rated – indeed it is highly recommended that each site customise the factors – however an initial set of factors is offered. These factors, identified generically as 'potential organisational weaknesses' include typical contributing factors seen in event investigations but sought *before* they contribute to an event. Examples from maintenance activities in nuclear power include: Communication; Goals and Priorities; Interfaces; Organisational Structure; Planning and Scheduling; Policies; Procedures and Work Control Documents; Roles and Responsibilities; Task Structure; Training & Experience; and Norms. In light of initial feedback, an organisation may decide to delve into one factor in more detail; additionally the factors can be customised to reflect current concerns, e.g., human performance factors that have been frequently observed in event root-cause investigations might be a useful starting point.

Based on initial trial applications, some of the more popular features of PAOWF include:

1. The ability to provide real-time data about what is happening in the workplace (asking the people doing the work today to assess the problems, today).
2. Immediate analysis and feedback capability (no need to wait a month or more to discover the source of problems).
3. Burden-free (an average PAOWF rating takes no more than 90 seconds to complete).
4. Additionally, PAOWF can be modified to serve a multitude of purposes, including:
   a. *Workplace observations* (the software can be installed on hand-held equipment, such that observers directly record their ratings of workplace conditions into a data base in real time as they tour the site).
   b. Short form *culture surveys* (rather than an annual all-in survey of all employees, the tool can be used to collect ongoing, stratified samples of opinions).

PAOWF Applications in Aviation

Behavioural observation is a popular tool in aviation as witnessed by the existence of check airmen, ghost passengers, and more recently LOSA and even FOQA (a mechanical observation of the flight's behaviour and deviations). The real benefit of PAOWF is that it provides another perspective – that of the person actually doing the work. How does the world look from where they are sitting? The front-line operators alone can answer this question. The largest source of information about working conditions and problems are the workers themselves, and an organisation that has both a just and learning culture is able to tap into that knowledge. This is not to challenge behavioural observations, but simply to say that another perspective can add richness to the information.

Once this first premise is accepted, the practical benefits of PAOWF are impressive. First, you are not paying someone to observe another person work. Second, the results are virtually instantaneous because they are entered directly into a database and the trending and summary features allow approved personnel to view that database and make sense of the data on a daily or weekly basis, allowing problems to be detected almost immediately. Third, it sends a message to front-line workers that they are the trusted custodians of safety (not children to be constantly observed) and that their understanding of the system is paramount to organisational learning and continued success. And finally, the data is more reliable than self-reported employee concerns alone because it is a systematic organised sampling across the workplace.

PAOWF can be used with ground crew, gate and boarding agents, cabin crew and yes, even pilots. Palm pilots or something equally convenient could be issued at different times to do a spot check or to investigate a potential problem. For example, the crews could rate all flights into a certain airport for two weeks or a month. Similarly, the introduction of new changes to the equipment or scheduling could be assessed and modified in light of real-time operational experience and feedback. These palm pilots could be dispatched with flight papers through crew scheduling.

There are many practical applications of PAOWF for the organisation that is interested in proactive management, and there are very sound reasons for incorporating the worker's perspective directly into an ongoing analysis of workplace and organisational factors. With appropriate direction, implementation and sampling, it is the fastest, most direct, and most accurate way to learn about the organisation's defences and barriers.

Other Tools

The nuclear industry, like aviation, is besieged with low-level incident reports that analysts are frankly unable to manage. Another tool under development is a database that generates options for corrective actions from a library of previous corrective actions, plus links to regulations, accident report recommendations, research, etc. It also includes the facility to track and check the effectiveness of the implemented actions. The emphasis is not on analysis (a reactive process), but on tracking the effectiveness of the corrective action. Did the 'solution' fix the problem? Have we learned, or are we implementing the same corrective actions repeatedly without success? Incidentally, because the number of repeated corrective actions can be quantified and tracked, it has been identified as a leading indicator for a learning culture.

Another tool under development is a hands-on training course for supervisors and managers in the intelligent use of data. We have observed that many believe the myth that 'data is good, data is god' but that few seem able to make sense of the mountain of data that comes to them. Turning data into useful information is a vital management skill, and while it may not be intuitive, it is something that can be learned. At this point in our technological evolution, we do not need more data, we need to know how to make sense of the data we already have!

Conclusion

Our work with the EPRI in the USA to develop leading indicators of organisational health is based on the industry's conclusion that the environment for individual workers' and teams' activities can be improved, and that doing so will improve the effectiveness of human activity and reduce the likelihood (and costs) of events caused by human error. Along with individual and team training in managing human performance, the tools discussed here provide added value, yet none by itself is the 'magic bullet'. Trying to find a single 'magic bullet' tends to create the syndrome of 'management whiplash,' going from one technique to the next, with the workforce, supervisors, and middle management being driven in one direction, then another and another. However using a set of principles developed from a systems perspective, we believe we have identified some areas of potential improvement in the overall management of human performance that could be equally important in the aviation environment.

Recall the challenges in accomplishing excellence in human (and overall) performance outlined at the beginning of this paper: no events to learn from; the cost of even 'non-events'; the continuing high contribution of human performance to overall performance problems; and the role of technical expertise rather than 'people skills' in selecting supervisor and management candidates. How do the concepts summarised in this paper help?

First the concepts of *proactive tools* provide a basis for action without waiting for events. These tools comprise one example of what Westrum has called 'faint signals' (Westrum, 1999), which if paid attention to, can provide a basis for identification of potential problems. This requires a willingness to listen, however. In order to accomplish this willingness to listen, education of the management is needed. Most management in technologically driven industries like the nuclear industry are technically oriented. Such people are helped by a technological framework to understand the issues and the tools – Figure 15.3 provides such a framework. We have already heard comments that one of the major accomplishments of this project with EPRI has been to put some of the underlying concepts into the dialogue of power-plant management. As the psychologist co-author of this paper has been known to observe, to communicate with engineers and technologists, you must draw boxes and arrows with labels. The engineering co-author has no disagreement with this observation.

References

Boeing. (2000). *Statistical Summary of Commercial Jet Airplane Accidents Worldwide Operations 1959-1999*. Seattle, WA: Boeing Commercial Airplane Group.

Marx, D. A. (1998). *Learning from our Mistakes: A Review of Maintenance Error Investigation and Analysis Systems*. Washington, DC: U.S. Federal Aviation Administration.

NRC. (1999). *The NRC Annual Report* (NUREG-1145, Vol. 15). Rockville, MD: US Nuclear Regulatory Commission.

Reason, J. (1998). *Managing the Risks of Organizational Accidents*. Brookfield, VT: Ashgate Publishing Company.

Weick, K. E. (1994). Organizational culture as a source of high reliability. In H. Tsoukas (Ed.), *New Thinking in Organizational Behaviour: From Social Engineering to Reflective Action*. Boston: Butterworth-Heinemann.

Westrum, R. (1999). *Faint hearts and faint signals – How organizations manage signs of trouble*. Paper presented at the 1999 Workshop of the Center for Human Performance in Complex Systems, Madison, WI.

Chapter 16

Artificial Seasoning: Enhancing Experience-Based Training for Aviation

David O'Hare
University of Otago, New Zealand

Introduction

Many fields of human endeavour require lengthy periods of training and practice to achieve high-levels of performance. Indeed, expertise is generally defined as 'having or showing special skill or knowledge derived from training or experience' (Allen, 2000). Much attention has been paid to the field of training in aviation and elsewhere. Enormous investments of time and resources have been made in developing training systems and improving training technology. The world's civil airlines and military forces have been at the forefront of simulator use in training. An extensive scientific literature on training in general, and simulator use in particular, has now been built up.

Completion of the prescribed training has traditionally been dependent on the accumulation of a defined quantity of training hours (e.g., 5 hours night flying, etc.) and a pass/fail check ride. Fulfilling the quantity of training was an immutable requirement even if competency in the task was reached earlier. Once the required hours had been completed and the check rides passed, the trainee was then deemed ready to begin the 'real job' of acquiring the experience necessary to become a seasoned performer. This process has been wonderfully described by Gann (1961) in his classic account of the early days of commercial air transport in the US.

Gann (1961) describes his arrival at the airline's six-week ground school: 'Thus, unlearned, credulous, and bewildered, certain of us emerged from the lower strata of aerial society' (p. 29). Assigned to his first base at Newark, Gann is checked out in the DC-2 and DC-3. In contrast to his high-flying compatriots, Gann finds himself assigned to a 'grasshopper' existence flying the AM-21 route between Newark and Cleveland. This involves exposure to a wide range of airports and weather conditions in a relatively short space of time: 'So began my true apprenticeship' (p. 62).

The process of seasoning described by Gann involved the accumulation of knowledge and skill through encounters with many and varied situations. As well as the ability to 'read' the weather, Gann frequently refers to the honing of mental skills such as planning and anticipation. And then there are knife-edge encounters with extremes of weather: 'The seasoning: where the mind is honed and sweat is found to mix with ice' (p. 78). These critical 'cases' may play an important role in the development of skilled performance.

The argument of the present chapter is that this process of seasoning, which is so essential to expert performance, can be enhanced by artificial means. Just as training in aviation has long been enhanced by technological innovations such as the flight simulator, so the acquisition of experience can be accelerated by the appropriate use of technologies from computer-based learning to PC-based flight simulation. In this chapter I will outline a rationale for such an approach, based on the cognitive science literature. I will then briefly outline some of the results from our on-going empirical studies of pilot performance that address the question of the role of case-based reasoning in skilled performance in aviation.

Natural Learning

Schank has made a number of important suggestions about learning and intelligence in real-world environments (Schank & Cleary, 1995). Three key principles underlie the ability of people to learn vast amounts about all manner of subjects from physics to football. The first is that people are motivated to learn when this is pertinent to their goals.

Much classroom teaching fails to have a significant effect because the recipients have no goals related to geometry or history, etc. In aviation, people are generally highly motivated to learn but it might be that providing information under the heading of 'safety' is not the best way of connecting with their current goals. Goals such as 'proficiency' or 'professionalism', for example, might provide a better context for motivating those in aviation. Having a goal inspires interest which generates enquiry or questioning about the domain. This is the second principle of natural learning.

The activity of generating questions and reflecting or musing on the outcomes of one's experiences is pivotal in learning. The important role of reflection in learning in aviation has been emphasised by Henley, Anderson, and Wiggins (1999). According to Schank and Cleary (1995), reflection on experiences serves two purposes. The first is to derive generalisations that can be applied to future situations. The second is to provide labels or indices that allow the original experiences to be brought to mind at a later date. In Schank's view, the use of previous experiences or cases is a fundamental component of human intelligence: 'Case-based reasoning is the predominant way in which people think about their worlds' (Schank & Cleary, 1995, p. 124). Recent theories of decision making (e.g., Klein, 1998) also emphasise the importance of matching the current situation to memories of previously encountered examples. If a good match can be achieved, the decision maker automatically knows what to expect and what course of action is likely to be most appropriate.

The process of answering the questions raised as a result of experiences thus provides the basis for subsequent performance. This third principle is covered in more detail in the concept of the 'understanding cycle', which Schank (1999) describes as 'the cornerstone of the human intelligence process' (p. 278).

These three basic steps in natural learning form the basis of what it means to be an experienced or seasoned performer. The key ingredient is that the learner is exposed to experiences which then provide the basis for reflection, generalisation and the development of a knowledge base consisting of cases that support the generalisations as well as exceptional cases which run counter to them (see Figure 16.1). In contrast, much aviation training is devoted to the idea of instilling generalisations in the form of rules for which the learner has no experiential basis. Once equipped with these teacher-provided rules, the learner is permitted to venture forth to acquire the necessary supporting experience. Unfortunately, as the aviation accident record shows, novices are frequently unable to recall or apply the rules learnt in training at the appropriate moments.

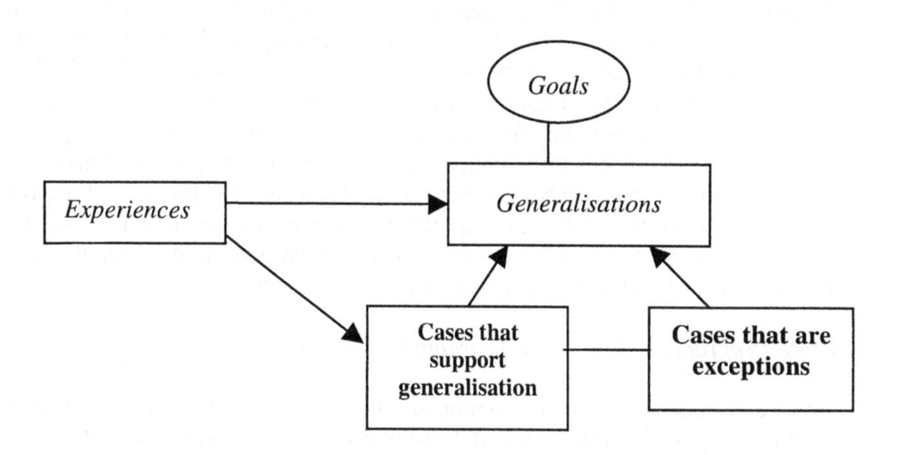

Figure 16.1 The process of natural learning

The Understanding Cycle

Schank (1999) places considerable emphasis on the role of failure in promoting learning. This refers to failure of one's expectations to match current reality, rather than to the experience of disaster or negative outcomes. The formation of generalisations from experience provides the learner with expectations. In general, when subsequent experience conforms to our expectations there is little further learning. However, when experience disconfirms our expectations this provides a valuable learning opportunity. These failures lead to a re-examination of the generalisations, resulting in more-refined or better-tuned generalisations, as well as an extended library of cases that are exceptions to the generalisation. It is important to note that it is the failure of expectations, rather than the learner that is important. Expectations may fail in a positive way. For example, the expectation that one will not enjoy aerobatics may be disconfirmed by a few gentle low-g manoeuvres. The learner has experienced a failure of expectations whilst also enjoying a positive experience.

The role of the teacher or instructor is, therefore, to provide the initial experience either through existing cases or by putting trainees in situations where they will experience failure of their expectations. These may be actual flight experiences or simulated flight experiences using desktop simulations. The instructor should help the trainee to draw the appropriate generalisations from these experiences. Research has shown that the critical factor in determining the usefulness of prior experiences is the ability to access those experiences at the appropriate time. This, in turn, depends on how those experiences have been indexed or labeled in the trainee's memory.

Schank and Cleary (1995) suggest that it is crucial that the learner attach functionally useful labels to experiences. For example, the label 'Tenerife air disaster' is less useful than 'crash caused by miscommunication and lack of CRM'. The question of how to design instruction so that learners draw the most appropriate generalisations and label their experiences in the most useful ways is a critical one. Research needs to be undertaken to determine the processes involved in learning from various kinds of experiences. Klein and Hoffman (1993) have drawn attention to the potential importance of vicarious experiences in the development of expertise. We have begun a research program to examine this and related issues based on the theoretical approach outlined above.

Research Program on Case-based Learning in Aviation

The most consequential errors in aviation are those associated with judgement and decision making. Klein (1993) studied decision errors by various real-world groups (including pilots) and found that the largest category was due to 'lack of experience'. Real-world experience is hard to gain – pilots may move quickly through the aviation system from *ab initio* to commercial pilot without any of the seasoning experiences described by Gann (1962). At the same time, because of the role played by air transport in modern society, air crashes and aviation safety are matters of intense public concern. Crashes are widely publicised and, in most parts of the world, subject to detailed investigation. In contrast, despite a vastly inferior safety record, very little is heard about safety in the marine transport industry.

The aviation industry is, therefore, well equipped to provide vicarious or second-hand experiences in the form of accident investigation reports, as well as published and spoken stories and cases describing a wide range of events. Little is known about the extent to which pilots assimilate this information and whether or not such vicariously acquired experiences play any significant role in their flight decision making. With the aid of a grant from NASA, we began a program of empirical research designed to throw some light upon these issues. The eventual aim of this research program is to provide insight and guidance into the development of tools to accelerate the acquisition of experience from vicarious sources. In other words, to provide the seasoning, formerly obtained through a long process of trial-and-error, by synthetic or artificial means.

Do Pilots utilise Information from Safety Materials during In-flight Decision Making?

Over a thousand pilots, ranging in flight experience from 4 to 33,000 hours, from all over the globe completed an on-line survey about the role of case recall in in-flight decision making. As part of the survey, respondents were asked to indicate whether they read any aviation safety material. The overwhelming majority (97.9 percent) indicated that they did so, although holders of commercial pilot licences were significantly less likely to read safety material than were private or airline transport pilot licence holders. This confirms the view that aviation safety is indeed a widely covered subject from official accident investigation reports to all manner of safety magazines and other aviation publications. Almost all participants in the aviation system are exposed to safety material in one form or another. As will be seen shortly, cases cited in these safety materials play an important role in aiding in-flight decision making.

The critical question is whether or not case-based recall plays a significant role in aiding in-flight decision making. Traditionally, decision making has been seen as a process of weighing up alternatives and opting for the one which yields the greatest expected benefits for the pilot (O'Hare, 2002). Normative models, such as subjective expected utility theory, provide the framework for evaluating decision making performance. An alternative perspective on decision making and problem solving is provided by case-based reasoning (CBR). According to Kolodner (1993), most problems are solved and decisions made by recourse to memories for previous problems and their solutions. Sometimes, use is made of analogical reasoning where a comparison is made with a case or example from an entirely different domain. Recent models of decision making, such as Klein's (1989) recognition-primed decision (RPD) model also emphasise the role of case comparisons in weighing up situations so as to decide on an appropriate course of action.

Survey respondents were asked to indicate whether they had ever recalled a previous case when dealing with a critical flight event. From previous surveys (Hunter, 1995; O'Hare & Chalmers, 1999), we know that almost all pilots have encountered critical flight events at some time. Just over half the respondents (52.5 percent) indicated that case-based recall played a part in responding to a critical flight event. This provides striking evidence for the role of case-based reasoning processes in aeronautical decision making. However, a substantial number of pilots were unable to recall using a case when making an in-flight decision. Whether this is due to an inability to recall such instances as may have occurred, or whether this is entirely due to these pilots resolving problems by other means, remains to be determined.

The majority (88 percent) of pilots agreed that the likelihood of a case being useful in decision making increased as a function of the amount of time available for the decision. The instances provided by the respondents of case-based recall in dealing with in-flight problems were strongly in accordance with this view. The most commonly cited situation where case-based recall was used involved weather-related problems (46 percent). Cruise was the most frequently cited phase of flight.

Consistent with the recognition-primed model of decision making (Klein, 1989), cases were more commonly used in the early stages of decision making (e.g., 58 percent were used to evaluate the current situation), rather than the later stages (only 28 percent were used to evaluate intended actions). Most interestingly, the source of exactly 50 percent of the cases used during in-flight problem solving was from written materials. Safety magazines provided 29 percent of the cases used, and accident reports 21 percent. It appears that the high exposure of pilots to written safety materials can result in information being retained and used during in-flight decision making well into the future.

A good many of the cases used during in-flight decision making were read or heard about some time previously – nearly 31 percent were more than five years old. There was a not-so-surprising tendency for older pilots to recall cases from further back in the past than younger pilots but even the oldest pilots (aged over 65 years) recalled cases from the relatively recent past – only 5-10 years previously.

The evidence from this survey strongly suggests that case-based reasoning does play a significant role in aeronautical decision making. These data show that the recall of a previous case or example is more likely if time allows for more deliberative decision making. However, in accordance with the recognition-primed decision model, cases are more useful in problem recognition and identification than in choosing a course of action. Cases are frequently drawn from the aviation safety literature.

Two caveats to these observations should be noted. Firstly, because of the nature of such a survey, it is impossible to know how many other pilots might have used case-based reasoning in their in-flight problem solving at the time but subsequently failed to remember doing so. The probable impact of this reporting bias would be to underplay the role of case-based recall in aeronautical decision making. Secondly, the sample of pilots who responded to the survey may have been unusually aware of aviation safety matters and more likely than the general pilot population to read and retain aviation safety information. As such, the probable impact on the present findings would be to overstate the impact of written safety materials.

Clearly, more experimentally controlled research is required on the role of case-based reasoning during in-flight decision making. It would be desirable to control participants' exposure to case-based material and to assess the subsequent impact (if any) on decision making in appropriately constructed scenarios. Such a framework would allow for the investigation of important issues such as the extent of generalisation from a case to related flight events. How closely does the in-flight event need to match the stored case in order for it to be used in decision making? What sorts of features are used to make the match between in-flight event and stored examples?

Both field and laboratory studies could be used to further address the learning that takes place from aviation safety materials, particularly those detailing accidents and incidents. What kinds of reflective activities (Henley, Anderson, & Wiggins, 1999) need to be undertaken for the appropriate lessons to be learnt in a form that allows them to be applied to future events? Is it beneficial to read about a large number of events or is it more beneficial to store the details of a smaller number of prototypical examples? What is the difference between expert and novice knowledge of aviation cases? The case for further research of these issues is clear and compelling.

Acknowledgements

The research reported here was made possible by the support of a NASA-Ames research grant (NAG-2-1395). I am grateful for this support and for the support and encouragement of Dr Immanuel Barshi at NASA-Ames. I am also grateful to Adele Arnold, Rachel Avery, Nadia Mullen, and Karen Walsh for their work on this project

References

Allen, R. (Ed.). (2000). *New Penguin English Dictionary*. Harmondsworth, UK: Penguin.

Gann, E. (1961). *Fate is the Hunter*. London: Hodder & Stoughton.

Henley, I., Anderson, P., & Wiggins, M. (1999). Integrating human factors education in general aviation: Issues and teaching strategies. In D. O'Hare (Ed.), *Human Performance in General Aviation* (pp. 89-117). Aldershot: Ashgate.

Hunter, D. (1995). *Airman Research Questionnaire: Methodology and Overall Results* (Rep. No. DOT/FAA/AM-95/27). Washington, DC: Federal Aviation Administration Office of Aviation Medicine.

Klein, G. A. (1989). Recognition-primed decisions. In W. B. Rouse (Ed.), *Advances in Man-machine Systems Research: Vol. 5* (pp. 47-92). Greenwich, CT: JAI Press.

Klein, G. A. (1998). *Sources of Power*. Cambridge, MA: MIT Press.

Klein, G. A., & Hoffman. (1993). Seeing the invisible: Perceptual-cognitive aspects of expertise. In M. Rabinowitz (Ed.), *Cognitive Science Foundations of Instruction* (pp. 203-226). Hillsdale, NJ: Erlbaum.

Kolodner, J. (1993). *Case-based Reasoning*. San Mateo, CA: Morgan Kaufmann.

O'Hare, D. (2002). Aeronautical decision making: Metaphors, models, and methods. In P. Tsang & M. Vidulich (Eds.), *Principles and Practice of Aviation Psychology* (pp. 201-237). Mahwah, NJ: Lawrence Erlbaum.

O'Hare, D., & Chalmers, D. (1999). The incidence of incidents: A nationwide study of flight experience and exposure to accidents and incidents. *The International Journal of Aviation Psychology, 9*, 1-18.

Schank, R. (1999). *Dynamic Memory Revisited.* Cambridge: Cambridge University Press.

Schank, R., & Cleary, C. (1995). *Engines for Education*. Mahwah, NJ: Lawrence Erlbaum.

Chapter 17

Decision Skills Training for the Aviation Community

Gary Klein
Klein Associates Inc, Fairborn, Ohio, USA

Introduction

How should pilots make decisions? By determining the preferred decision strategies that pilots should use in handling difficult situations, we can develop training programs that reduce error and improve performance.

The original answer was that pilots should make decisions by considering a range of options, evaluating these options on a set of evaluation categories, judging how each option scored on each evaluation category, perhaps assigning different weights to the categories to reflect their importance, totalling up the results, and finding the option with the highest score. This can be referred to as a Rational Choice strategy (Klein, 1998). It is widely taught in business schools and schools of engineering, and for many years it has been taught to pilots. One popular version is the DECIDE model (Benner, 1975). The six steps of the prescriptive model of decision making are to:

1. Detect that an important change has occurred.
2. Estimate how that change will impact the mission.
3. Choose a safe outcome.
4. Identify plausible courses of action.
5. Do (execute) the action that was chosen.
6. Evaluate the effect of the action.

The rationale for using DECIDE to train pilots was to encourage pilots to consider all the relevant options, consider the relevant evidence, and find the best. The Rational Choice method can be useful, and it has a number of strengths. It is a general method, so pilots can use it with different types of decisions. It is reliable-pilots can rely on getting the same answer if they run the method several times. It is comprehensive. The method requires pilots to consider a wide range of evidence and issues.

However, as researchers have studied decision making in natural settings (Klein, 1998; Klein, Orasanu, Calderwood, & Zsambok, 1993; Zsambok & Klein, 1997), we have also learned that this Rational Choice method has some limitations. It takes a lot of time to perform the analysis, and in settings such as aviation, a pilot may not have that time. An aborted takeoff does not permit careful consideration of alternative courses of action. The Rational Choice method requires a great deal of data, and pilots often have to wrestle with uncertainty about weather conditions, the nature of a malfunction, the possibility of a go-around, and so forth. (A go-around is an aborted landing, followed by a route to attempt the landing a second time.) Another limitation is that a Rational Choice method is most helpful when the options are close together, rather than having one option clearly good and the others clearly bad. However, the closer together the options are, the less it matters to find the best. In fact, under time pressure, pilots often look for the first workable option, and do not care about finding the absolute best. Klein (1989) found that in a variety of natural settings, skilled decision makers rarely compared options. Even a watered-down version of the Rational Choice method may be used less than 5 percent to 10 percent of the time for critical decisions (Klein, 1998). If pilots are not comparing options, then how are they making decisions? We have found that skilled decision makers can use their experience to generate an effective option as the first one they consider (Klein, Wolf, Militello, & Zsambok, 1995).

The Recognition-primed Decision Model

The Recognition-primed Decision (RPD) model (Klein, Calderwood, & Clinton-Cirocco, 1986) attempts to explain how people can make decisions in field settings, under conditions that make it difficult to generate and compare multiple options. It must be emphasised that the RPD model is only one of many Naturalistic Decision Making (NDM) models. Naturalistic Decision Making is the study of how people use their experience to make decisions in field settings, examining the strategies that people actually employ rather than studying their deviations from mathematical or statistical formulations. Lipshitz (1993) has reviewed nine different NDM models. The basic assertions of the RPD model are that:

People are able to use their experience to size up a situation as familiar. By recognising a situation, people can generate a reasonable option as typical. Usually, this typical option is the first one they consider. If they are unsure of the situation, they may build stories to explain different events (although this may take a little time). If they are unsure of the option they generated, and have the time, they may build stories to evaluate how the option will be carried out in context.

The RPD model consists of three variations. In the simplest case, the decision maker judges the situation to be functionally identical to a prototype. A prototype is a typical, or representative case. The match to a prototype allows a determination of the relevant cues, reasonable goals, expected developments, and typical course of action. This is a form of rule-following behaviour. However, we should not treat this as mechanistic. Skill is needed to judge whether the antecedent condition of the rule has been satisfied. Experience is needed to recognise a situation as typical, and to make an appropriate prototype match. All pilots know the rule not to fly into turbulent weather. It is a simple and important rule. But merely telling a novice pilot the rule is no guarantee of safety. Skill is needed to determine what counts as turbulent weather. Many rules sound simple, but turn out to be complex under field conditions.

Sometimes, pilots may be unsure about what is going on in a situation. They do not try to represent their uncertainty by estimating probabilities of different explanations. Rather, they try to build stories to explain the evidence, and they judge the plausibility of each story. Thus, when faced with a strange reading that one of the fuel tanks seems to be running lower than the others, a pilot might explain this away as a failure of the measuring instruments. An alternative explanation is that there might be a leak in that fuel tank, but this is more rare than a fault in a sensor. The Pilot does not attempt to estimate the probability of each, but rather compares the plausibility of the two. It is easier to imagine a sensor going bad than a fuel tank suddenly developing a leak for no apparent reason. The assessment relies on the heuristic of the availability of the two stories. This is described in Variation 2 of the RPD model. It takes experience to be able to build coherent stories (Cohen & Freeman, 1996; Klein & Crandall, 1995). Novices are unable to do this, particularly when the evidence is complex. This type of story building is often seen in diagnosis of equipment malfunctions (De Keyser & Woods, 1993). It is more difficult to build stories in time-pressured situations.

Pilots may need to evaluate a course of action, once they recognise it. How can you evaluate an option without comparing it to others? We have found that decision makers again build stories, to see how the option will be carried out. This is Variation 3 of the RPD model. We call this a process of mental simulation, thinking out how the action will proceed. If they do not find any difficulties, then they are fine. If they do spot a problem, they may find a way to improve the action plan. If they cannot find a way to improve the action, then they may reject the option and look for a better one. For example, a Captain preparing to land with engine problems might rehearse the sequence with the First Officer. In so doing, they might identify points at which nonessential power requirements could be cut, to ensure sufficient power in case a go-around was necessary. If the uncertainty is too high (e.g., if the airport has been experiencing wind shear), then the pilots may reject the option of landing at that field, and select another.

Clearly, there are times when pilots do need to compare several options rather than select the first that is satisfactory. For example, if a malfunction requires an airplane to divert (i.e., abandon the plan to land at a given airport in favour of an alternative), the Pilot will have to consider which airports are in the vicinity, and to pick one of them. Here, we might see some sort of Rational Choice comparison. The RPD model was not developed to explain cases in which multiple options are compared. However, I believe that in many cases the evaluation of multiple options is done by mentally simulating each option, to see if there are any difficulties, rather than by setting up evaluation criteria for judging each of the options. By relying on mental simulation to examine each option, some airports will be dropped from consideration. The remaining ones, all acceptable, may be compared at a global level, depending on which one provides the greatest comfort level. This explanation follows the work of de Groot (1946/1978), who referred to the process of progressive deepening (essentially the same as mental simulation) whereby chess players imagine how a line of play might develop, and determine a global reaction to the type of position they might expect, then progressively deepen another move, form an impression of it, and eventually select the move that triggers the strongest emotional reaction. In this way, different moves are compared. However, the comparison does not take place along a common set of evaluation dimensions (e.g., how much centre control does each move provide, how strong is the king side attacking potential, and so forth). This type of strategy is not the same as non-compensatory strategies such as elimination by aspects (Tversky, 1972), in which all options are evaluated along the most important evaluation dimension, and the inadequate options are dropped, then they are rated on the next most important dimension, until one option remains. De Groot's work showed that the chess players were imagining each move in the context of the game, not along common dimensions.

I suspect the same is true of pilots. It is possible that pilots who need to divert are using an elimination by aspects strategy (Tversky, 1972) in comparing different airports on common dimensions: which airports are within 100 miles, then, if more than one remains, which airport has the longest runways, then which airport has the fewest terrain risks, etc. But I hypothesise that the process is more along the lines of de Groot's concept of progressive deepening. A pilot might first identify the airports that are close enough to consider. Then, for each one, a progressive deepening would either cull it out (too much of a chance that they will close that airport by the time I get there), or continue it (I can live with it, but I wish it had better connections for my passengers), or select it (I am not crazy about flying in there because of the mountains, but the winds seem to be light enough so that I should not have much trouble).

Decision Skills Training: An Approach for Improving the Decisions of Pilots

We developed the Decision Skills Training (DST) method in response to a request from the US Marine Corps, who wanted to provide a program to improve decision making in squad leaders (Klein, McCloskey, Pliske, & Schmitt, 1997).

The Decision Skills Training program (Klein, 1997; Klein, in press) was in keeping with the NDM assertion that effective decision making depends on experience, rather than on following procedures or normative strategies. Therefore, the approach we adopted was to help the squad leaders rapidly boost their experience level with regard to making difficult decisions. Instead of trying to teach them to make difficult discriminations, or to distinguish different types of situational dynamics, we instead attempted to provide them with the tools to boost their experience level. That included ways of preparing better for the training they received, ways of augmenting their training, and ways of reflecting on the lessons they learned during training.

To help them *prepare* better for training, we helped them identify the decision requirements of their missions: the difficult and critical judgements and decisions, along with the reasons for the difficulty, types of errors that are often made, and strategies that seem helpful. The squad leaders' missions were primarily reconnaissance and surveillance activities in tracking enemy forces. The decision requirements included being able to estimate the time it would take them to get into their positions once they were inserted by helicopter, the ability to identify good landing sites for helicopters sent to extract them, the ability to infer enemy force disposition from glimpses of a few vehicles, and so forth. This exercise helped them define the decision skills that they needed, and helped them determine how to get practice and feedback on these specific decision skills during their field training exercises. We also taught them to use a mental simulation strategy prior to beginning a field exercise in which they would imagine that their plans had fallen apart, and to try to discover what might have gone wrong. This helped them identify the weaknesses in their plans prior to execution. The key to successful practice includes the availability of feedback that is timely, accurate, and diagnostic. In addition, experts in many fields engage in deliberate practice, such as setting goals for themselves in each session, rather than just putting in their time. Therefore, careful preparation prior to field exercises or training exercises can result in a greater level of learning during those exercises.

To help *augment* their training, we prepared tactical decision games (e.g., Schmitt, 1994) to provide low-level simulations. The tactical decision games offered high degrees of uncertainty in order to allow them to practice making decisions with key data elements missing or ambiguous. The games were also conducted under extreme time pressure, to prepare them to handle this type of stressor. The intent behind these tactical decision games was to provide them with a conditioning process in which they practiced making difficult decisions several times a week in preparation for their culminating field exercise. The program also taught methods for personnel to build their own tactical decision games in response to decision requirements that they needed to practice further.

To help them *reflect* on the training, we provided them with a decision checklist to help them debrief themselves after field exercises (and also after tactical decision games). The checklist asked them to: identify areas of uncertainty in the exercise and how they handled the uncertainty; identify difficult judgements and subtle cues and patterns, and how they discriminated these; reflect on the difficult judgements and decisions that they had just encountered.

The qualitative evaluations of the Decision Skills Training were very positive. Both attitude surveys and interviews showed enthusiasm for the training and its perceived benefit. As a result, the US Marine Corps has requested that we further develop this program for use in different courses at the Marine Corps University. In one version, we were asked to include a component on coaching, to help the more experienced officers articulate what they are seeing in order to do a better job of providing on-the-job training to officers with less experience. Zsambok, Crandall, and Militello (1994) have found in a variety of domains that on-the-job training accounts for the majority of skill development. However, most organisations do not have any program for performing on-the-job training, and most of the people who wind up providing on-the-job training have little idea of how to do it. They have a very small repertoire of training methods. Worse still, they tend to think they are doing a decent job because they do not know any better and because their own supervisors are not able to give them feedback. Aviation offers many effective opportunities for using on-the-job training, and the advantage is that it is not expensive since it can be provided during flying operations. Furthermore, the training generalises to the operational setting because it is presented in that setting. Currently, many trainers provide feedback about the actions taken, but not about the decision strategies used. We have found that trainers can be shown how to pay attention to the processes of decision making, along with the actions. They can check what pilots were seeing, what cues and patterns they were noticing, which options they were considering, and so forth. Decision training can, and should, occur in the context of regular training opportunities. This helps to ensure that the learning will generalise, and it also keeps the costs down. By preparing instructors to provide feedback about processes, the regular training can become more valuable.

It would appear that many aspects of this program could be useful for aviation training. In the past, instructors noted that pilots needed to do a good job of making decisions, and without differentiating types of decisions, were content to teach generic strategies. There is little evidence for the effectiveness of this approach. Instead, we may be able to do a much better job by:

1. Helping pilots identify the specific judgements and decisions that they need to practice.
2. Providing pilots with tools to use, and to build decision scenarios that allow practice in making decisions under uncertainty and time pressure.
3. Designing debriefing checklists to enable pilots to identify the decision requirements they encounter during challenging situations, so that they can use these challenges to direct their own learning.
4. Showing pilots that there are coaching methods for accelerating skill development.

In defining Naturalistic Decision Making as the study of how people use their experience to make decisions in field settings, the spotlight is placed on expertise and the way it develops. Instead of emphasising generic decision strategies, NDM applications tend toward specifying the types of decisions that are of concern, and the training is designed to build up expertise for these types of decisions. This brings us to the linkage between NDM and the field of expertise. The NDM community needs to take advantage of research on expertise, such as the framework presented by Shanteau (1992) for different aspects of expertise. It appears that we can differentiate experts from novices in several different ways. These include the ability of the expert to adapt and improvise more freely than novices, to project ahead more skilfully than novices, as well as to focus attention more effectively than novices.

Ericsson and Charness (1994) have taken the strong position that high levels of expertise in many fields can be attained through practice. To become a chess grandmaster, Charness (1989) estimated that it takes approximately 32,000 hours, or 4 hours of practice a day, every day, for about 20 years. The Decision Skills Training program is an attempt to improve decision making through the accelerated development of expertise. For the past few years, we have used the DST program to train pilots and maintenance technicians in the US Navy's S-3B program, as a means of improving safety by improving decision quality.

Conclusions

The questions posed by NDM are different from the questions raised in the past. As long as there was a belief in generic decision strategies, it made sense to ask how to teach pilots to make better decisions. Once that belief was challenged, the limitations of the question became clear. NDM is radical in rejecting the question about how to train pilots to make better decisions. It is radical in insisting that we substitute better questions, by specifying which decisions are the source of difficulty, why they are difficult, what types of errors are being made, and how proficient pilots are avoiding these errors. Rather than seeking to substitute new and improved decision strategies, we should be tunnelling down into the types of judgements and decisions that are difficult, in order to help pilots make better use of their existing strategies. NDM posits the use of decision requirements in place of global and unfocussed questions. NDM claims that in many cases, the magic is in the question, not in the solution. Once the right question is posed, at the right level, the solution may appear mundane. Rather than being disappointed, we should acknowledge the value of a straightforward answer.

The solutions posed by NDM are different from the training approaches offered in the past. One prominent decision researcher was once asked how he would train pilots to make better decisions and, without hesitation, he answered, 'I would teach them probability theory'. This is not an answer that would be given by NDM researchers. Generic strategies such as Rational Choice methods, Bayesian statistics, and de-biasing procedures are all weak because generic methods must be weak, and because they have a limited role for expertise. Even worse, as Erev, Bornstein, and Wallsten (1993), Wilson and Schooler (1991), and others have shown, when people apply analytical strategies, their performance may deteriorate, perhaps because pattern matching processes are interfered with. Sometimes, the generic strategies can be helpful, and NDM researchers are trying to learn the boundary conditions for the quantitative methods, such as multi-attribute utility analysis, that decompose tasks, as well as the boundary conditions for holistic methods that depend on heuristics and expertise. Sometimes, NDM researchers do offer generic approaches, such as Cohen, Freeman, and Thompson's (1998) Critical Thinking program. But this is done with caution and with acknowledgement that the program will not always be appropriate. Part of the program includes practice in determining when not to apply the Critical Thinking strategy. Yet another difference is in the way NDM researchers treat uncertainty, not as a judgement that informs the rating of a course of action, but as a dynamic feature of the environment and the decision maker that affects the strategies used.

In short, NDM is radical in rejecting the multi-attribute utility analysis, Rational Choice strategy, that has been the mainstay of training programs in Aviation Decision Making and in many other applied domains.

Author Note

This paper is a concise version of: Klein, G. (2000). How can we train pilots to make better decisions? In H. O'Neil & D. Andrews (Eds.), *Aircrew Training and Assessment* (pp. 165-195). Mahwah, NJ: Lawrence Erlbaum Associates.

Robert Hutton provided very useful comments and suggestions on the original manuscript.

References

Benner, L. (1975). D.E.C.I.D.E. in the hazardous materials emergencies. *Fire Journal, 69*(4), 13-18.

Charness, N. (1989). Expertise in chess and bridge. In D. Klahr & K. Kotovsky (Eds.), *Complex Information Processing: The Impact of Herbert A. Simon* (pp. 183-208). Mahwah, NJ: Lawrence Erlbaum Associates.

Cohen, M. S., & Freeman, J. T. (1996). Thinking naturally about uncertainty. *Proceedings of the 40th Human Factors & Ergonomics Society,* 179-183.

Cohen, M. S., Freeman, J. T., & Thompson, B. (1998). Critical thinking skills in tactical decision making: A model and a training method. In J. Cannon-Bowers & E. Salas (Eds.), *Making Decisions Under Stress: Implications for Individual and Team Training* (pp. 155-189). Washington, DC: APA Press.

de Groot, A. D. (1946/1978). *Thought and Choice in Chess.* New York, NY: Mouton.

De Keyser, V., & Woods, D. D. (1993). Fixation errors: Failures to revise situation assessment in dynamic and risky systems. In A. G. Colombo & A. Saiz de Bustamente (Eds.), *Advanced Systems in Reliability Modeling.* Norwell, MA: Kluwer Academic.

Erev, I., Bornstein, G., & Wallsten, T. S. (1993, June). The negative effect of probability assessments on decision quality. *Organizational Behaviour and Human Decision Processes, 51*(1), 79-94.

Ericsson, K. A., & Charness, N. (1994). Expert performance: Its structure and acquisition. *American Psychologist, 49*(8), 725-747.

Klein, G. (1997). *Developing Expertise in Decision Making, Thinking, and Reasoning: Vol. 3* (pp. 337-352). East Sussex, UK: Psychology Press Ltd.

Klein, G. (1998). *Sources of Power: How People Make Decisions.* Cambridge, MA: MIT Press.

Klein, G. (in press). *Intuition at Work.* New York: Doubleday.

Klein, G., McCloskey, M. J., Pliske, R. M., & Schmitt, J. (1997). Decision skills training. *Proceedings of the Human Factors and Ergonomics Society 41st Annual Meeting* (pp. 182-185). Santa Monica, CA: HFES.

Klein, G., Wolf, S., Militello, L., & Zsambok, C. (1995). Characteristics of skilled option generation in chess. *Organizational Behaviour and Human Decision Processes, 62*(1), 63-69.

Klein, G. A. (1989). Strategies of decision making. *Military Review,* 56-64.

Klein, G. A., Calderwood, R., & Clinton-Cirocco, A. (1986). Rapid decision making on the fireground. *Proceedings of the 30th Annual Human Factors Society: Vol. 1* (pp. 576-580). Santa Monica, CA: The Human Factors Society.

Klein, G. A., & Crandall, B. W. (1995). The role of mental simulation in naturalistic decision making. In P. Hancock, J. Flach, J. Caird, & K. Vicente (Eds.), *Local Applications of the Ecological Approach to Human-machine Systems: Vol. 2* (pp. 324-358). Mahwah, NJ: Lawrence Erlbaum Associates.

Klein, G. A., Orasanu, J., Calderwood, R., & Zsambok, C. E. (Eds.). (1993). *Decision Making in Action: Models and Methods.* Norwood, NJ: Ablex.

Lipshitz, R. (1993). Converging themes in the study of decision making in realistic settings. In G. A. Klein, J. Orasanu, R. Calderwood, & C. E. Zsambok (Eds.), *Decision Making in Action: Models and Methods* (pp. 103-137). Norwood, NJ: Ablex.

Schmitt, J. F. (1994). *Mastering Tactics.* Quantico, VA: Marine Corps Association.

Shanteau, J. (1992). Competence in experts: The role of task characteristics. *Organizational Behaviour and Human Decision Processes, 53,* 252-266.

Tversky, A. (1972). Elimination by aspects: A theory of choice. *Psychological Review, 79*(4), 281-299.

Wilson, T. D., & Schooler, J. W. (1991). Thinking too much: Introspection can reduce the quality of preferences and decisions. *Journal of Personality and Social Psychology, 60,* 181-192.

Zsambok, C. E., Crandall, B., & Militello, L. (1994). *OJT: Models, Programs, and Related Issues* (Contract MDA903-93-C-0092 for the US Army Research Institute for the Behavioral and Social Sciences, Alexandria, VA). Fairborn, OH: Klein Associates Inc.

Zsambok, C. E., & Klein, G. (Eds.). (1997). *Naturalistic Decision Making.* Mahwah, NJ: Lawrence Erlbaum Associates.

Chapter 18

Ab-initio Flight Training and Airline Pilot Performance: Differences in Job Requirements

Peter Maschke and Klaus-Martin Goeters
DLR, Germany

Introduction

A knowledge of the basic job requirements is an important basis whenever a selection system has to be established. In ab-initio pilot selection the question arises as to whether the requirements in both initial training and line duty are identical or differ in level or profile. A similar question comes up regarding the validation of already existing selection systems (Hörmann & Maschke, 1996). Concerning the relevant criteria, whether those from the initial training should be selected or from the airline operation. Alternatively, there may be no difference because the requirements are both the same. These issues call for a review of the basic personal capabilities needed by pilots in initial training and later profession and a closer look at any possible differences in requirement profiles.

Method

Instrument

The method used to evaluate different requirements for student pilots and airline pilots was the Job Analysis Survey F-JAS by Fleishman (1992). With the F-JAS, job holders are asked to rate their jobs on 7-point rating scales with respect to the level of the ability required. Application of the Fleishman system is particularly appropriate for selection issues because it provides a list of tests to measure the identified abilities. Thus it is possible to transfer the results of the job analysis directly into assessment methods (Fleishman & Reilly, 1992). Another advantage is that it is easy and economical to administer. Although the F-JAS is a general job analysis tool and was therefore developed for all kinds of professions, it has been applied with great success in different areas of aviation (Eissfeldt, 1997; Maschke, Goeters, & Klamm, 1998).

The F-JAS consists of 72 behaviourally-anchored rating scales, mainly focussing on aptitudes, knowledge, and skills. There are also nine interactive/social scales, which were supplemented in our study by nine additional scales developed by the German Aerospace Centre (DLR) in order to cover this area more comprehensively. This was done in order to take into account prior research indicating the importance of attitude and personality for a successful career as an airline pilot (Chidester et al., 1991; Hörmann & Maschke, 1996). Five knowledge/skills scales were not used for this study, because they were clearly not related to the professional activities of pilots, e.g., 'Typing'. This was done in order to increase the acceptability of the survey by the pilots. In the applied version the survey consisted of 76 scales: 21 cognitive, 10 psychomotor, 9 physical, 12 sensory, 18 interactive/social, and 6 knowledge/skills scales.

Participants

Two different subject groups were selected as samples. Sample 1 consisted of 25 student pilots from Lufthansa Pilot School who were in the last phase of the 2-year initial training scheme. Students in this phase were chosen because they had overall experience in all the initial training requirements. They had almost finished the theoretical and practical training, with 180 to 190 flight hours. Their ages varied from 21 to 31 with a mean of 25.

In sample 2 experienced airline pilots served as a reference group. Participants were 141 airline pilots from Lufthansa German Airlines (49 captains and 92 first officers flying on the jet aircraft types B737, B747, A320, A340, A310, or A300). The mean age was 37 years (standard deviation: 8.5 years), the average number of flight hours flown was 6,900 in total, and on current type 2,100.

Procedure

The study was conducted by the German Aerospace Center (DLR). The modified version of the Fleishman survey was administered to the student pilot sample under classroom conditions and to the active pilot sample in fleet meetings. This guaranteed standardised instruction and application. F-JAS data was collected anonymously in two steps: In a first step the participants were asked to rate the required level of a particular factor for airline pilots in *general*. In a second step they were asked to rate the level of each factor required for their *specific* initial training at the flight school (sample 1) or for their *specific* position and type of aircraft within the airline operation (sample 2). This method of calling special attention to the specific requirements of flight students and airline pilots was chosen because this study was mainly focussed on these differences.

Results

Student Pilots: Specific Requirements

As the Fleishman system consists of 7-point rating scales (scale-mean=4) and the rating scores in our study were distributed with a standard deviation of about 1, abilities rated with a mean of more than 5 were interpreted as 'important' because they clearly exceed the scale average.

Table 18.1 shows the results of the student pilots sample regarding the specific ratings in order of importance. Most of the scales rated as important (>5) were cognitive (9 out of the 21), psychomotor (4 out of 10) and sensory abilities (6 out of 12) as well as interactive/social capabilities (10 out of 18). Although 'Map Reading' received the highest score, the majority of the knowledge/skills scales and all physical abilities were rated as less important for ab-initio training.

Student Pilots and Airline Pilots: Different Requirements

Table 18.2 shows the differences regarding important requirements between student pilots and active airline pilots. In all cases of significant difference the airline pilots' ratings were higher than the student pilots'. With one exception (Control Precision) the *general* ratings do not differ significantly. In contrast to the general ratings, many significant differences were found due to the *specific* ratings. The majority of these differences are related to the interactive/social capabilities (8 out of 10 important scales) and the psychomotor abilities (3 out of 4 important scales) supplemented by some cognitive abilities (2 out of 9 important scales). No significant differences were found in important sensory abilities and knowledge/skill scales.

Discussion

The results of the specific student pilot ratings emphasise the importance of most of the classical ab-initio pilot selection criteria in the cognitive, psychomotor, and sensory areas. Factors such as 'Time Sharing', 'Spatial Orientation', or 'Auditory Attention' are still basic requirements in pilot training. However, two results are unexpected: Cognitive and knowledge/skills related job demands involving mathematical/technical elements ('Mathematical Reasoning', 'Mechanical Knowledge', 'Electrical Knowledge') are rated surprisingly low (and are therefore not mentioned in Table 18.1). These classical pilot selection criteria seem to have declined in value. Even more surprising is the high importance of the interactive/social factors, which can be seen as a blend of personality components and attitudes. The initial training scheme already contains factors as 'Stress Resistance', 'Motivation', and 'Cooperation', and these are as important as the classical cognitive and psychomotor factors and should therefore be equally weighted within the selection.

Table 18.1 Job analysis of student pilots: Factors rated as important (M>5), N=25 student pilots, ordered by magnitude

Scale	Area	Mean Rating M
Map Reading	Knowledge/Skills	6.56
Stress Resistance	Interactive/Social	5.76
Time Sharing	Interactive/Social	5.72
Motivation	Interactive/Social	5.64
Cooperation	Interactive/Social	5.60
Spatial Orientation	Cognitive	5.56
Behaviour Flexibility	Interactive/Social	5.48
Rate Control	Psychomotor	5.48
Resilience	Interactive/Social	5.48
Selective Attention	Cognitive	5.48
Decision Making	Interactive/Social	5.44
Memorisation	Cognitive	5.44
Auditory Attention	Sensory	5.40
Self Awareness	Interactive/Social	5.28
Situation Awareness	Interactive/Social	5.28
Speech Recognition	Sensory	5.25
Glare Sensitivity	Sensory	5.24
Perceptual Speed	Cognitive	5.24
Written Comprehension	Cognitive	5.24
Depth Perception	Sensory	5.20
Reading Plans	Knowledge/Skills	5.20
Resistance to Premature Judgement	Interactive/Social	5.20
Communication	Interactive/Social	5.16
Night Vision	Sensory	5.16
Number Facility	Cognitive	5.12
Oral Comprehension	Cognitive	5.12
Visualisation	Cognitive	5.12
Response Orientation	Psychomotor	5.08
Control Precision	Psychomotor	5.04
Far Vision	Sensory	5.04
Multi-limb Coordination	Psychomotor	5.04

Hardly any significant differences between the student pilot and the airline pilot ratings could be found on the basis of the general ratings. This result shows that the job requirements of airline pilots are generally perceived in the same way by the two groups. On the other hand significant differences are found in the specific ratings between student and airline pilots: In many important scales the requirements are rated significantly higher by active pilots than by student pilots. This is especially the case in the interactive/social area. Although these personality

Table 18.2 Differences of average F-JAS-ratings (general and specific) between student pilots (N=25) and airline pilots (N=141). Only scales with substantial average ratings (see Table 18.1) are listed. In all cases of significance the airline pilots' average is higher

Scale Difference	Area	Mean Scale	
		General Rating	Specific Rating
Map Reading	Knowledge/Skills	ns	ns
Stress Resistance	Interactive/Social	ns	.67 ***
Time Sharing	Interactive/Social	ns	.56***
Motivation	Interactive Social	ns	ns
Cooperation	Interactive/Social	ns	.83***
Spatial Orientation	Cognitive	ns	.53**
Behaviour Flexibility	Cognitive	ns	.39*
Rate Control	Psychomotor	ns	.56**
Resilience	Interactive/Social	ns	ns
Selective Attention	Cognitive	ns	ns
Decision Making	Interactive/Social	ns	.72***
Memorisation	Cognitive	ns	ns
Auditory Attention	Sensory	ns	ns
Self Awareness	Interactive/Social	ns	.63***
Situational Awareness	Interactive/Social	ns	.71***
Speech Recognition	Sensory	ns	ns
Glare Sensitivity	Sensory	ns	ns
Perceptual Speed	Cognitive	ns	ns
Written Comprehension	Cognitive	ns	ns
Depth Perception	Sensory	ns	ns
Reading Plans	Knowledge/Skills	ns	ns
Resistance to Premature Judgement	Interactive/Social	ns	.74***
Communication	Interactive/Social	ns	1.22***
Night Vision	Sensory	ns	ns
Number Facility	Cognitive	ns	ns
Oral Comprehension	Cognitive	ns	ns
Visualisation	Cognitive	ns	ns
Response Orientation	Psychomotor	ns	.59**
Control Precision	Psychomotor	.39*	.62**
Far Vision	Sensory	ns	ns
Multi-limb Coordination	Psychomotor	ns	ns

(ns – non significant; *p<.05; **p<.01; p<.001).

factors are already of surprisingly high importance during the initial training, they become even more important during active service. These results may also explain the different outcomes of validity studies based on the criteria used: When based on flying school criteria, personality factors had lower validities than when criteria data from airline operation were used (see Maschke & Hörmann, 1988; Hörmann & Maschke, 1996).

Contrary to expectation, the student pilots did not record higher requirements than the active pilots in any of the important scales. Successfully completing a flying school is clearly not the most difficult part of the entire career development: the contrary seems to be the case.

Which conclusions should be drawn for ab-initio airline pilot selection? The test profile should consider not only traditional aptitude traits but also personality factors. As personality evaluation still seems to be under-represented in most current selection systems, more activities should be carried out in developing reliable and valid personality assessment instruments (Goeters et al., 1993; Hörmann et al., 1997).

The differences in requirements between student and airline pilots indicate that the level of acceptance criteria should be applied more strictly in the selection methods, if they are to predict success not only in flying school but also in a subsequent airline career. In this case it is also recommended, that airline validity criteria are used if possible. Flying school data could show an incomplete picture of the airline profile.

The study can also be seen as an explanation for negative experience some airlines have had in employing already licensed pilots (direct entries) by taking flying school results instead of psychological tests: Due to the higher airline requirements, a pass in the flying school exam is a necessary precondition but not a guarantee of a successful airline career.

References

Chidester, T. R., Helmreich, R. L., Gregorich, S. E., & Geis, C. E. (1991). Pilot personality and crew coordination: Implications for training and selection. *The International Journal of Aviation Psychology, 1,* 25-44.

Eissfeldt, H. (1997). Ability requirements for different ATC positions. *Proceedings of The Ninth International Symposium on Aviation Psychology*, Columbus, Ohio.

Fleishman, E. A. (1992). *The Fleishman Job Analysis Survey (F-JAS).* Palo Alto: Consulting Psychologists Press, Inc.

Fleishman, E. A., & Reilly, M. E. (1992). *Handbook of Human Abilities: Definitions, Measurements, and Job Task Requirements.* Palo Alto: Consulting Psychologists Press, Inc.

Goeters, K. M., Timmermann, B., & Maschke, P. (1993). The construction of personality questionnaires for selection of aviation personnel. *The International Journal of Aviation Psychology, 3,* 123-141.

Hörmann, H. J., & Maschke, P. (1996). On the relation between personality and job performance of airline pilots. *The International Journal of Aviation Psychology, 6,* 171-178.

Hörmann, H. J., Manzey, D., Maschke, P., & Pecena, Y. (1997). Behavior-oriented assessment of interpersonal skills in pilot selection: Concepts, methods, and empirical findings. *Proceedings of The Ninth International Symposium on Aviation Psychology,* Columbus, Ohio.

Maschke P., Goeters, K. M., & Klamm, A. (1998). Job requirements of airline pilots: Results of a job analysis. *Proceedings of the Fourth Australian Aviation Psychology Symposium,* Sydney.

Maschke, P., & Hörmann, H. J. (1988). Zur Bewährung Psychologischer Auswahlverfahren für Operationelle Berufe in der Luft und Raumfahrt. *Zeitschrift für Flugwissenschaften und Weltraumforschung, 12,* 181-186.

Chapter 19

Validating a Computer-Based Training Tool for In-Flight Weather-Related Decision-Making

Mark Wiggins, *University of Western Sydney, Australia*
David O'Hare, *University of Otago, New Zealand*
and Oliver Lods, *University of Western Sydney, Australia*

Introduction

Inadvertent visual flight into Instrument Meteorological Conditions (IMC) remains one of the most significant causes of fatalities within the US general aviation industry. Between 1983 and 1992, so-called VFR (Visual Flight Rules) into IMC accidents comprised 9.8 percent of the total number of weather-related general aviation accidents in the United States, but accounted for 27 percent of the fatalities that occurred within this period (Aircraft Owners and Pilots Association, 1996).

Strategies for the prevention of VFR into IMC accidents can be classified loosely into those that focus upon the motivation component of decision-making and those that focus upon the cognitive component (Jensen, 1992). In the case of the former, it has been assumed that visual pilots persist in flying into IMC due to a misplaced motivation to reach a destination. This approach is encapsulated within the hazardous thoughts training program in which pilots are taught to recognise and respond to the onset of inappropriate motivators (Buch & Diehl, 1984).

While there is undoubtedly some validity associated with the motivational approach to weather-related decision-making, skilled performance also requires the capacity to acquire, interpret, process, and respond to situations involving a deterioration in weather-related conditions during flight. Therefore, these cognitive elements of the weather-related decision-making process also need some consideration if improvements are expected in reducing the incidence of decision errors (Wiggins & O'Hare, 1995). This study describes the evaluation of a computer-based Pilot Judgement Tutoring System (PJTS) that was developed to target some of the cognitive features associated with in-flight weather-related decision-making. Referred to as 'Weatherwise', the program was designed to facilitate the acquisition of the skills necessary to recognise the onset of deteriorating weather conditions and identify the most appropriate alternate destination under the conditions.

Program Development

'Weatherwise' was developed on the basis of the assumption that expert decision-making is highly individualised, and is a product of the particular experiences acquired by a decision-maker (Sternberg, 1998). Therefore, rather than prescribe an 'optimal' decision-strategy, 'Weatherwise' was designed to provide an environment that simulated the cognitive demands that are associated with the operational environment. Exposure to this environment was expected to provide an opportunity for individuals to develop their own decision strategies and receive task-oriented feedback concerning their performance.

A secondary assumption that underscored the development of the program concerned the differences that appear to exist between experts and novices exposed to the decision-making environment. In particular, novices appear to require more time than experts to acquire and process task-related information, and develop an appropriate strategy (Klein, 1989). From an operational perspective, this information processing strategy needs to be accomplished while maintaining physical control of the aircraft.

As a consequence of the distinction between the performance of experts and novices, the program was designed to provide a series of cues that would enable less experienced pilots to recognise a deterioration in the weather conditions at the earliest stage of the flight. The aim was to ensure that novice pilots both recognise that a deterioration has occurred, and maximise the time available for the decision-making process.

The weather-related cues were identified following a cognitive interview with expert pilots who were asked to describe a situation in which they had made what they considered to be an inappropriate weather-related decision. As part of the interview process, experts were asked to identify those cues within the environment that would have indicated that the conditions were deteriorating (Wiggins, 1999). Nine cues were identified and these were evaluated using an internet-based survey in which participants were asked to make a series of judgements concerning 10 images of in-flight weather conditions (Wiggins & O'Hare, in press). Having validated the nine cues as indicators of deteriorating weather conditions during flight, the cues were integrated into 'Weatherwise' and formed the basis of the initial stage of the program.

Program Structure

'Weatherwise' is divided into three parallel streams, each of which is designed to develop knowledge and skills pertaining to in-flight weather-related decision-making. The first of these streams is designed to facilitate the development of an understanding of the cues necessary to recognise and respond appropriately to a deterioration in the weather conditions during flight (problem recognition; see Figure 19.1). The second stream is designed to facilitate the development of an awareness of decision-making styles amongst users (problem-solving). The final stream involves a simulated flight during which users have the opportunity to apply the skills developed within the problem recognition and problem solving streams.

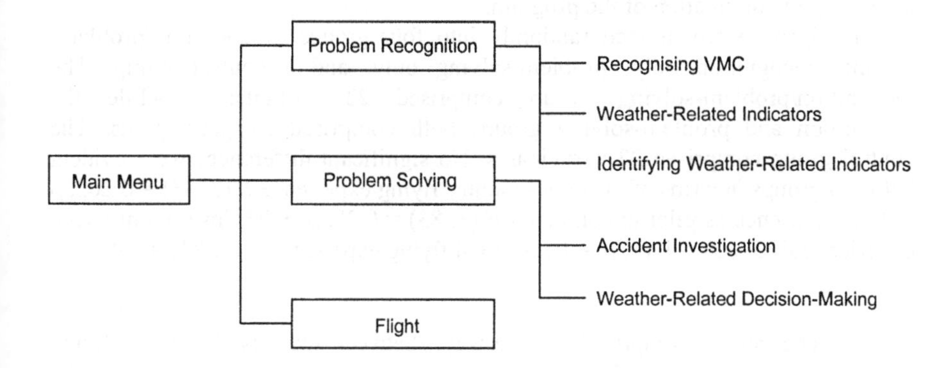

Figure 19.1 A diagram of the structure of the pilot judgement tutoring system 'Weatherwise' illustrating the main components and the sub-components of the system

Study Aims

From a theoretical perspective, it was assumed that the timely identification and recognition of a change in the weather-related cues would enable inexperienced pilots to recognise, at a relatively early stage of the flight, that a deterioration in the weather conditions had occurred. This capability was expected to increase the time available for pilots to engage a decision-making strategy that was appropriate under the circumstances.

The validation of 'Weatherwise' occurred during the flight phase of the computer program. During this stage, the timeliness and the appropriateness of the decision could be recorded. Therefore, it was hypothesised that exposure to both the problem recognition and the problem solving phases of the program would improve the timeliness and the appropriateness of the decision during the flight phase, above that recorded for the problem recognition and problem solving phases presented in isolation. It was further hypothesised that the timeliness and the appropriateness of these decisions would exceed that recorded for a control group who were not exposed to either of the training phases prior to undertaking the flight.

Methodology

Participants The participants in this study comprised 87 licensed private pilots recruited from throughout Australia. Although the computer program has been developed primarily for a United States audience, the principles discussed were presumed to be universal, and Australian pilots were not expected to be influenced unduly by the orientation of the program.

Participants were divided randomly into four groups: Recognition/problem-solving; recognition only; problem-solving only; and a control group. The recognition/problem-solving group comprised 23 participants, while the recognition and problem-solving groups both comprised 21 participants. The control group comprised 22 participants. No significant differences were evident between groups in terms of the cross-country flying experience $F(3, 83) = 0.262, p > .05$, experience as pilot in command $F(3, 83) = 0.28, p > .05$, instrument flying experience $F(3, 83) = 1.18, p > .05$, or total flying experience $F(3, 83) = 0.44, p > .05$.

Design The study comprised a one-way, between-subjects factorial design incorporating four levels. These levels were consistent with the extent of the exposure to the various phases of the computer program. Participants in the 'recognition' group only had access to the recognition phase of the program prior to completing the flight phase. Similarly, participants in the 'problem-solving' group only had access to the problem-solving phase of the program prior to completing the flight phase. Those participants in the 'recognition/problem-solving' group were exposed to all aspects of the computer program prior to completing the flight phase, while the control group did not complete any of the preliminary phases of the program prior to completing the flight phase.

Stimuli and Materials The stimuli comprised the Pilot Judgement Tutoring System 'Weatherwise' that was developed with the assistance of the Federal Aviation Administration. Once a 'beta' version of the program had been developed in-house, professional designers were engaged to produce the computer program using Macromedia Authorware™ and Macromedia Director™ (see Figure 19.2).

The materials comprised a data evaluation sheet on which participants were asked to indicate their level of operational experience, their use of the weather-related decision-making cues prior to, and following the program, their assessment of the applicability of the material, and their perceptions of the usability features of the computer-program.

Procedure

Participants were tested individually, either in a laboratory or in a quiet room within the operational environment. They were recruited by approaching individual flying schools, and through advertisements placed in aviation-related newsletters and magazines. Each participant was provided with an information sheet and a consent form and began the study by completing the first stage of the data evaluation sheet.

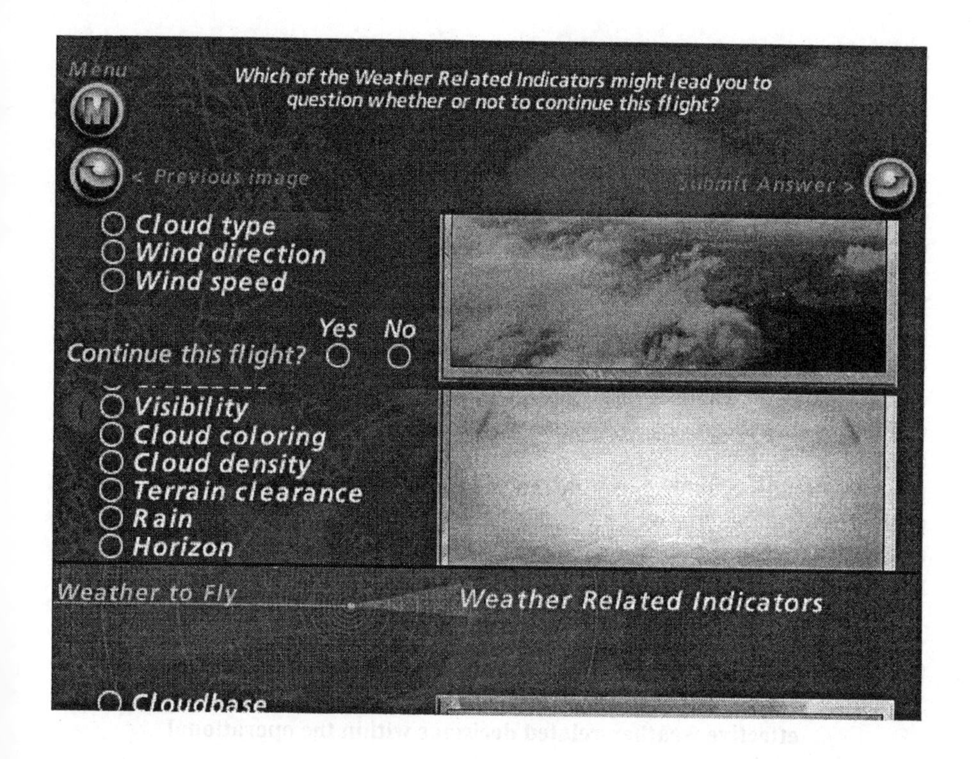

Figure 19.2 An example of the weather-indicators interface from the PJTS

Depending upon the group to which the participant had been allocated, the program was commenced at the appropriate stage. There was no communication or direction provided to the participant during any of the phases, since one of the aims of the study was to establish the overall usability of the program. There was no time limit imposed on those participants who were engaged in the problem recognition or problem solving phases of the program. Participants were encouraged to progress only when they felt comfortable with the material provided.

Results

The results were considered on the basis of subjective and objective assessments of the computer-based training. In the absence of a validation of the program within the operational environment, the combination of subjective and objective assessment served to increase the validity of any conclusions drawn concerning the application of program within the operational environment (Wiggins, 1996).

Subjective Assessment

At the lowest level of system assessment, participants were asked to indicate, using a Likert Scale, the extent to which they considered that the program would assist them to formulate effective weather-related decisions within the operational environment. A oneway ANOVA, incorporating the testing group as the manipulated variable, revealed a statistically significant main effect F (3, 83) = 9.17, $p < .01$, and a summary of the mean responses is provided in Table 19.1. Post-hoc tests using a Bonferroni adjustment revealed that the differences lay between the recognition group and the control group, and the recognition/problem-solving group and the control group ($p <.02$).

The results in Table 19.1 suggest that those groups in which participants were exposed to the recognition section of the computer-based training system provided a significantly higher rating of the extent to which the information from the program could be applied within the operational environment. Consistent with expectations, the control group yielded the lowest response of the four groups, although simply exposing participants to a simulated flight appears to invoke a perception of a learning experience amongst participants.

Table 19.1 A summary of mean responses from participants indicating the extent to which the program would assist them to formulate effective weather-related decisions within the operational environment

Group	Mean Response (SD)
Recognition	3.95 (0.67)
Problem-Solving	3.29 (1.01)
Recognition/ Problem-Solving	3.65 (0.83)
Control	2.59 (1.05)

Objective Assessment

In the case of the evaluation of 'Weatherwise', the main objective assessment criterion was performance during a simulated flight at the conclusion of the program. The simulated flight involved a series of five decision points at which participants could elect either to continue the flight or divert to an alternate destination. Performance during the flight was assessed in terms of the decision point at which the participants decided to divert to an alternate. Of the five decision points, decision point three was designed as the optimal stage at which to divert and thereby remain clear of cloud. At decision point three, a number of weather-related indicators had deteriorated to the point such that a diversion would be necessary.

A chi-square analysis was used to determine the extent to which differences occurred between the frequency with which members of each of the groups would divert at, or before, the optimal decision point. Due to the sample size, the cells were collapsed such that responses before the optimal decision point were coded in one group, while responses at, or after, the optimal decision point were coded as a second and third group respectively. A subsequent chi-square analysis revealed a statistically significant relationship between the frequency with which pilots in the different groups continued beyond the optimal decision point $\chi^2(6, N = 87) = 15.33, p < .02$. An inspection of the contingency table indicated that 54 percent of participants allocated to the control group continued beyond the optimal decision point in comparison to 14 percent of participants in the recognition group and 23 percent of participants in the problem-solving group. In comparison, only 8.5 percent of participants allocated to the recognition/problem-solving group continued beyond the optimal decision point (see Table 19.2). These results suggest that exposure to the recognition and problem-solving elements of the computer program may have the capacity to improve performance in terms of preventing pilots from continuing beyond the optimal decision point during simulated in-flight weather-related decision making.

Table 19.2 Frequency with which participants decided to divert prior to the optimal decision point, at the decision point or beyond the decision point during the evaluation flight

Group	Pre-optimal	Optimal	Post-optimal
Recognition	5	13	3
Problem-Solving	4	12	5
Recognition/ Problem-Solving	7	14	2
Control	2	8	12

In terms of the choice of airport, each of the options was allocated a rating, depending upon the distance of the airport from the decision point, the nature of the terrain to be traversed to reach the airport, and the nature of the weather conditions that were likely upon arrival. These ratings comprised a score, whereby 100 was the optimal choice. Since the data were non-parametric, a Kruskal-Wallis test, using four independent samples based upon the groups was used to assess the relative efficacy of the choice. The results failed to reveal any differences between the groups $\chi^2(3) = 4.70, p>.05$, although the median ratings for the recognition and recognition/problem-solving groups exceeded the median ratings for the problem-solving and control groups (see Table 19.3).

Table 19.3 Median rankings for the choice of airport for each of the four groups during the flight stage of 'Weatherwise'

Group	Median Response
Recognition	46.33
Problem-Solving	39.52
Recognition/Problem-Solving	52.00
Control	37.68

Discussion

'Weatherwise' was developed as a computer-based training system to facilitate the acquisition of the skills necessary to develop effective and efficient in-flight weather-related decision-making amongst inexperienced pilots. The validity of this training system was examined on the basis of a combination of subjective assessment, and objective assessments of performance during a simulated flight.

In terms of the subjective assessment of 'Weatherwise', the results indicate that differences do exist, particularly in terms of the subjective assessment of the extent to which the program will facilitate effective weather-related decisions within the operational environment. More specifically, the results differentiated between those participants who were exposed to the recognition section of the program, and those who were either allocated to the control group or who were allocated exclusively to the problem-solving section of the program.

The differences that emerged between participants on the basis of subjective assessments were consistent with the results that emerged on the basis of objective assessments of the program. The objective assessment was undertaken using a simulated in-flight scenario in which participants 'flew' a flight and made a series of decisions concerning whether to divert or continue the flight in the face of deteriorating weather conditions. The frequency with which participants continued beyond the optimal decision point was compared, and the results indicated that those participants who were allocated to the control group were more likely than the recognition or problem-solving groups to continue beyond the optimal decision point. The group least likely to continue beyond the optimal decision point was the group that had been exposed to both the recognition and the problem-solving stages of the program. On the basis of the statistical differences evident between groups, it might be argued that exposure to the training phases of 'Weatherwise' appears to result in some level of improvement in the timeliness of decisions, albeit within a simulated environment.

In terms of the choice of airport, no statistically significant differences were evident between groups. However, consistent with the results pertaining to the timeliness of decisions, the results associated with the choice of airport did provide some encouraging indications, particularly in terms of the performance of the recognition and recognition/problem-solving groups. In particular, the pattern of responses in terms of the choice of airport was surprisingly similar to the pattern of responses associated with the decision to divert. Nevertheless, it must be recognised that the efficacy of 'Weatherwise' in relation to the selection between alternates has yet to be established empirically.

Combined with the subjective assessments of 'Weatherwise', the objective assessment of the program suggests that improvements in performance may be evident following exposure to the program, particularly over the longer term. Moreover, since no post-hoc differences were evident between the subjective assessments of system usability, it might be argued that any differences evident were unlikely to be due to differences associated with the usability features associated with the different features of the program.

Limitations of the Study

Although 'Weatherwise' has yielded a number of positive outcomes, the assessment of the program was based upon a simulated assessment of in-flight decision-making. The extent to which performance in a simulated environment reflects performance within the operational environment is matter of debate. However, it should be noted that the simulated flight within 'Weatherwise' was based upon actual weather data and followed an actual route within the United States. Furthermore, a time limit was imposed on the decision-making process to reflect the limited time available during which to formulate actual weather-related decisions.

While it might be possible, from an experimental perspective, to examine the decision-making performance of pilots during actual flight, ethical considerations preclude the exposure of pilots to the significant level of risk that an in-flight assessment of weather-related decision-making might impose. However, O'Hare and Batt (1999) suggest that Personal Computer Aviation Training Devices (PCATD) may provide a useful alternative in this regard, particularly in terms of the levels of realism that can now be achieved in terms of the simulated weather conditions.

Implications

The aim of this study was to test the validity of 'Weatherwise' in an environment that simulated the cognitive demands that are associated with in-flight weather-related decision-making within the operational environment. Although the results arising from the study were not conclusive, there was evidence to suggest that 'Weatherwise' may have the capacity to improve pilot performance during in-flight weather-related decision-making. Therefore, it might be argued that the theoretical and empirical basis for the development of the computer program was also justified and has some level of validity.

The differences that emerged between the different features of the program were such that the recognition stage, incorporating the weather-related cues, was perceived by participants as the most useful aspect of the program. This outcome was also reflected in the objective data, suggesting that the integration of the weather-related cues within the program was justified and has the potential to improve human performance within the operational environment.

Although the evaluation of the program comprised a simulated flight, the results have broader implications, particularly in terms of the extent to which computer-based training systems can facilitate the acquisition of cognitive skills within complex technological environments. Previous attempts to integrate computer-based training systems within aviation have generally focussed upon the development of procedural (Koonce, Moore, & Benton, 1995; Taylor et al., 1997) and/or perceptual-motor skills (Koonce & Bramble, 1998; Mattoon, 1994) amongst users. However, computer-based training systems also have the capacity to develop cognitive skills amongst users (Marinelli, 1994). The results arising from the present study provide support for this assertion and suggest that greater effort needs to be directed towards the development of interactive computer-based training systems that focus upon the development of task-specific cognitive skills.

References

Aircraft Owners and Pilots Association. (1996). *Safety Review: General Aviation Weather Accidents.* Frederick, MD: Aircraft Owners and Pilots Association: Air Safety Foundation.

Buch, G., & Diehl, A. (1984). An investigation of the effectiveness of pilot judgement Training. *Human Factors, 26,* 557-564.

Jensen, R. S. (1992). Do pilots need human factors training. *CSERIAC Gateway, 3,* 1-4.

Klein, G. A. (1989). Recognition-primed Decisions (RPD). *Advances in Man-Machine Systems, 5,* 47-92.

Koonce, J. M., & Bramble, W. J. (1998). Personal computer-based flight training devices. *The International Journal of Aviation Psychology, 8,* 277-292.

Koonce, J. M., Moore, S. L., & Benton, C. J. (1995). Initial validation of a basic flight instruction tutoring systems (BFITS). In R. S. Jensen & L. A. Rakovan (Eds.), *Proceedings of the Eighth International Symposium on Aviation Psychology* (pp. 1037-1040). Columbus, OH: Ohio State University Press.

Marinelli, M. (1994). *Multimedia Systems and Cognitive Aspects in the Training of Pilots.* Bologna, Italy: Baskerville Communications Centre.

Mattoon, J. S. (1994). Designing instructional simulations: Effects of instructional control and type of training task on developing display-interpretation skills. *The International Journal of Aviation Psychology, 4,* 189-210.

O'Hare, D., & Batt, R. (1999). A pilot for all seasons: Beyond simulation. In D. O'Hare (Ed.), *Human Performance in General Aviation* (pp. 173-192). Aldershot, UK: Ashgate.

Sternberg, R. J. (1998). Abilities are forms of developing expertise. *Educational Researcher, 27,* 11-20.

Taylor, H. L., Lintern, G., Hulin, C. L., Talleur, D., Emanuel, T., & Phillips, S. (1997). *Transfer of Training Effectiveness of Personal Computer-based Aviation Training Devices* (NTIS DOT/FAA/AM-97/11). Washington, DC: Federal Aviation Administration.

Wiggins, M. (1996). A computer-based approach to human factors education. In B. J. Hayward & A. R. Lowe (Eds.), *Applied Aviation Psychology: Achievement, Change, and Challenge* (pp. 201-208). Aldershot, UK: Ashgate.

Wiggins, M. W. (1999). The development of computer-assisted learning systems for general aviation. In D. O'Hare (Ed.), *Human Performance in General Aviation* (pp. 153-172). Aldershot, UK: Avebury Aviation.

Wiggins, M. W., & O'Hare, D. (1995). Expertise in aeronautical weather-related decision-making: A Cross-sectional analysis of general aviation pilots. *Journal of Experimental Psychology: Applied, 1,* 304-319.

Wiggins, M. W., & O'Hare, D. (In press). Expert and novice pilot perceptions of in-flight weather conditions. *The International Journal of Aviation Psychology.*

Chapter 20

Learning to Land: The Role of Perceptual Feedback and Individual Differences

Susannah J. Tiller, Peter Pfister, Allen T. G. Lansdowne, Stephen C. Provost
Human Factors Group, The University of Newcastle, Australia
and
David J. Allerton
Cranfield Aeronautical University, Bedford, UK

Introduction

Over 50 years ago, Langewiesche (1944/1972) commented that learning to land was the most difficult part of flying training. This was largely due to the difficulty of perceiving the aircraft's position relative to the glideslope angle (GSA). However, Langewiesche felt that this judgement was a skill that could be taught, and detailed several perceptual strategies that pilots could use to determine their position vis-à-vis the GSA. The most important of these he termed 'perspective', known in the psychological literature as form ratio.

This strategy relies on the shape of the runway to maintain a constant GSA. When approaching a runway at an angle, it will appear trapezoidal. If a constant GSA is maintained, the trapezium will not change shape, only size. However, deviations from the GSA will result in a changed shape to the trapezium. According to Langewiesche (1944/1972), an experienced pilot can use these perceptual changes to the form of the runway to determine whether the aircraft is too high, too low, or on the ideal GSA.

Langewiesche (1944/1972) made the ambitious claim that if novice pilots "would consciously practice the visual tricks described here [i.e., the form ratio], it would probably save … many expensive hours of just shooting approaches and landing" (Langewiesche, 1944/1972, p. 265). However, these claims were not substantiated by Langewiesche, and do not appear to have been tested by subsequent research.

Electronic guidance systems can be used to provide feedback on performance during an approach and landing. These generally take the form of symbols indicating the path to be followed to maintain an optimum GSA. They can be set so they are always present, never present, or appear only when the aircraft exceeds certain horizontal and vertical deviations from the GSA. Studies (e.g., Lintern & Koonce, 1990; Lintern, Roscoe, Koonce, & Segal, 1990) have shown that these devices can significantly reduce horizontal and vertical deviation from the GSA, irrespective of whether the guidance is constant or adaptive. In these studies, feedback is provided during each trial. Within the aviation context, there are few – if any – studies examining the effect of feedback when it is withheld to the end of each trial. Theories and results from other experimental situations (e.g., Buekers, Magill, & Hall, 1992) suggest that under these conditions, feedback will also improve performance.

Aviation is a male-dominated industry. Approximately 95 percent of pilots are male, yet there is almost no empirical evidence to suggest that males are better pilots than females, as the majority of studies testing performance factors do not measure gender differences. In one of the few studies testing gender differences (Lintern, Taylor, Koonce, Kaiser, & Morrison, 1997), results showed a small, but significant tendency for males to perform better than females on a simulated landing task. However, it is possible that these results were artificially altered by sample characteristics, as in the study, males outnumbered females 6:1.

This study aimed to investigate the role played by perceptual characteristics and instruction in the development of landing skills. A secondary aim was to use a more balanced sample to examine gender differences in performance.

Method

Participants

Thirty-two undergraduate students at The University of Newcastle participated in this study. The sample comprised 15 males and 17 females, with a mean age of 24.8 years (*SD* = 6.99 years). All were flight-naïve.

Apparatus

The flight simulator used was a fixed-based NovaSim, with enhanced graphics and data capture capabilities. For this study, the NovaSim was configured to simulate a multi-engine Piper PA-30 aircraft. The same simulated approach to landing was used for all trials. The aircraft was at a point 2.05 nautical miles from the runway, at an altitude of 600 feet, about to intercept a 3° glideslope. Aircraft settings such as power, flaps, and rudder were standardised for all trials.

Procedure

Participants were randomly allocated to experimental groups in a 2 x 2 between-subjects Factorial design. The two factors were feedback (present vs. absent) and form ratio (present vs. absent).

Participants were instructed in the experimental task using a standardised instructional video and training sessions on the flight simulator. Participants in the form ratio-present conditions received extra videotaped instruction in using the changing shape of the runway to determine their position relative to the glideslope.

All participants completed two practice trials, and six experimental trials. Participants in the feedback-present condition were given graphical feedback at the end of each trial. The graph plotted altitude over distance, with lines representing 2, 3, and 4° glideslope angles. The participant's altitude over distance was also displayed. The 3° glideslope was described to participants as the 'shortest distance to the runway'. Participants' performance was described relative to this line – too high, too low, or close to the correct approach.

Results

Participants' performance was measured in the degree deviation from a 3° GSA, known as the glideslope error (GSE). Due to software specifications, approaches below the GSA were recorded as positive deviations, while approaches above the GSA were recorded as negative deviations. Participants' GSE was recorded every 0.1 nautical miles, from 2.0 to 0.5 nautical miles. These scores were then averaged across trials and distances to give a mean GSE for each participant.

A 2 x 2 ANOVA was conducted on the mean GSE for each group. As shown in Figure 20.1, participants provided with form ratio information were closer to the GSA than participants not provided with this information. This effect was significant, $F(1, 28) = 10.32, p < .05$.

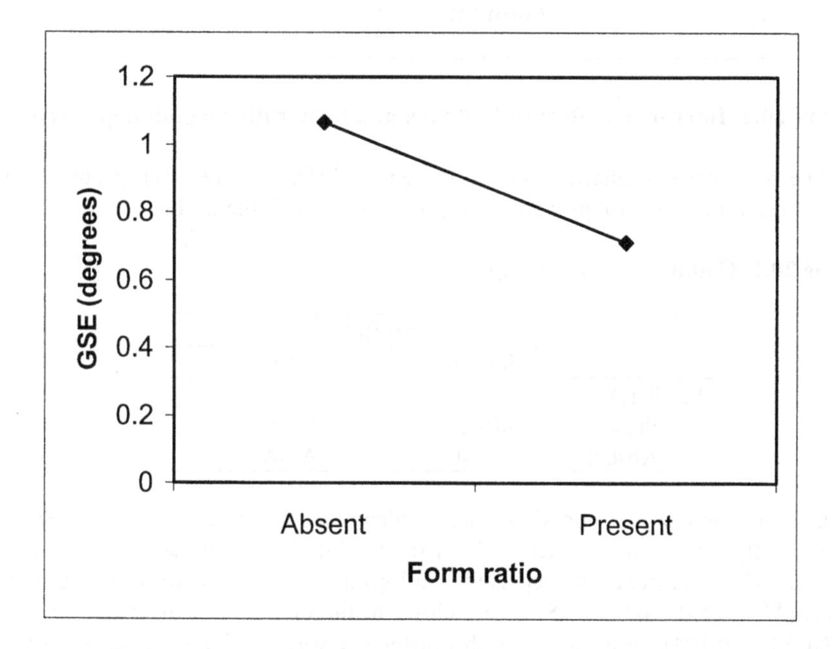

Figure 20.1 Main effect of form ratio on glideslope error

Participants provided with feedback were closer to the glideslope than participants not provided with feedback. However, this effect was not significant, $F(1, 28) = 2.68, p > .1$.

The interaction of feedback and form ratio is shown in Figure 20.2. Participants without feedback or form ratio showed greatest deviation from the 3° glideslope, while the performance of the remaining three groups was comparable. This interaction was significant, $F(1, 28) = 5.44, p < .05$.

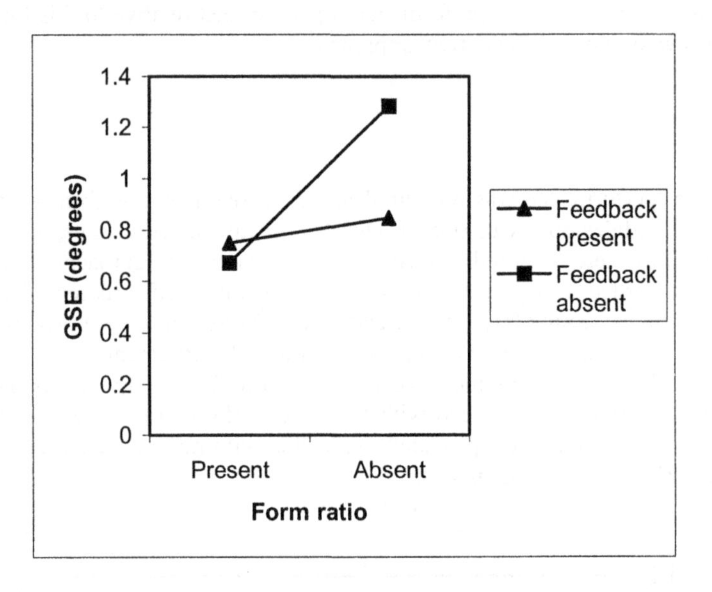

Figure 20.2 Interaction effect of feedback and form ratio on glideslope error

The next stage of analysis looked at gender differences between groups. For ease of reference, the groups were coded, as outlined in Table 20.1.

Table 20.1 Coding for each group

	Form Ratio	
	Present	Absent
Feedback		
Present	$_{FB}P_{FR}P$	$_{FB}P_{FR}A$
Absent	$_{FB}A_{FR}P$	$_{FB}A_{FR}A$

Overall, there was a non-significant tendency for females ($M = 0.883$, $SD = 0.331$) to fly simulated approaches closer to the glideslope than males ($M = 0.894$, $SD = 0.443$). This trend was repeated in all groups, except $_{FB}P_{FR}P$. In this group, males ($M = 0.535$, $SD = 0.281$) were closer to the glideslope than females ($M = 0.964$, SD = 0.104) were. However, this difference was not significant at $\alpha = .05$, $t(3) = -2.86, p = .06$. The only significant gender difference between groups was in the $_{FB}A_{FR}A$ group, where females (M = 1.153, SD = 0.118) were closer to the glideslope than males (M = 1.413, SD = 0.148) were. This difference was significant, $t\,(5) = 2.75, p < .05$.

The gender distribution is shown in Table 20.2. Two of the groups contained equal numbers of males and females; the remaining groups had unbalanced male:female ratios.

Table 20.2 Gender distribution for each group

	Males	Females
Group		
$_{FB}P_{FR}A$	2	6
$_{FB}P_{FR}P$	4	4
$_{FB}A_{FR}A$	4	4
$_{FB}A_{FR}P$	5	3
Total	15	17

Discussion

All participants flew simulated approaches below the glideslope. However, participants receiving the form ratio information recorded significantly smaller GSEs. This shows that providing novice pilots with information about the form ratio helps them to more accurately perceive their position relative to the glideslope.

Langewiesche's (1944/1972) anecdotal observations about the role of perceptual cues in the development of landing skills were supported by the data from this experiment. Across all trials and all distances, participants provided with form ratio information displayed significantly smaller glideslope errors (GSE) on the approach to landing than participants who did not receive this information. Although both groups (form ratio present vs. form ratio absent) flew below the glideslope, the group provided with form ratio information was significantly closer to it.

Contrary to the hypothesis, although there was a slight tendency for feedback to result in approaches closer to the GSA, this effect was non-significant. It is uncertain why this occurred, especially given previous research both within the aviation context (e.g., Lintern & Koonce, 1990) and more theoretical situations (Buekers, et al., 1993).

There are two possible explanations for the non-significant effect of feedback. Firstly, it may be that the nature of the feedback was unhelpful to participants. Although it did give some measure of their performance (e.g., too high), it did provide any practical information for improving performance. This may have been accentuated by the fact that, due to software specifications, feedback was provided at the end of each trial rather than during the trial itself. Therefore, any improvement, or decrement, in participants' performance seems more related to their interpretation of the feedback, rather than the provision of feedback itself.

The second possible explanation for the non-significant effect of feedback relates to the extent of participants' training and their subsequent knowledge levels. In previous studies testing the effect of feedback on novice pilot performance (e.g., Lintern & Koonce, 1990), while participants may be flight naïve, they have generally undergone a considerable period of theoretical pre-flight training. It is reasonable to assume that they have a higher level of knowledge than the participants in this experiment, who underwent a much shorter and less sophisticated period of instruction. Given the discrepancy in instruction and knowledge level, it is possible that participants in the Lintern and Koonce (1990) study were able to utilise the feedback more effectively than participants in this study.

Although there was no significant main effect of feedback, the interaction of feedback and form ratio was significant. Participants who were not provided with either feedback or form ratio information had the greatest GSE and poorest overall performance. Groups receiving feedback, form ratio information or both, performed significantly better than the group without additional visual information. This suggests that the provision of some form of visual cues provides the necessary feedback to improve performance.

The analysis of gender difference in performance revealed some unexpected results. In the ${}_{FB}P_{FR}P$ group (feedback present, form ratio present), males' performance was superior to females', although this did not reach significance (p = .06). However, overall, and in all other groups, females outperformed males. This was significant only for the ${}_{FB}A_{FR}A$ group (feedback absent, form ratio absent).

These results suggest that females perform better without feedback and form ratio information, with the opposite true for males. Although the superior performance of males in the ${}_{FB}P_{FB}P$ group is not significant, calculation of power for this group shows that there was a 48 percent possibility that this is due to Type II error.

The superior performance of females over all conditions may suggest that females are generally better at landing an aircraft. However, several factors need to be considered. In the ${}_{FB}P_{FR}A$ and ${}_{FB}A_{FR}P$ groups, the male:female ratios were 2:6 and 5:3 respectively. As unbalanced t-tests rely on pooled variances and weighted averages, it is possible that this may have artificially altered the results of the t tests.

The significant difference in the ${}_{FB}A_{FR}A$ may be a result of Type I error. As α = .05, there is a 5 percent possibility that this effect is due to chance, rather than some real difference.

It is intriguing to speculate whether these results show that females genuinely outperformed males in the flight task. In addition, finding gender-based differences in response to the level and type of instruction could have wide-reaching implications for training and selection programs in aviation.

This study confirmed the important role played by perceptual characteristics during the development of landing skills. The provision of feedback, form ratio information, or both, has a valuable role in helping novice pilots perceive their position relative to the ideal GSA.

This study produced several unexpected findings related to gender differences in performance. They may be due to statistical anomalies. Alternatively, and more optimistically, they may represent new directions for research in training and selection.

Author Note

The Human Factors Group at The University of Newcastle is a node of the ARC Funded Key Centre for Human Factors and Applied Cognitive Psychology, The University of Queensland, Australia. Susannah Tiller is at present a doctoral student at the Key Centre. Dr Allen Lansdowne and Dr Stephen Provost are currently at Southern Cross University. David Allerton is now Professor for Systems Engineering at the University of Sheffield.

References

Buekers, M. J. A., Magill, R. A., & Hall, K. G. (1992). The effect of erroneous knowledge of results on skill acquisition when augmented information is redundant. *Quarterly Journal of Experimental Psychology,* 44(A), 105-117.

Langewiesche, W. (1972). *Stick and Rudder: An Explanation of the Art of Flying* (Rev. ed.). New York: McGraw Hill.

Lintern, G., & Koonce, J. M. (1992). Visual augmentation and scene detail effects in flight training. *International Journal of Aviation Psychology, 2,* 281-301.

Lintern, G., Roscoe, S. N., Koonce, J. M., & Segal, L. D. (1990). Transfer of landing skills in beginning flight training. *Human Factors, 32,* 319-327.

Lintern, G., Taylor, H. L., Koonce, J. M., Kaiser, R. H., & Morrison, G. A. (1997). Transfer and quasi-transfer effects of scene detail and visual augmentation in landing training. *International Journal of Aviation Psychology, 7,* 149-169.

Chapter 21

A Cognitive Approach to the Development of Prescriptions for a New Flight Crew Licence: Psychological Perspective from a Regulator and Curriculum Designer

Graham J. F. Hunt
Massey University
and
Richard Macfarlane
New Zealand Civil Aviation Authority

Behavioural Approaches to Regulatory Prescriptions

Although systematic approaches to identifying and measuring learning and its outcomes can be traced back to over 400 years, most of the significant developments have occurred in only the last 50 years. Skinner's (1953) maxim that reinforcement was a necessary condition for learning was the catalyst for substantial research and development into programmed learning and its corollary the 'systems' approach to training. This model identified the behavioural components of stimulus control and the reinforcement processes of discrimination, generalisation, associations, and chaining into an analytical procedure for analysing tasks into constituent objectives with conditions and standards for prescribing acceptable performance. The fervour for the 'systems approach' in the late 1950s and early 1960s shared many parallels with today's messianic belief in the educative power of web-based learning.

The key tool in the systems approach arsenal was the behavioural objective. Although the name Mager (1962) and behavioural objectives are often seen to be synonymous with each other, it is to Tyler (1975) that most of the founding credit should be attributed. In 1934 he described objectives as '...terms which clarify the kind of behaviour which the course should help to develop' (Walbesser, 1972).

Later, as an integral component of the systems approach, behavioural (sometimes instructional) objectives were identified as the means for specifying learning outcomes by means of defining prior to instruction, the requisite behaviour to be demonstrated, the conditions supporting that behaviour and the standards of acceptable behaviour.

In education, the behavioural approach to learning provided the theoretical basis for a number of major innovations in curriculum and instructional development. One of the earliest of these was the Keller Plan (Keller, 1974) in which Keller identified five distinguishing features of his system: (a) unit mastery of the material taught, (b) student self-paced learning, (c) learning support through the use of student proctors, (d) reliance on written instruction, (e) a significant de-emphasis of lectures. An application of the Keller system formed a basis for the development of New Zealand's Human Resource Development in Aviation (HURDA) student learning materials in the late 1980s.

Although the systems approach has had a well-chartered history in education since the 1950s its application to aviation has been much slower. It may be argued that its lack of impact in aviation training is less a reflection of any inherent difficulty in applying such a science to the content of professional aviation than has been the regulator's role in proscribing innovations in curricula and assessment.

Traditional Approaches: The Inspector-Regulator

Since 1944 and even earlier, most civil aviation administrations have had teams of airworthiness surveyors, general aviation and airline inspectors whose duty it has been to 'inspect' maintenance works and the performance of operating personnel. The origin of the concept of inspection has probably been drawn from other traditionally state regulated activities such as education. Here, inspectors have universally had the right to visit schools, watch teachers at work, interpolate the quality of their professional performance by examining representative samples of student work and examinations, and ensuring by such means that the curriculum as articulated in government publications was being met. In New Zealand, prior to 1987 the role of the 'flight crew licensing' examiner was mainly to ensure that adequate entry levels of flight standards were being maintained (through the conduct of Flight Tests) and that the security of the examination system was not being compromised – individuals putting themselves forward to be examined were not finding out in advance about the content of the tests. As Stevens (1999) has said of the system, 'the [aviation] industry was often frustrated by the inconsistencies generated by such a system, although some exploited this inconsistency to their own advantage'. However, even these inspectorial checks were largely confined to so called 'approved schools' while the majority of flight training organisations lay outside the range of the inspector's gaze.

A Cognitive Approach to Competency Specifications: The HURDA Model

One example of an attempt to specify aviation standards in terms of cognitively-derived competencies was that developed in a project for the New Zealand Civil Aviation Authority by Hunt and Crook (Hunt & Crook, 1988) called HURDA. The focus of this project was to identify flight crew competencies as the basis for developing valid and reliable assessment tools for licensing. A cognitive knowledge structures hierarchy model (Hunt, 1986) was used to define performances and their related abilities as means for determining broad licensing-relevant competencies.

In this model, from an analysis of a regulatory authority's personnel licensing mission – the purpose to which all personnel licensing activities are focussed, broad functional capabilities can be derived. These were described as professional accomplishments. Accomplishments provide the focus for the identification and analysis of domain specific knowledge (Gagné, 1987). For example, the pilot accomplishment of command management defines a capacity to exercise formal, legal power and authority over an aircraft's crew, passengers and load while in transition from one location to another and in ensuring the effective and efficient management of all the resources requisite to achieving the organisation's objectives relevant to that goal.

Each accomplishment is in turn defined by two or more performances. A performance is a procedurally based group of intellectual skills which may be summarised as 'knowing how' to do a major aspect of a job as identified in the accomplishment. This entity is the application of the concept developed by Newall and Simon (Newall & Simon, 1972) when they proposed the notion of a complex entity as a production, which entered into more complex production systems. Such an entity comprised a rule of procedural knowledge, which was composed of a condition and action (Gagné, 1987). In this knowledge structures hierarchy model performances provide the intellectual skill definitions that relate to individual accomplishments.

The base of the hierarchy is provided by the identification of specific abilities providing expositions of the superordinate performances. These abilities are the individual cognitive, affective (attitude), attribute (culture, personality, and motivation), or motor skills, which can be taught or shaped through learning, training, and education. For example, for the accomplishment of command management and its performance crew interacting, specific abilities within a particular context may include the pilot in command assessing, monitoring, and decision-making with a corresponding need for that pilot to demonstrate an accepting attitude and a high level of achievement motivation (attributional abilities; Hunt 1997.

In the HURDA study the results of a cognitive task analysis of flight crew licenses requirements were matched to assessment scenarios forming the contextual basis of a set of licensing-based competency specifications. Each competency specification contained two dimensions: a cognitively derived generic knowledge base (individual accomplishments, performances, and related abilities) and the specific assessment contexts in which the competencies would be measured. Thus, each licensing-related competency specification was constructed from a generic and 'constant' knowledge base and applied to the 'variable' contexts from which the individual's competencies were to be measured. By changing the context – the aircraft type (single engine to multi engine piston; single pilot to multi crew) the required application of the knowledge would also change. However, the generic nature of that knowledge or set of specific abilities would remain unchanged.

The challenge from this approach to competency specification and assessment is to develop measurement tools, which can answer in an operational context the question 'how good/accurate/professional should this performance be?' Traditional paper and pencil multi-choice or essay styled questions are not likely to generate valid predictors of successful performance. Nor are current methods in practical flight-testing. These conventional 'content' measures may in fact provide more relevant information on 'perseverance' or other attributional characteristics than they do about answering the related question 'do they know how?' A more useful approach may well be to research and develop 'holistic' performance measures designed to capture high-level problem solving and decision making in carefully structured but operationally 'real' assessment contexts.

The Regulator's Role: In Transition

A global feature of civil aviation regulatory systems still demonstrates a fundamental lack of internal correction, particularly at the lower end of the regulatory spectrum. Thus, most systems still rely on external supervision and intervention by the regulator with safety appended as a 'clip-on' requirement. But change is underway. Increasing numbers of authorities are moving towards requiring aviation organisation's to demonstrate internal systems which emphasise quality assurance, self correction and procedures which inculcate the attributes of organisational safety culture.

High accident rates, frequent audit demand and sometimes prohibitive compliance costs have given way to much lower accident rates and a commensurate reduction in audit requirements and therefore attendant compliance costs. The detailed inspection by the regulator has become a voluntary return to compliance by transgressing operators and there has been a definitive shift from 'zero error tolerance' to error management: that is, regulators have moved from a position of penalising mistakes to allowing operator control of acknowledging mistakes and taking the appropriate corrective actions (Macfarlane, Batchelor, & Crosthwaite, 1999). Refer to Figure 21.1.

It is fair to say that this international trend from constant regulatory scrutiny to a monitoring and surveillance role has been differentially applied even within single regulatory systems. This trend needs to be accelerated even though some authorities and individuals will remain unconvinced as to the need for change. The easy option of not changing is supported by an element of conservative inertia, which is based on the current high safety esteem in which aviation is held, whether deservedly or not. 'If it ain't broke, don't fix it' (Maurino, 1995) summarises the current situation in many cases and the financial reality is that in most cases, a cost-benefit analysis could not quantify the necessity for change strongly enough to overcome that conservative inertia.

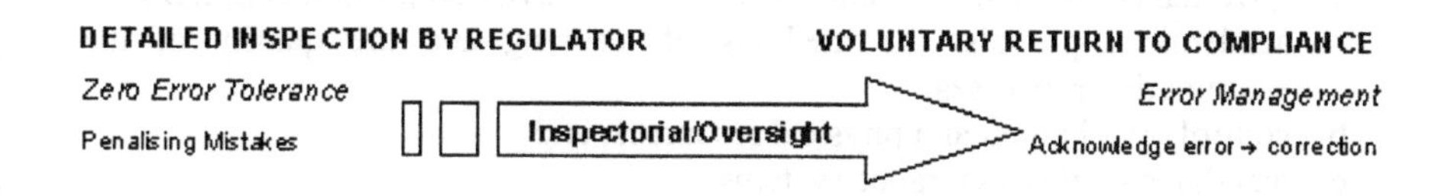

Figure 21.1 Inspectorial-oversight continuum

The mandate for such regulatory change has been largely championed through academic endeavour in the search for the cause of human error and likely strategies for reducing it. Reason (1990, 1997) has provided the catalyst for much of the recent thinking on the subject. Safety trends within the industry are also likely to have a significant impact on the demands placed on any regulatory authority. Current accident and incident rates combined with the projected increase in air traffic would see a major hull loss occurring on a weekly basis (Wells, 1997). This would clearly not be acceptable to the travelling public and it is expected that the CAA, at least in New Zealand, will play an integral part in the development of strategies that target a reduction in the present accident rate. There is a possibility that these requirements will become all consuming leaving little resources for either the maintenance of present activities, no matter how crucial to organisational and industry needs, or the development of the organisation as a whole (adapted from Capper, 1996). For example, in New Zealand's case, proactive policy on issues such as adventure aviation, the ageing of the country's general aviation fleet, the automated environment and the removal of statutory monopolies for air traffic services and examination delivery, may need to be slowed or postponed indefinitely. The pragmatics of these constraints and trends need to be taken into account when considering the options available to a regulatory authority in terms of strategic direction.

Licensing 'Equivalence': A First Step to a New Air Transport Licence?

In 1992 the New Zealand Civil Aviation Authority agreed to a major regulatory innovation by recognising Massey University's Bachelor of Aviation Flight Crew degree as providing an equivalent means of meeting the examination requirements for pilot licensing as prescribed in Civil Aviation Rules Part 61, flight crew licensing. The equivalence process has been defined as 'a moderation and surveillance system between the regulator and the tertiary institution such that the achievement outcomes from an agreed curriculum, with integrated flight practicum components, are recognised as meeting the requirements for the issuance of aviation documents'.

The characteristics of this model are:

1. Additional requirements for meeting Civil Aviation Rules Part 141 (Aviation Training Organisation's certification) including:
   a. governance of the organisation's professional aviation programs in terms of academic approvals, the methods of teaching and the quality of the examination processes;
   b. control over human and physical resources; and
   c. mandated quality assurance systems.
2. Additional requirements for meeting Civil Aviation Rules Part 61 (Pilot Licenses and Ratings) including:
   a. curriculum development (including the integration of academic, technical and practicum knowledge in the contexts of multi-crew performance in single and multi-engine aircraft;
   b. application of advanced aircraft and instructional simulation tools; and
   c. assessment and evaluation.

History may judge that the New Zealand Civil Aviation Authority approval of a system of equivalence to the traditional licensing system through Massey University's tertiary level Bachelor of Aviation provided a curriculum and quality assurance model from which a future ab-initio air transport pilot licence (AATPL) might be based.

ICAOs Initiatives in Proposing Regulatory Direction

The International Civil Aviation Organisation (ICAO) at an informal Air Navigation Commission meeting with industry in September 1997 expressed the concern of participants about the level of flight crew competency in many member states. Evidence from ICAO Safety Oversight Assessments indicated that few countries had established formal standardisation of the performance criteria expected for the examinations for the issuance of licences or for the demonstration of maintenance of competency required by Annex 6. As a result, the flight tests, which in most cases are the ultimate step before a licence is issued or renewed, are left to the sole judgement of the examiner. As a result, the potential for significant inconsistencies in the competency of flight crew members world-wide exists. In most countries, national licensing regulations are essentially inventories of knowledge, skill, and experience requirements and lack specific performance criteria.

It was to this situation that in October 2000, the Air Navigation Commission proposed an informal meeting on the future of international personnel licensing and training standards. The meeting held in Madrid almost unanimously agreed that the current licensing system needed to be revised to meet the flight crew competency demands of the air transport industry today, particularly in the preparation of ab-initio pilots for initial air transport employment. Deficiencies in the current national licensing systems derived from Annex 1 included:

1. the wide spread practice of providing flight instruction sequences independent of theoretical knowledge which should underpin the practice;
2. no standardisation of the flight experience;
3. single pilot orientation;
4. primary focus on visual navigation;
5. little requirement for procedural experience;
6. little requirement for experience in advanced flight and navigation technologies; and
7. significant gap between the 'CPL' entry level and competencies required for initial transition to air transport operations.

In contrast, the meeting recommended that ICAO give consideration to:

1. defining a licensing structure and curriculum processes which would facilitate the transition of competency from novice to air transport 'first officer';
2. identifying the critical competency elements of knowledge, attributes, and practical performance relevant to transitioning to professional flight crew roles;
3. defining training syllabi relevant to the privileges of licenses and ratings;
4. assessing the integration of competency-based licensing standards with current standards;
5. re-assessing the application of instructional resources such as synthetic part-task training devices in creating or maintaining competency and providing appropriate credits for such functionalities; and
6. reviewing current instructor selection and training standards and identifying the level and range of competencies which might be appropriate to an ab-initio air transport pilot system.

The Future: Tripartite Partnership – Regulator, Operator, and Educational Provider

Should these proposals be developed as a set of amendments to Annex 1 and Annex 6, the result is likely to be increased recognition of the need to develop tripartite partnerships between the regulator, operator, and education-training provider (currently, ICAO makes no representation provision for education-training providers). For the regulator, new strategies will need to be adopted. These strategies may result in a significantly reduced regulatory involvement in the future. Robust error management systems could provide an authority with valid and reliable organisational oversight without compromising individual entry control standards. To meet these needs though, current aspects of licensing standards, that are not listed as part of compliancy checks, will need to be identified and stated as assessable outcomes.

The challenge facing the regulatory authorities is to apply safety culture/just culture principles not only to their organisational behaviour but also to the licensing processes themselves. There are two aspects to this proposal. First, it will necessitate a redefining of the organisational aspects of the system to accommodate this new perspective. Second, it will be necessary to define the elements that would form the basis of measuring individual performance and behaviour commensurate with a public safety perspective. Key aspects of behaviour would include those referred to as 'just' culture, specifically those currently unmeasured characteristics that might be incorporated into a more detailed 'fit and proper person' criterion relating to issues such as risky or reckless factors. The tools for measuring these attributes at the initial licensing stage are still to be developed. However, the expertise for developing such already exists. The final assessment of an individual's competence then will include their attitudes to safety, effectiveness, and efficiency as well as the traditional knowledge and skills. At a time when reporters are still attributing more that 70 percent of accidents to Human Error, this may well have a tangible improvement on our current accident rates.

This paper proposes the need to design alternative ways of meeting air transport licensing standards. Cognitive instructional system design models have already demonstrated their utility in empirically identifying knowledge, attributes, and practicum dimensions of professional flight crew competency. Operators are in the best position to know what structures and characteristics they are looking for, especially when it comes to intangibles. These organisations are best placed to know whether individual applicants measure up to these structures and characteristics and they will need to provide effective measures of those current intangibles to satisfy the accreditation of the operators. The regulator can then revert to a program of safety monitoring and surveillance through information gathering from both training organisations and operators (see Figure 21.2) (Macfarlane, 1999). In this respect the regulatory processes needs to allow training organisations to evaluate the attributes of individuals wishing to enter the system, especially those attributes which are currently seen to be intangible. This approach will require the regulator to revise their monitoring regime to guard against the application of bias, discrimination and the idiosyncratic application of the 'right stuff' criteria. It is our contention that most human factor characteristics are measurable in a system such as this and will provide the essence of the new regulatory environment through the application of individual assessment along the lines of the 'just culture' principles. The suggestion here is to propose a refocus of the regulatory authority mandate from that which is seen to be technically based and output orientated, to one that is demonstrated to be governance based and outcome orientated (Reason, 1997), especially with respect to licensing protocols.

A Future Licensing Structure

It may be reasonable to predict that the new tripartite relationships will create industry-based 'Standards Councils'. The purpose of these councils would be to establish and regulate standards of professional competency for flight crew and potentially, cabin crew and maintenance personnel as a means of standardising performance and assuring public demands for safety and quality assurance in relation to the input variables (training) and outcomes (operational performance) (Hunt, 2000). Such a body might set standards beyond national regulatory criteria and eventually replace some of the traditional regulatory activities, which are applied to air transport operators. These could include:

1. 'certifying' providers;
2. agreeing to the characteristics of acceptable professional competency;
3. establishing curriculum, training, and assessment criteria;
4. developing agreed standardised methods for reviewing and assessing the maintenance of professional competency; and
5. independently monitoring and assessment of all professional groups within the alliance in terms of competency specifications which might be established at higher levels of compliance than those which would be the minimum requirements defined by ICAO or national authorities. These specifications for 'good professional practice' would probably, but not necessarily require Standards Council's to have the power (acknowledged by ICAO and national authorities) to monitor compliance and in cases of non-compliance, to sanction and ultimately withdraw individual licensed person's 'right to practice'.

These and other criteria would need to be agreed to by national regulatory authorities as meeting 'alternative means' of achieving professional licensing requirements under 'equivalent' prescriptions for the issuance of aviation documents. Whilst not challenging the sovereignty of the documents themselves, the processes and standards for meeting the issuance requirements might be very different from those processes and standards, which have been traditionally defined by national aviation authorities. This will require international recognition agreements between participating States in the first instance, but these may in turn result in international harmonisation of Standards and Recommended Practices.

References

Capper, P. (1996). *System Safety Issues in the Wake of the Cave Creek Disaster.* Australasian Evaluation Society.

Gagné. (1987). *Instructional Technology: Foundations.* Hillsdale, New Jersey: Lawrence Erlbaum Associates.

Hunt, G. J. F. (1986). Needs assessment in adult education: Tactical and strategic considerations. *Instructional Science, 15,* 287-296.

Hunt, G. J. F. (1997). Instruction and evaluation: Design principles in instructional design. In G. J. F. Hunt (Ed.), *Designing Instruction for Human Factors Training in Aviation* (pp. 3-16). Aldershot: Ashgate.

Hunt, G. J. F. (2000). *Future licensing for air transport pilots.* Paper presented at World Air Transport Conference, Frankfurt, Germany.

Hunt, G. J. F., & Crook, C. (1988). *Competent Flight Crew Licensing II: The Terms of Reference for the HURDA Programme.* Palmerston North: Instructional Systems Programme, Massey University.

Keller, F. S. (1974). Ten years of personalized instruction. *Teaching of Psychology, 1,* 4-9.

Macfarlane, R., Batchelor, O., & Crosthwaite, R. (1999). The role of the regulator in the future aviation system. *ICAO Human Factors Digest No. 14*.
Mager, R. F. (1962). *Preparing Objectives for Programmed Instruction*. Belmont, CA: Fearon.
Maurino, D. E., Reason, J., Johnstone, N., & Lee, R. B. (1995). *Beyond Aviation Human Factors: Safety in High Technology Systems*. Aldershot: Avebury.
Newall, A. A., & Simon, H. A. (1972). *Human Problem Solving*. Inglewood Cliffs, New Jersey: Prentice Hall.
Reason, J. (1990). *Human Error*. Cambridge: Cambridge University Press.
Reason, J. (1997). *Managing the Risks of Organizational Accidents*. Aldershot: Ashgate.
Skinner, B. F. (1953). *Science and Human Behaviour*. New York: Macmillan.
Stevens, M. (1999). *A decade of change in aviation safety regulation has prepared New Zealand for the 21st Century*. Paper presented at the Joint International Meeting, FSF, IFA, IATA; 52nd Annual International Air Safety Seminar: Enhancing Safety in the 21st Century, Rio de Janeiro, Brazil.
Tyler, R. W. (1975). Educational benchmarks in retrospect: Educational change since 1915. *Viewpoints, 51*(2), 11-31.
Walbesser, H. H., & Eisenberg, T. A. (1972). *A Review of Research on Behavioural Objectives and Learning Hierarchies*. Columbus, OH: The Ohio State University, Center for Science and Mathematics Education.
Wells, A. T. (1997). *Commercial Aviation Safety*. New York: McGraw-Hill.

Chapter 22

Legislating Behaviour: The Regulator's Dilemma

Mick Toller
Civil Aviation Safety Authority, Canberra, Australia

Introduction

The key question I want to address is how we, as the regulator, can embrace and incorporate Human Factors issues into the regulatory environment.

First we need to look at the core role of the safety regulator. At the most general level, the Regulator's role is to represent the interests of the public by making mandatory the application of accumulated knowledge in the pursuit and maintenance of their safety. The regulator is the instrument through which cumulative knowledge on maintaining safety is applied. This role involves two direct requirements: the Civil Aviation Safety Authority (CASA) must set safety standards, and must enforce those standards. The role also calls up two obligations: CASA must educate the aviation industry about standards and safe or recommended practices; and we must encourage the creation of safety cultures that engender self-compliance and the adoption of higher standards of safety. So CASA is involved in setting and enforcing safety standards and encouraging the adoption of higher standards of safety through promotion and education. As a result of these functions, the Civil Aviation Safety Authority has three main divisions: standards, compliance and safety promotion.

The dilemma is how do you balance the encouragement role with the 'stick' you need to use in compliance at all times?

On the sanctions side we use the stick and we sometimes must use the stick hard. On the encouragement side we use education and safety promotion through seminars, forums, guides, a magazine and a website.

Both encouragement and sanctions work, and both are necessary. We have to get ourselves to the point where the two are in balance in order to serve the best interests of safety. It would be nice to have it all on the absolution or safety promotion and education side, but that does not take account of the real world. A classic case is air rage where there is no point in not using some stick. Between the blame or culpability and blamelessness or absolution is a large grey area. There is a bit of grey in the balance between blame (sanctions) and absolution (education and promotion). But pilots and engineers are very comfortable with black and white. However, as a regulator I live almost entirely in the grey area. Punishment on its own is not a particularly effective tool. Our main tool is administrative action aimed at maintaining the safety of the travelling public. Yes, there is a role for zero tolerance, and that is where we target the marginal operators, the bottom feeders who spend little time or effort on safety while undercutting other operators.

Overall, aviation has an outstanding reputation in safety. But how do we make it safer? I will take some observations from Rene Amalberti: High risk operations where the risk of a serious accident is 10-3 (e.g., hang gliding) respond to tailored regulations. Here the regulator has a supervisory role to ensure that sports aviators regulate themselves properly. Regulated systems where the risk of a serious accident is 10-3-10-5 (e.g., general aviation) include regulations and procedures, better designs, reporting systems, professional safety officers. Ultra safe systems where the risk of a serious accident is 10-6 (e.g., regular public transport flights) are fully mature regulated systems. A different approach is needed if we are to move safety to the next order of magnitude (i.e., 10-7).

If people are buying a ticket to go from A to B then they are entitled to assume the journey is safe. One strategy to improve the safety of ultra safe systems is to make sure that commercial operators adopt safety management programs. A safety management program is how an organisation manages the safety of its operations. The critical elements of a safety management program are:

1. Top management sets and commits to safety standards. This is really critical, because if you do not have top management commitment you are going to have problems throughout the organisation. If top management is not seen to back the safety management program, how can you expect it to be adopted by a worker?
2. The company follows and maintains safety standards. That is, there is some form of recognising hazards and getting something done about them. CASA itself has recognised this, and has recently appointed a risk manager, whose task is to take an overview of the state of the industry.
3. Hazards are identified and reported in a timely manner.
4. A hazard resolution and management process is in place.

Through its safety trend indicator, CASA analyses each individual operator and maintenance organisation for operational risk and diverts resources to those exhibiting the highest risk. At the core of every safety management program is human error; in fact, human error is probably our single most serious threat to aviation safety. Therefore, error management is a key component of a safety management program. To improve safety results each operator must recognise the issue of human error. Error management has two parts – error prevention and error containment. Error prevention involves understanding why mistakes are made; addressing the factors that lead to errors; and modifying the workplace to reduce error rates. Error containment involves recognising that errors are inevitable;

avoiding the possibility of single-point failures; and establishing and maintaining robust defences.

So what is CASA Doing about This?

CASAs Human Factors goal is to become a centre of excellence in aviation Human Factors. We will do this by:

1. Incorporating Human Factors knowledge into our regulations and our educational materials.
2. Sponsoring and promoting basic and applied research on aviation Human Factors issues.
3. Providing all operational CASA staff with fundamental Human Factors knowledge. We are keen on ensuring that our staff get the same Human Factors training as the airlines.
4. Consulting, assisting and educating the aviation community on Human Factors issues.

When I first came to CASA we lacked a Human Factors group. In mid 1999 we set up our Human Factors group, with the appointment of CASAs first Manager of Human Factors (Michael Rodgers). This was followed by the appointment of a second Human Factors specialist in January 2001. One of the Group's first assignments was to establish an industry wide Human Factors advisory group. This was accomplished in late 1999 and comprises senior Human Factors managers from Qantas, Ansett, Airservices, the Australian Transport Safety Bureau, the Defence Science and Technology Organisation and others by invitation.

In addition we have established a PhD scholarship program, with the first awarded to the University of South Australia's Centre for Sleep Research in 2000. A second PhD scholarship was offered in 2001 to the University of Newcastle. And finally I have directed that our Human Factors Manager has sign off on Human Factors issues related to our rules, advisory circulars and so on. In the near future we will add two new Human Factors specialists to our team – a Human Factors in maintenance specialist and a Human Factors research officer.

CASAs current Human Factors activities include:

1. A joint CASA, Qantas, Ansett, Centre for Fatigue Research, fatigue and risk management project. CASA is looking at a whole new approach to flight and duty time limitations based on research. We believe this is a very important piece of work which may change the way we deal with flight time limitations. We have strong support from the government for this.
2. Error management strategies for our safety program for commercial operators.
3. The preparation of a Crew Resource Management discussion paper.
4. The completion of Human Factors training for CASA Flying Operations and other Inspectors, and CASA Standards officers.

5. The appointment of our Manager of Human Factors to the International Civil Aviation Organisation (ICAO) Human Factors and Flight Safety Panel.

In conclusion, aviation psychology provides much of the accumulated knowledge upon which we will base our Human Factors regulations and practices. Human Factors is particularly useful in looking at how to manage human error. We are looking at how to develop and encourage a positive safety culture. But can we achieve a perfect safety environment? To quote Mary Schaefer, 'Insisting on perfect safety is for those who do not have the [guts] to live in the real world'. Human error will never be eliminated; we must work towards systems and processes which are sufficiently resilient to maintain safety, and we must understand how people best manage error.

Chapter 23

Auditory Warnings in the Cockpit: An Evaluation of Potential Sound Types

Karen L. Stephan, Sean E. Smith, Simon P. A. Parker, Russell L. Martin and Ken I. McAnally
Air Operations Division,
Defence Science and Technology Organisation, Australia

Introduction

Warnings may be presented in several ways, the norm being via the visual or auditory modalities. The primary purpose of a warning is to alert and inform (Edworthy & Stanton, 1995; Wheale, 1981), as well as to guide the response. For a warning to be effective the signal used should be recognisable as a warning and the operator must be aware of both what the signal represents and the action that is required. Within the modern military cockpit, the visual channel is often overloaded, and auditory warnings provide a method to alleviate this. Due to the unique nature of the auditory system, auditory warnings can attract attention regardless of either the operator's head/eye position, or what task they are engaged in. In this way, our sense of hearing acts as a natural warning sense (Edworthy & Stanton, 1999).

Auditory warnings may be either verbal or non-verbal. Both have their relative advantages and disadvantages, which can be used as a guide for effective implementation. Verbal (or speech) signals have the advantage that negligible learning is required. They are also effective for conveying complex information. They can, however, take a long time to convey this information. In addition, they are not suitable for some situations, for example where loud verbal messages would be indiscreet (Edworthy & Stanton, 1995). Whilst all auditory warnings can be subject to masking by background noise, speech warnings have the additional problem of being susceptible to informational masking by other speech present in the cockpit.

Abstract sounds (e.g., simple tones) are commonly used as a non-verbal alternative in warning systems. They can convey information quickly, and they are not subject to informational masking by speech-rich environments as verbal warnings can be. The main disadvantages of abstract sounds lie in the difficulty that humans have in learning and remembering large sets of them (Patterson &

Milroy, 1980), and therefore in the limited amount of information they can convey. Due to this, it has been suggested that no more than five to eight items should make up a warning set if abstract sounds are to be used (Patterson, 1982; Sanders & McCormick, 1987).

Environmental sounds (or auditory icons) are another type of non-verbal sound with potential for use as auditory warnings. They share the advantages of abstract sounds with respect to masking and the speed of information transfer, but they are not as limited in set size. They also have the potential to convey more information. Humans enjoy a huge memory capacity for environmental sounds (Lawrence & Banks, 1973). Their association with real events means that they are easier to learn than abstract sounds. The use of environmental sounds as auditory warnings takes advantage of natural auditory processes through the use of the link between a sound signal and its semantic representation. They also share many attributes with speech and it is for this reason that Ballas and Howard (1987) conclude that they may be considered to be a form of language.

The following is a description of some of the research that Air Operations Division of DSTO has been conducting into several aspects of auditory warnings.

Comparative Research on Speech, Icons, and Abstract Sounds

Experiment 1: Learning and Retention

In choosing the most appropriate sound type to fulfil the role of an auditory warning, several issues need to be addressed. Environmental factors need to be taken into account, such as the ambient noise environment (Stanton & Edworthy, 1999). Psychological factors are also important, yet often neglected. For example, how easily can the relationship between the warning signal and the event it portrays be learnt and remembered?

An experiment was conducted to investigate subjects' ability to learn and retain three different warning types, i.e., speech, auditory icons, and abstract sounds (Leung, Smith, Parker & Martin, 1997). The speech warnings were generated by a speech synthesis software package (DECtalk 4.2, Digital Equipment Corporation). The auditory icons were digital recordings of environmental sounds, chosen to share strong relationships with the events they were to represent. For example the sound of guns firing was used to represent the event 'guns'. Patterson's (1982) 'attensons' were used as the abstract sound stimuli. These are synthetic complex sound pulses designed to be maximally discriminable.

The study measured the number of trials required for subjects to learn eight warning signal-event pairings, as well as the number of errors made. Subjects' retention of the warning-event pairings was tested after three days, and again after eight days. It was found that abstract sounds were significantly more difficult to learn and retain than speech and icons, as subjects required more trials to learn abstract sound-event pairings and made more errors doing so. No significant difference was found between subjects' ability to learn and retain speech and icons, although there was a trend in favour of speech warnings (Figure 23.1).

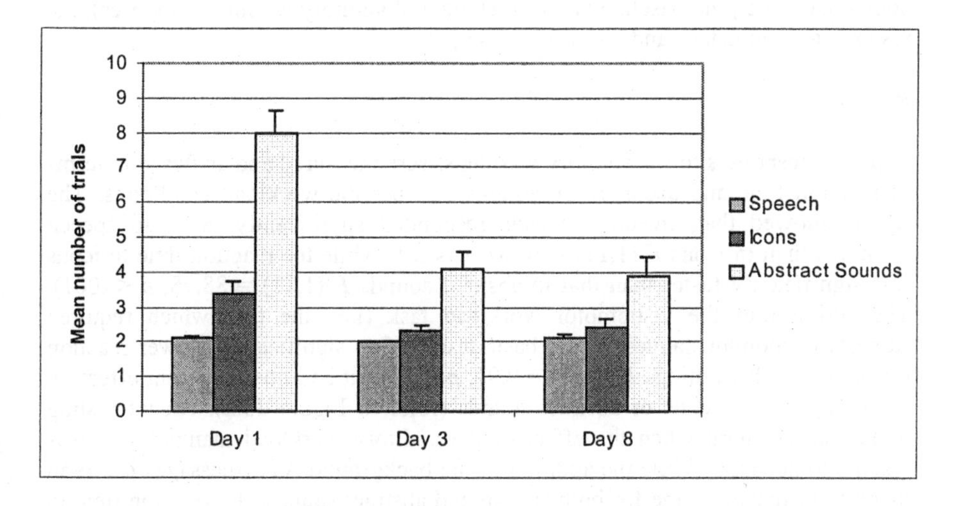

Figure 23.1 Mean number of trials required to learn each auditory warning type

Experiment 2: Effect of Workload

Auditory warnings must be usable in an environment where attention is divided and workload varies. It is possible that an auditory warning that is effective in normal circumstances may lose its effectiveness when workload is high, and the operator's attentional capacity is decreased. To investigate this issue, an experiment was performed to assess subjects' performance using speech, icon and abstract warnings under varying workload conditions (Smith, Parker, Martin, McAnally & Stephan, 1999).

Method

The methodology used was similar to that used in studies by Wheale (1981, 1982). There were four workload conditions. In the baseline condition, subjects were required to perform a tracking task. At various times throughout the trial, an auditory warning would occur, and the subject had to respond to the warning by pressing a corresponding key. Workload was manipulated by adding tasks to these baseline tasks. In one condition, subjects performed a visuomotor task in addition to the baseline tasks. The subjects were required to monitor gauges and respond when the gauges moved out of range, by pressing a button on the joystick used for the tracking task. The second condition involved an auditory attention task in addition to the baseline tasks. Air traffic control (ATC) messages were played to the subject throughout the trial, and subjects were required to attend to the messages and to remember any message which began with their call-sign (e.g., Condor, ascend 1000 feet). The fourth workload condition included all these components, i.e., the baseline tasks (tracking and auditory warning responses), the visuomotor gauge task and the ATC message task.

Results

Subjects' responses to the auditory warnings were measured and analysed in terms of reaction time and accuracy of response, across the workload conditions. The results showed that, overall, subjects responded significantly faster to speech warnings than to icons $F\ (1,11) = 16.67$, $p < .01$, while the reaction time to icons was significantly faster than that to abstract sounds $F\ (1,11) = 88.78$, $p < .001$). The addition of the visuomotor workload task (i.e., the task which required subjects to monitor gauges) to the basic tasks led to significantly slower reaction times to speech warnings $F\ (1,11) = 9.97$, $p <.01$, but had no significant effect on reaction times to icons or abstract sounds (refer to Figure 23.2). An interesting trend was apparent when the effect of the auditory workload manipulation was taken into account. While the addition of the background ATC messages led to an increase in reaction time for both speech and abstract sounds, the reaction time to icons remained more stable. This increase in reaction time for speech and abstract sounds, however, did not reach significance (Figure 23.3). Note that further data are required to establish whether this effect is robust.

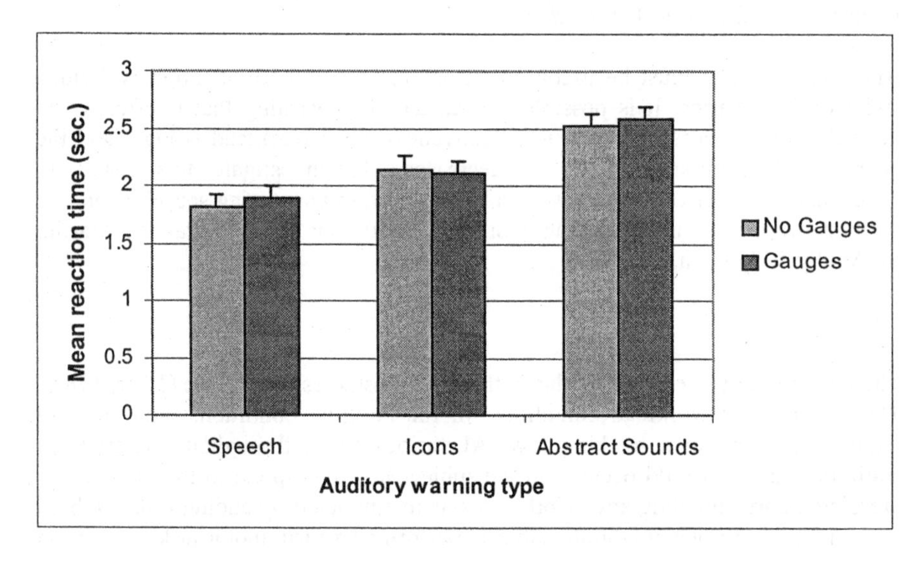

Figure 23.2 Mean reaction time to auditory warnings by visuomotor gauge task

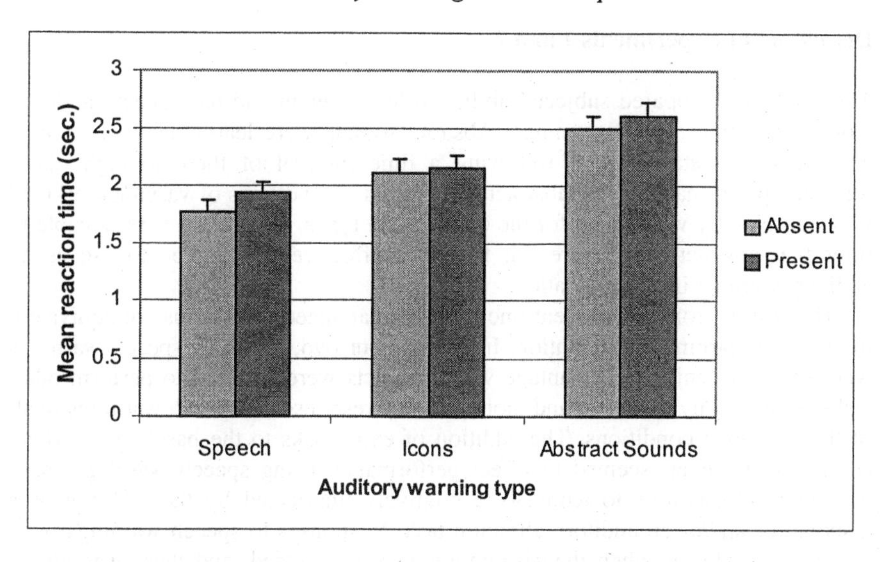

Figure 23.3 Mean reaction time to auditory warnings by auditory attention task

The analysis of mean response accuracy indicated that speech warnings were identified significantly more accurately than both icons and abstract sounds. There was no significant difference between the mean accuracy scores for icons and abstract sounds (Figure 23.4).

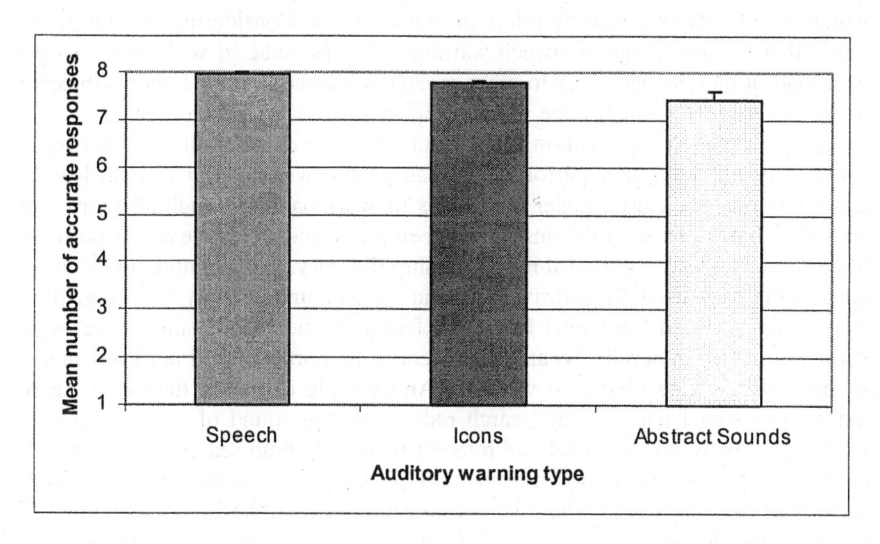

Figure 23.4 Mean number of accurate responses per trial to auditory warnings

Discussion of Experiments 1 and 2

These studies compared subjects' ability to learn, retain and use speech, auditory icon and abstract sound warnings. Abstract sounds were learnt relatively slowly and were not retained well following a time interval of three to eight days. Subjects' performance using abstract sounds under conditions of varying workload was significantly worse than for the other sound types. As such, it is reasonable to infer that abstract sounds are not the most effective sound type to use as an auditory warning in these situations.

The results from Experiment one indicate that speech and icons are equivalent in terms of learning and retention. In Experiment two, however, speech warnings were shown to enjoy an advantage when subjects were required to perform other tasks, as shown by the faster and more accurate responses to speech warnings in all of the workload conditions. The addition of extra tasks to the baseline workload condition, however, seemed to affect performance using speech warnings more than icons. Responses to icons were relatively unaffected by the addition of a visuomotor and/or an auditory attention task. Responses to speech warnings were significantly slower when the visuomotor task was added, and there was also a trend towards slower reaction times for speech warnings with the addition of the auditory attention task.

It is important to note that while the overall level of workload in Experiment two was designed to be demanding, subjects had little difficulty with the baseline workload condition, and it was not until both the visuomotor task and the auditory attention task were added that the subjects reported any difficulties. It is likely that the workload resulting from this laboratory-based experiment did not approach that which may be experienced by pilots in real cockpits. Considering that the results imply that the advantage of speech warnings may be reduced with an increase in workload, it may be necessary to place greater workload demands upon subjects in future experiments to determine whether the advantage for speech remains.

A more detailed examination of the results of these experiments established that some icons elicited better performance than others, with respect to both learning and retention, when used under conditions of workload. Although all icons were chosen to have a strong relationship between the sound and the event it portrayed (the referent), the strength of this relationship did vary. For example, the sound of guns firing was used to portray the event 'guns', undoubtedly a strong, direct association between the signal and the referent. Some relationships were not as strong, however, generally because the event to be portrayed did not itself make a sound that could be used as a warning. An example of this is the event 'search radar'. The sound paired with 'search radar' was the sound of a sonar ping, the relationship between the signal and referent being that both search radar and sonar perform searching functions. An inspection of the results revealed that the subjects performed extremely well when the relationship between the sound signal and the event appeared to be a strong one. This observation has led to recognition of the need to systematically investigate issues relating to signal-referent associations.

Auditory Icons: Signal – Referent Associations

Experiment 3

In order to further investigate the role of signal-referent association strength, a learning and retention study was conducted using different types of auditory icons. Three types of signal-referent associations were identified and evaluated in the experiment. Direct signal-referent relationships were defined as existing when the referent, or event, was signalled by a sound produced by the referent. For example, the sound of a dog barking (the signal) was used to represent the referent 'dog'. Indirect signal-referent associations were those where the object producing the sound signal and the referent had some sort of existing, but indirect, relationship, e.g., the sound of a cat miaowing to represent the referent 'dog', the relationship being that both cats and dogs are typical pets. Arbitrary relationships were defined as sound-event pairings that had no pre-existing relationship, for example the sound of a cork popping to represent the referent 'dog'.

Method

Subjects were assigned to one of the three signal-referent association conditions; direct, indirect or arbitrary. There were 21 subjects per condition. Each subject participated in two experimental sessions, an initial session in which they learnt eight sound-word (i.e., signal-referent) pairs, and a second session after four weeks to test retention.

The sounds used in the experiment were chosen to be easily recognisable. They were selected on the basis of a pilot study, in which a number of sounds were presented to 15 people who were asked to identify them. The sounds used in the experiment had a mean identification accuracy of 93.33 percent. To ensure that subjects recognised the sounds in the actual experiment, a questionnaire was administered at the conclusion asking them to identify each sound. Subjects were also asked to define the relationship between the sound and the event that it represented, and to rate the strength of that relationship (on a scale of 1=not at all related, to 7=absolutely related). This was to ensure that the subjects in the indirect condition recognised the pre-existing relationship between the sounds they heard and the events that the sounds portrayed. It also provided a subjective measure of the strength of the relationship between the sound signal and its referent for all association conditions. Any subject who misidentified a sound was not included in the experimental analysis. Likewise, any subject in the indirect association condition who did not recognise a pre-existing relationship between the signal and the referent was not included.

Results

The results were analysed in terms of the number of training sessions required to learn the icon-event relationships and the number of errors made (Figure 23.5). The analysis revealed that direct associations were learnt and retained with little or no difficulty, and few errors were made. Overall, the learning and retention performance of the subjects in the indirect association condition was not significantly different from that of the subjects in the direct association condition, although there was a trend in favour of direct associations. There was no significant difference between the number of errors made for the direct and indirect association conditions, either at the initial test $F\,(1,60) = 2.62$, $p>.05$, or the re-test $Z = 8.12$, $p>.05$. There was also no significant difference in the number of training sessions required for the direct and indirect conditions at the initial test $F\,(1,60) = 2.26$, $p>.05$ (Figure 23.5). Subjects did, however, require significantly more training sessions in the indirect compared with the direct condition at the re-test session $F\,(1,60) = 6.93$, $p<.05$ (Figure 23.5).

Arbitrary sound-event associations were significantly more difficult to learn and retain than both indirect and direct associations. Subjects required more training sessions to learn the arbitrary associations (session 1: $F\,(1,60) = 32.7$, $p<.001$; session 2: $F\,(1,60) = 25.05$, $p<.001$), and made more incorrect responses (session 1: $F\,(1,60) = 42.47$, $p<.001$; session 2: $Z = 24.69$, $p<.05$).

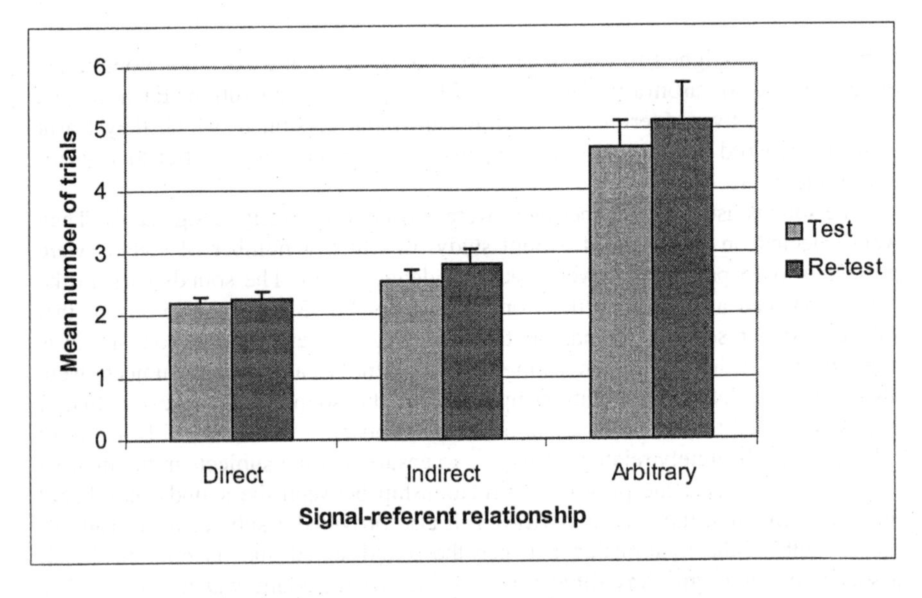

Figure 23.5 Mean number of trials required to learn the signal-referent pairs per association condition

The subjective ratings of the relationship strengths between the signals and their referents were also analysed (Figure 23.6). Signal-referent pairs with a direct association were rated as very strong, with a mean rating of 6.5 out of 7 (where 7=absolutely related). Indirect associations were also rated as strong (mean rating=5.4), however they were rated as being significantly weaker than the direct associations $F\,(1{,}69) = 42.64, p<.001$. Arbitrary signal-referent associations, with a mean rating of 2.5, were rated as being significantly weaker than indirect associations $F\,(1{,}69) = 260.52, p<.001$.

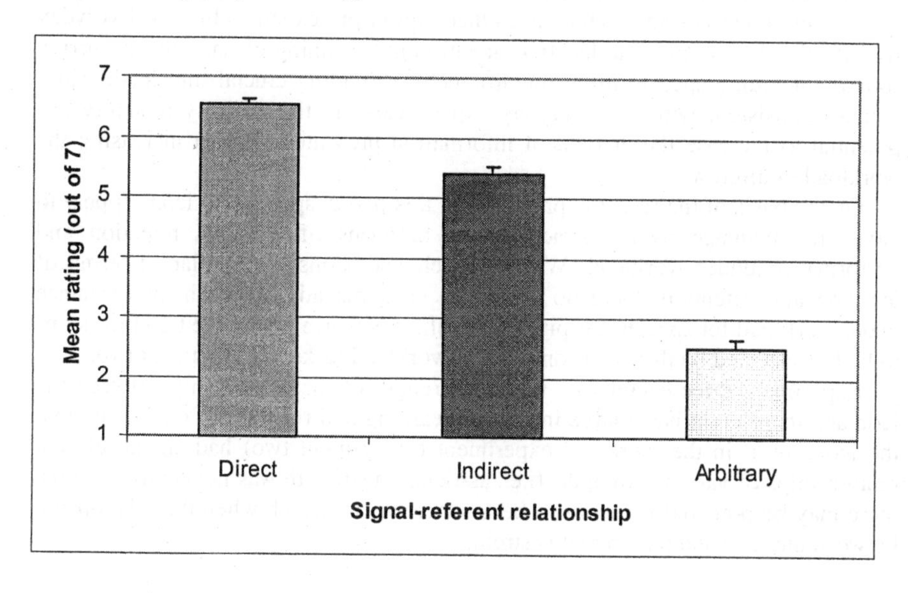

Figure 23.6 Mean ratings for strength of relationship between signal and referent per association condition

Discussion

The subjects learnt both direct and indirect signal-referent associations easily. Direct associations were retained almost perfectly after a four-week interval, with indirect associations requiring minimal retraining. Arbitrary associations, however, were difficult to both learn and retain. Presumably this reflects the cognitive processes underlying each of the association types. Direct associations require minimal processing, and generally rely solely on a correct identification of the sound. Indirect associations also take advantage of existing relationships but these relationships are not as strong, as they require both correct identification of the sound and recognition of the relationship between the sound and the referent. Arbitrary associations are the most difficult to learn, as the subject must form a new relationship between the sound and the referent.

It is particularly encouraging that indirect signal-referent relationships are learnt and retained almost as easily as direct signal-referent relationships. This supports the possibility of finding effective auditory icons to represent events/concepts that do not themselves make a sound, by using a related sound as the warning.

Overall Discussion and Conclusions

The research outlined in this paper reflects the authors' design philosophy of developing interfaces and technologies that exploit pre-existing abilities. Everyday we effortlessly interpret the location, identity and meaning of naturally occurring sounds, including speech, in our environment. Providing crucial information in a manner consistent with our everyday experiences via the auditory modality has potential to increase the amount of information presented without increasing the workload of aircrew.

On the basis of the research presented in this paper, speech and icons appear to have an advantage over abstract sounds in terms of learning, retention and performance under workload. Whilst speech and icons are similar in terms of learning and retention, there do appear to be some advantages in performance under workload for speech. Despite this result, it is worth noting the trend for icons to be less affected by the inclusion of extra workload tasks in Experiment two.

Experiment three established that the strength of the relationship between the icon and its referent has a large impact on learning and retention. Considering that the icons used in the workload experiment (experiment two) had signal-referent relationships of varying strengths (i.e., association strength was not controlled for), there may be potential for icons to be as effective as speech when the relationship between the icon and its referent is strong.

References

Ballas, J. A., & Howard, Jr., J. H. (1987). Interpreting the language of environmental sounds. *Environment and Behavior, 19 (1),* 91-114.

Edworthy, J., & Stanton, N. (1995). A user centred approach to the design and evaluation of auditory warning signals: 1. Methodology. *Ergonomics, 38(11),* 2262-2280.

Edworthy, J., & Stanton, N. A. (1999). Auditory warnings and displays: An overview. In J. Stanton & N. A. Edworthy (Eds.), *Human Factors in Auditory Warnings.* Aldershot: Ashgate.

Lawrence, D. M., & Banks, W. P. (1973). Accuracy of recognition memory for common sounds. *Bulletin of the Psychonomic Society, 1,* 298-300.

Leung, Y. K., Smith, S., Parker, S. A. & Martin, R. L. (1997). Learning and retention of auditory warnings. *Proceedings of the Fourth International Conference on Auditory Displays,* 129-133.

Patterson, R. D. (1982). *Guidelines for Auditory Warning Systems on Civil Aircraft* (CAA Paper 82017). London: Civil Aviation Authority.

Patterson, R. D., & Milroy, R. (1980). *Auditory Warnings on Civil Aircraft: The Learning and Retention of Warnings* (CAA Paper 7D/S/0142). London: Civil Aviation Authority.

Sanders, M. S., & McCormick, E. J. (1987). *Human Factors in Engineering and Design* (6th ed.). New York: McGraw-Hill.

Smith, S., Parker, S. A., Martin, R. L., McAnally, K. L., & Stephan, K. (1999). Evaluation of sound types for use as auditory warnings. *Australian Journal of Psychology, 51 (Supp),* 42.

Wheale, J. L. (1981). The speed of response to synthesized voice messages. *British Journal of Audiology, 15*, 205-212.

Wheale, J. L. (1982). Decrements in performance on a primary task associated with the reaction to voice warning messages. *British Journal of Audiology, 16*, 265-272.

Chapter 24

Dealing with Conflicting Information: Will Crews Rely on Automation?

Kathleen L. Mosier, Jeffrey Keys and Robert Bernhard
San Francisco State University, USA

Introduction

The availability of sophisticated automated aids in modern aircraft feeds into a general human tendency to travel the road of least cognitive effort (Fiske & Taylor, 1994). Omission and commission errors resulting from automation bias, the tendency to utilise automated cues as a heuristic replacement for vigilant information seeking and processing, have been documented in professional pilots and students, in one and two-person crews. Underlying causes of automation-related omission errors have been traced in part to vigilance issues. Crews count on automation to provide the most salient and reliable information about flight progress and system status, and often 'miss' events that are not pointed out to them by automated systems. This tendency is exacerbated by the fact that operators often have incomplete or fuzzy mental models of how various modes of automation work (Norman, 1990; Sarter & Woods, 1992). Additionally, the opaque interfaces of many automated systems, which provide only limited information about actual status and criteria for conclusions, make it difficult for even vigilant decision makers to detect errors.

Pilots also rely on automation, particularly automated warning systems, to guide their actions, as is the case with Ground Proximity Warning System (GPWS) or Traffic Collision Avoidance System (TCAS) aural and visual displays. Automation-related commission errors result when pilots follow automated information or directives inappropriately (e.g., when other information in the environment contradicts or is inconsistent with the automated cue). Automation bias and resultant errors have been documented in current flight crew Aviation Safety Reporting System (ASRS) incident reports (Mosier, Skitka, & Korte, 1994), as well as in controlled studies using students and professional pilots as participants in solo or team configurations. In these studies, professional pilots were sensitive to the importance of correctness for critical flight tasks, and made fewer errors on events involving altitude and heading errors than frequency discrepancies (Mosier, Skitka, Heers, & Burdick, 1998; Mosier, Skitka, Dunbar, &

McDonnell, in press). Training for automation bias reduced commission errors for students, suggesting the importance of early intervention and training on this issue (Skitka, Mosier, & Burdick, 2000; Skitka, Mosier, Burdick, & Rosenblatt, 2000). In another study, participants in a non-automated condition out-performed those using an automated aid during equivalent failure events (Skitka, Mosier, & Burdick, 1999).

Commission Errors

Preference for Action?

It has been hypothesised that commission errors may be related to a tendency of pilots to take action in a given situation as a way of maintaining control. This notion was supported in an exploration of pilots' use of conflict probe displays (Cashion, Mackintosh, McGann, & Lozito, 1997). Several of the crews in that study (6 out of 20) chose to manoeuvre during an event in which the 'rules of the road' suggested that action was not required of them, but rather of the other aircraft involved in the conflict. The tendency toward action makes sense for pilots, as pro-activity has typically been associated with superior crew performance (e.g., Fischer, Orasanu, & Montalvo, 1993; Mosier & Chidester, 1991). Operators may feel that taking an action gives them more control over the situation, or they may be taking action as a way of dealing with a potential situation immediately to eliminate the risk it may develop into a problem and the workload involved in monitoring it.

Making Information Sources Equally Salient

Most of the studies mentioned above utilised low, medium, or high-fidelity flying tasks, and features of the information display may have fostered the tendency to rely heavily on automated information. For example, automated information is typically designed to be more available, compelling, and salient than other kinds of information. What happens when other sources of information are made equally salient to automated cues?

A series of paper-and-pencil scenario studies was conducted with student participants (Skitka, 1999). Scenario 1 was modelled after a simulator event used by Mosier, Palmer, and Degani (1992), and involved a decision about which one of two engines was more damaged and needed to be shut down. Automated monitoring devices indicated that one engine was severely damaged, whereas traditional system indices suggested that the other engine was the source of the problem. Reliability of each of the indicators as well as relative risk associated with making a mistake were systematically varied. In this situation, people on average responded in very rational ways. Under both high and low risk, they opted to go with the recommendation that had the highest probability of being correct; and when both sources of information had an equal probability of being right, they showed no systematic preference for one source over the other. In sum, when other sources of information were presented on a par, and with equal salience to automated recommendations, automation bias did not emerge.

Scenario 2 involved a car with the latest in on-board computers, and presented participants with conflicting information from the computer and the engine gauges. In this study, the source of an 'action' recommendation was varied, as was the reliability of each of the sources and the risk involved in choosing each action. Results indicated significant main effects for action and reliability, and an action by risk interaction. In this case, in addition to a general tendency to choose action and follow reliable information, people were particularly likely to exhibit a preference for action – whatever the information source and whatever its reliability – when low risk was involved.

Scenario 3, a nuclear power plant scenario, was constructed so that there were risks associated with failing to act, rather than with acting. Again, the reliability of each information source (computer or gauges) and the source of the action recommendation were systematically varied. The results of this study conceptually replicated the findings of the airplane and car scenario studies, in that no systematic preference for automated information was observed. When information from non-automated sources was presented with equal salience to automated information, automation bias effects did not emerge. When risk varied as a function of inaction, rather than action, people again acted to minimise the potential hazard associated with their choices. Although there was a preference for action under both conditions of manipulated risk, the effect was much stronger when high risk was associated with inaction.

The Present Studies

The scenario studies described above suggest that the tendency toward automation bias may be at least in part a function of the way automated information is displayed. When individuals were made equally aware of conflicting information from all sources, they tended to be more analytical in their decisions, and displayed no systematic tendency to prefer automated cues over others. These findings suggest that in decision-making contexts such as the glass cockpit, it may be the display features of automated information, such as its salience and opaque interface, that causes it to be used as a heuristic, rather than the quality of the information itself. If all information in the cockpit was displayed equally compellingly, the phenomenon of automation bias, at least with respect to commission errors, might be mitigated – or even disappear. The present studies, which focussed on regional, or Part 135 operations (i.e., commercially-operated aircraft with fewer than 50 passenger seats), were conducted as an initial exploration of this possibility.

An ASRS analysis of Part 135 reported incidents was conducted to gather preliminary information. It was intended to accomplish three tasks: (a) to document the existence of issues or problems related to the use of automated and other information in Part 135 operations; (b) to trace the patterns of action that pilots take in various types of incidents, especially when information conflicts are present; and (c) to identify scenarios that would be good candidates for the scenario study which was to follow. In the scenario study, pilots were asked to choose between two sources of information that conflicted with each other. The information was presented in a paper-and-pencil format, to give both sources equal salience. The intent of this study was to determine whether a tendency toward automation bias would be a factor in pilot decisions when data were presented in a manner that gave information from all sources equal salience. We also wanted to examine the impact of perceived risk and action vs. inaction on their choices.

ASRS Study

Data for the ASRS study were obtained from the ASRS incident database, and covered reports submitted between 1994 and 1998. Using several broad search queries, such as aircraft type, passenger operations, and automated displays or instruments (e.g., EFIS, integrated navigation), we created a preliminary sample of 1,200 reports that were submitted by Part 135 pilots. Each of these was screened for appropriateness for our study of conflicting information and/or automation-related errors, and candidate reports were subject to a second scrutiny. We were particularly interested in incidents involving conflicting information from different sources. Through this procedure, we ultimately created a sample of 189 ASRS reports involving some form of automation and/or information conflict. These reports were coded in two ways. First, the variable fields present in the ASRS database were entered directly into a data file to collate information on aircraft type, phase of flight, and incident type and resolution as coded by ASRS personnel. Second, the narratives entered by the pilots were coded to identify incidents in which information conflicts were present, and to track the actions that were taken in each case. Narrative coders cross-checked each others' work until 90 percent inter-rater agreement was attained. Incidents were coded with respect to the sources of information that were cited concerning the critical event, whether the sources provided consistent information or were in conflict, and how the incident was resolved. Factors such as fatigue, distraction, and risk level were coded whenever mentioned in the report.

Results and Discussion

Descriptive Information

The final sample of 189 reports represents approximately 16 percent of incidents reported by Part 135 pilots, demonstrating the presence of automation-related issues and incidents in regional operations. That number can be expected to increase as regional operators upgrade to more sophisticated aircraft types. Reports came from pilots of at least 13 different aircraft types. Breakdown of the data by aircraft type showed that the categories represented by the largest numbers of reports were 'commercial, fixed wing' (n = 71), followed by Brasilia EMB-120 (n = 29), SF-340A/B (n = 29), and Beech 1900 (n = 20). With respect to phase of flight, the largest number of incidents occurred during climb (n = 64), cruise (n = 40), descent (n = 31), and approach (n = 28).

Narrative Coding

The richest source of data for this study was found in the narratives, which offered candid information about the incidents in the words of the reporting pilots. From the narratives, incidents were categorised as: (a) system/mechanical failures or faulty warnings (n = 67; 35.4 percent); (b) traffic incidents (n = 71; 37.04 percent); and (c) altitude, heading, or route failures or errors (n = 62; 27.51percent). The most commonly reported cause for traffic incidents was ATC error (n = 20), and crew error and/or incorrect system set-up was cited most often as the source of altitude, heading or route failures (n = 47). Twelve incidents involved automation-related omission errors, and nine involved automation-related commission errors.

Whenever possible, we coded factors that pilots noted had contributed to the incident they were reporting. Distraction was cited as a contributing factor in 23.81 percent of the incident reports. Failure to double-check indicators and/or fatigue were occasionally cited as contributing to incidents (15.9 percent and 7.41 percent respectively). Distraction and failure to double-check indicators were most often associated with altitude, heading or route errors. Risk was mentioned in 9 percent of the reports. In 103 reports (55 percent), pilots reported dealing with conflicting information from two or more sources. Of these, 70 involved some form of automated information that conflicted with another source.

Sources coded included:

1. physical – visual or auditory cue;
2. other human source (ATC, passenger, etc.);
3. checklists/handbooks;
4. 'steam gauges' – not electronic data;
5. electric warning system (e.g., master caution warning lights, fire lights, oil lights, etc.);
6. electronic warning system;
7. computer system (e.g., EFIS, EICAS, FMS, FMC);
8. autopilot;
9. TCAS I;
10. TCAS II;
11. auto feathering system;
12. weather detection system;
13. GPWS; and
14. unknown/unable to determine.

Most conflicts involved instrument or electronic data that was contradicted by some human element, such as ATC information or a crew member's physical sensations (i.e., sees, hears, or feels something). Pilots tended to trust TCAS and their own senses over instruments or ATC information. A few incidents (n = 7) involved conflicts between traditional sources and electronic data (not including TCAS), and in all but one of these, crews followed the traditional source. Many of the incidents involving conflicting information were traffic incidents, and involved a conflict between TCAS and some other source (e.g., ATC or visual cues; n = 33). High risk was most often cited as a factor in conjunction with traffic incidents. Analyses indicated that, when TCAS information entailed taking evasive action, crews typically followed TCAS recommendations – even when visual information contradicted the need for the manoeuvre.

These incident reports document the existence of conflicts among information sources in the cockpit. It seems that any preference for automated information, however, is impacted by the action orientation of the information. In many cases, crews tended to follow their own senses or traditional sources of information rather than preferring electronic data. When TCAS was involved, however, crews generally followed its action directives, regardless of contradictory information. This strategy may be related at least in part to procedures prescribed in conjunction with TCAS directives, and also supports the notion of a 'take action' tendency.

Scenario Study

Scenario Development

Scenarios were created using incidents from the ASRS analysis and from previous research studies (Fischer, Orasanu, & Wich, 1995). Care was taken to ensure that scenarios were representative enough that they could be responded to by pilots of several different aircraft types. Each scenario conveyed a situation involving conflicting information from two sources: an automated source plus either a human source or a traditional indicator. In each scenario, information from one source suggested making some change (action); information from the other source suggested maintaining status quo. Each scenario was followed by two decision options – for example:

> You are the pilot flying on approach into your destination in VMC. You would really like to expedite your arrival, because you are already late and many of the passengers are in danger of missing their connections. You are being vectored in for a landing behind a 757, which you know is notorious for causing wake turbulence problems for aircraft following it. Air traffic control has told you that you are presently 5 miles behind the 757, in no danger of encountering wake turbulence, and to maintain your present speed of 200 knots to stay in sequence for landing. You look at the TCAS display, and it shows you only 3 miles behind the 757.
>
> Given this information, what would be your decision?

- Hold present speed and distance from the 757.

- Get ATC clearance to slow down to increase the distance from the 757.

The above scenario contains information from an automated source (TCAS) and a human source (Air Traffic Controller). In this version, the information from the human source suggests that the pilot maintain status quo; information from the automated source suggests a change. Pilots were asked to choose one of the options, and to report their level of confidence in the decision (not confident à very confident) as well as the risk involved in the scenario (minimal risk à high risk) on 1-9 scales. Pilots were told in a cover letter that we realised the scenarios might not contain all of the information or decision options that they would like to have, but asked them to make a choice based on what was available in the scenarios.

Procedures

Two different packets of 10 scenarios each were created. Seven of the scenarios were matched between packets – that is, the same scenario was manipulated so that, in Packet 1, the information from the automated source suggested action, and in Packet 2, the information from the other source suggested the same action. Pilots saw only one version of each scenario. One scenario contained conflicting action recommendations – an automated source suggested that one of two engines was on fire and should be shut down; traditional indications suggested that it was actually the other engine that was damaged. Two additional scenarios were added to each packet to even out the number of human and traditional indicators contained in the scenarios. Demographic information solicited included flight hours, years with current airline, and experience by aircraft type.

Approximately 700 packets were distributed to the mailboxes of pilots of a US regional carrier. Pilots were asked to place completed packets into a collection box in their operations office. One hundred twenty-five packets with usable data were returned to us. These were roughly divided between Packets 1 and 2 (54 and 71, respectively).

Results and Discussion

Descriptive Information

Pilot respondents ranged in age from 22-55 years (M = 34), and had total hours of flight experience ranging from 1,000-23,000 hours (M = 5382; SD = 3787). Glass cockpit hours varied from 0-10,400 hours (M = 1,355; SD = 2,047). It should be noted that this sample represents a broad range of flight experience, particularly with respect to glass cockpit experience.

Decision Choices

The nature of the data did not lend itself to traditional statistical analyses of all scenarios against each other. However, in looking at scenario pairs, we found no systematic evidence of a preference for automated information in pilot decisions – in fact, in none of the scenario pairs was automated information followed across packets. Rather, we saw a pronounced scenario effect; that is, in most scenarios there was high agreement across packets on the preferred option, the risk level of the scenario, and the confidence with which pilots chose an option. The most dramatic scenario response splits exhibited either a clear source effect (e.g., following a particular source), or a response effect (e.g., taking action vs. maintaining status quo). We did not find evidence of a preference for action across all scenarios (which was, in most cases, the more conservative option), although the higher the estimated risk of a scenario, the more likely pilots were to choose action, and the more confident they were in their choice. A more complete picture can be gained by looking specifically at the scenario pairs, as displayed in Table 24.1. The table shows pilot responses, by source of information and by action/status quo options. Numbers for the predominant predictors by scenario, source, or response are in bold type.

Table 24.1 Decision choice totals by scenario

Information Conflict	Source Effect		Response Effect	
	Automation	Human/ Traditional	Action	Status Quo
TCAS vs. ATC	61	64	110	15
TCAS vs. ATC	65	58	98	25
Warning light vs. human/indirect	74	50	119	5
TCAS vs. PNF	34	90	74	50
Computer vs. PNF	35	86	56	65
FMS vs. traditional (VHF nav)	31	99	71	51
Ambiguous engine fire – warning system vs. engine gauges	70	54	6	118
Engine fire – which engine? warning system vs. engine gauges	9	94		

When automated information conflicted with information from a human source, for example, the nature of the human information impacted decisions. If the 'human' was the air traffic controller, or when the human offered indirect information (e.g., remembered a similar incident being a false alarm), pilots exhibited a response effect, and tended to follow whichever source recommended action. However, when the PNF (pilot-not-flying) was the source of direct information, pilots made the decision suggested by the human (source) rather than the automation. In automation vs. traditional indicator conflicts, we observed a tendency to follow traditional rather than automated indicators (source). Two scenarios contained engine fire scenarios. In one of these, the dilemma was whether or not an engine was actually on fire, and should be shut down. In this conflict, pilots responded conservatively, and this was the only scenario in which an inaction response was prevalent (response). For the pair of scenarios that contained conflicting engine fire indications (which engine was on fire), pilots most often believed traditional, rather than automated indicators (source).

Summary

These results roughly parallel the findings of the ASRS report analysis, and again suggest a tendency toward action and a trust of direct human information. When other data are made to be equally salient with automated information, pilots do not exhibit a systematic preference for automated information. In fact, they seem, at least in this study, to trust traditional indicators over automated information. Other factors, such as the perceived validity of conflicting information, also impacted whether or not automated cues were trusted. Pilots seem to assume high validity when information comes from fellow crewmembers, and less validity when the reporting human is an air traffic controller.

Results of this study are encouraging in several ways. Regional airline pilots are typically less experienced with automated aircraft than the commercial, B-737/747/767 pilots of previous studies. These results suggest that we may be able to impact automation bias if we train pilots early enough in their careers to evaluate automated cues in context with other cues. As a caveat, however, we need to be very cautious about generalising from the paper-and-pencil venue to the glass cockpit. This format provides information differently than it is shown within the cockpit, and allows the information to be processed in a less biased and more analytical way. Additionally, we have previous evidence that, when encountering a situation in an actual or simulated aircraft environment, pilots do not always do what they say they would do.

One important implication of these findings is that making automated and other information equally salient may go a long way toward promoting a more analytical and less biased decision-making process in the cockpit. Input from other crewmembers, for example, proved to be just as important in evaluating situations as automated information. Follow-up studies will be required to determine if results of the paper-and-pencil study will hold in other venues.

Acknowledgements

We are grateful for the valued input from Dr. Beth Lyall, Co-investigator in this research program, and her colleagues at Research Integrations, Inc. This work was funded by the NASA Aviation Safety Program, NASA Grant # NAG2-1285. Dr. Judith Orasanu at NASA Ames Research Center is our Technical Monitor.

References

Cashion, P., Mackintosh, M. A., McGann, A., & Lozito, S. (1997, October). A study of commercial flight crew self-separation. *Proceedings of the Sixteenth American Institute of Aeronautics and Astronautics Digital Avionics Technical Committee* (pp. 6-18).

Fischer, U., Orasanu, J., & Montalvo, M. (1993). Efficient decision strategies on the flight deck. *Proceedings of the 7th International Symposium on Aviation Psychology*, Columbus, OH.

Fischer, U., Orasanu, J., & Wich, M. (1995). Expert pilots' perceptions of problem situations. *Proceedings of the 8th International Symposium on Aviation Psychology*, Columbus, OH.

Fiske , S. T., & Taylor, S. E. (1994). *Social Cognition* (2nd ed.). NY: MacGraw Hill.

Mosier, K. L., & Chidester, T. R. (1991, July). Situation assessment and situation awareness in a team setting: Designing for everyone. *Proceedings of the 11th Congress of the International Ergonomics Association* (pp. 798-800). Paris.

Mosier, K. L., Palmer, E. A., & Degani, A. (1992). Electronic checklists: Implications for decision making. *Proceedings of the Human Factors Society 36th Annual Meeting* (pp. 7-11). Santa Monica, CA: Human Factors Society.

Mosier, K. L., Skitka, L. J., Dunbar, M., & McDonnell, L. (in press). Air crews and automation bias: The advantages of teamwork? *International Journal of Aviation Psychology*.

Mosier, K. L., Skitka, L. J., Heers, S., & Burdick, M. D. (1998). Automation bias: Decision making and performance in high-tech cockpits. *International Journal of Aviation Psychology, 8*(1), 47 - 63.

Mosier, K. L., Skitka, L. J., & Korte, K. J. (1994). Cognitive and social psychological issues in flight crew/automation interaction. In M. Mouloua & R. Parasuraman (Eds.), *Human Performance in Automated Systems: Current Research and Trends* (pp. 191-197). Hillsdale, NJ: Lawrence Erlbaum Associates, Inc.

Norman, D. A. (1990). *The 'problem' with automation: Inappropriate feedback and interaction, not 'over-automation'* (B 327). Philosophical Transactions of the Royal Society of London.

Sarter, N., & Woods, D. (1992). Pilot interaction with cockpit automation: Operational experiences with the flight management system. *International Journal of Aviation Psychology, 2*(4), 303-321.

Skitka, L. J. (1999). *Automation: Decision aid or decision maker?* (NASA Contractor Report). Moffett Field, CA: National Aeronautics and Space Administration.

Skitka, L. J., Mosier, K. L., & Burdick, M. D. (1999). Does automation bias decision making? *International Journal of Human-Computer Studies, 50,* 991-1006.

Skitka, L. J., Mosier, K. L., & Burdick, M. D. (2000). Accountability and automation bias. *International Journal of Human-Computer Studies, 52,* 701-717.

Skitka, L. J., Mosier, K. L., Burdick, M. D., & Rosenblatt, B. (2000). Automation bias and errors: Are crews better than individuals? *International Journal of Aviation Psychology, 10*(1), 83-95.

Chapter 25

Development of Hazard Analysis Techniques for Human-Computer Systems

Andrew Neal, Michael Humphreys, David Leadbetter and Peter Lindsay
The University of Queensland, Australia

Introduction

Human error is known to be responsible for approximately 80 percent of all system failures within industries such as aviation, power generation, and mining (Hollnagel, 1993). Many of these errors can be traced back to the design of the human-computer or human-machine system. For example, the London Ambulance Service installed a new computerised dispatch system in 1992 resulting in lengthy delays in the dispatch of ambulances to emergencies (Finklestein & Dowell, 1996). A number of the errors were caused by a slow human-computer interface in which exception messages were not prioritised, queues scrolled off the screen with no means of retrieval and duplicated calls were not identified. In order to overcome these types of design problems, a range of techniques have been developed to analyse the potential for human error within safety-critical systems, and to examine the consequences of errors for the system as a whole.

It is interesting to compare the types of techniques that are used for analysing human error, with the hazard analysis techniques that are used for the design and evaluation of hardware and software. International system safety standards – such as in the defence, railways, and process industries – mandate or highly recommend formal (mathematical) modelling of safety-critical aspects of hardware and software functionality (Commonwealth of Australia, 1998; European Committee for Electrotechnical Standardisation, 1995; International Electrotechnical Commission, 1997). Formal models are used for safety assurance with software and hardware systems, because they are precise, systematic, reproducible, and auditable.

By contrast, the techniques currently used for modelling and analysing the safety of Human-Computer Interface (HCI) designs, and operator error rates, are informal. One of the most commonly used methods for safety analysis is Failure-Modes and Effects Analysis (FMEA). Two examples are Systematic Human Error Reduction and Prediction Approach (SHERPA) and Technique for

Human Error Rate Prediction (THERP; Kirwan, 1994). In such an FMEA, the designer inspects components of the system and identifies possible human failure modes and their potential effects using a 'checklist' of common human failure-modes (Hussey, 1998). These approaches require a subject matter expert to estimate the likelihood of different types of errors occurring. Such judgements are frequently difficult to make, and are inherently subjective. Empirical data regarding the frequency of different types of errors is often not available, or is difficult to collect, particularly for systems that are under development.

There are a number of reasons why formal models are not currently used for modelling the performance of human operators within safety-critical systems. These include:

- difficulty formally modelling the interaction between operators and the computer;
- lack of understanding of the psychological processes responsible for operator error;
- inability to formally specify the antecedent conditions that trigger those processes, and to estimate the resulting likelihood of errors; and
- lack of precise methods for determining system risk due to operator errors.

The aim of the current paper is to describe the first stages of the development of a new methodology for safety assurance. This method includes a formal model of operator performance, and incorporates this model into a formal model of the computer system. The method is designed to be used as a risk analysis tool, allowing the user to estimate the probability of operator errors under different operational scenarios, and evaluate the effect of those errors on the performance of the system as a whole.

The Safe HCI Methodology

The safety assurance methodology involves the following steps:

1. Definition of system concept and scope. This involves identifying the functions that the human-computer system performs, the range of conditions ('operational profiles') that it operates under, and possible system hazards.
2. Modelling of the safety-critical aspects of the system. This involves:
   a) A system risk model, to categorise system failure modes and identify the mechanisms that give rise to system failures and safety hazards and mishaps;
   b) Cognitive models of human error. These models identify the characteristics of the system's operational profile and HCI that enhance the likelihood of error; and

c) A model of human-computer interaction that incorporates the cognitive models of human error within a formal specification of the human-computer system. This model identifies the protective features of the HCI and ways of detecting whether interactions are diverging from safe operation of the system.

3. A series of experiments to collect data for calibrating the risk model and fine-tuning the cognitive models, by varying the operational profiles and measuring the rates of human error, and rates of system hazards.
4. A series of experiments to validate the model of human-computer interaction. This involves using the models to (a) predict error rates and system hazards under a new set of operational profiles, and a range of new HCI design configurations; and (b) empirically testing these hypotheses.

The output of the methodology is a 'formula' for calculating system safety risk under varying operational profiles, and recommendations for improving HCI design to reduce risk. One of the key advantages of this type of model is that it allows the analyst to simulate the performance of a system under a wide range of different operational profiles, and design configurations. Given the large variety of potential conditions that any moderately complex system can operate under, it is generally not possible to empirically evaluate the performance of the system under all conditions. Simulation, therefore, provides a powerful and cost-effective tool for risk analysis.

A prototype of this methodology has been developed at The University of Queensland, using a highly simplified air traffic control task. In the following sections, we illustrate the cognitive model that was developed for this task, show how this model can be used to analyse the safety hazards stemming from operator error within this system.

An ATC Case Study

The ATC system

The ATC task that we are using in this research program runs on a personal computer, and is simplified to the point that naïve participants can learn to perform the task with an adequate level of proficiency within a two hour experimental period. The task involves routing aircraft through a series of waypoints, and detecting and preventing conflicts by controlling the speed of the aircraft. The aircraft fly on a small number of fixed routes in straight lines in two dimensions, and the HCI functionality is very simple. Altitude is not represented in this task, and participants do not have control over the route of the aircraft. Participants have to ensure that no aircraft ever violate separation standards. A five nautical mile separation standard is used in this task. Participants are informed that their primary goal is to ensure that no aircraft pass within five nautical miles of each other. In reality, air traffic controllers are concerned with the orderliness and efficiency of traffic flow, as well as safety. We do not consider these outcomes in the current analysis.

A formal model of this human-computer system has been developed, representing the key functions that are performed by the system as a whole, the states that the system can enter, the nature of the human-computer interface, and the mechanisms by which the operator interacts with the system (see Hussey, Leadbetter, Lindsay, Neal & Humphreys, 2000).

The Cognitive Model

A simplified version of the cognitive model is shown below in Figure 25.1. Operator performance is modelled as a cyclic process, involving scanning for potential conflicts between aircraft, projecting potential conflicts forward in time to assess whether there will be a violation of separation, prioritising potential conflicts, making a decision, and performing the intended action.

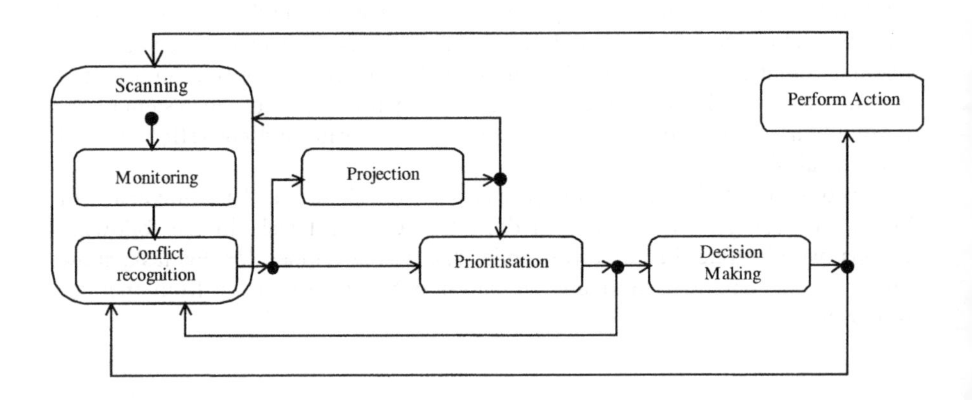

Figure 25.1 The ATC cognitive model

The scanning function involves two sub-processes: monitoring and conflict recognition. The participants are assumed to monitor the aircraft within their sector by considering the attributes of pairs of aircraft. These attributes can include the position, speed, and route of either or both aircraft. These attributes can cue conflict recognition in two ways: (a) by retrieving matching examples from memory, and (b) using rules or algorithms. Our participants, therefore, may recognise a conflict because it resembles previously seen conflicts, or because they have developed a set of rules that allow them to calculate whether a pair of aircraft will conflict.

If a potential conflict is recognised, then the participant may either project the event forward in time, or proceed direct to prioritisation. A potential conflict must be projected forward if the participant is unsure when the potential conflict will occur, or if the participant is unsure as to whether the potential conflict is simply a false alarm. By mentally projecting the event forward in time, the participant is able to provide a more accurate estimate as to when and where the conflict may occur. However, if the participant knows when and where the conflict will occur, or the event requires immediate action, the participant proceeds directly to prioritisation.

The prioritisation function assigns a priority to the conflict based on the time that is available for preventative action to be taken. The participant is then assumed to retrieve other current conflicts from short term memory, and to compare the priority of the currently attended conflict with the priorities of other conflicts. If the currently attended conflict has the highest priority, they continue to the decision process, otherwise they return to scanning in order to pick up the higher priority event.

There are two potential outcomes of the decision process: a decision is taken to change the speed of one of the aircraft immediately to prevent the conflict, or the decision is deferred until a later time. If the decision is deferred, then the participant returns to scanning. If the decision is taken, then the participant proceeds to perform the action. The actions are then taken through the HMI, and the participant returns to scanning.

The Error Model

The cognitive model provides a systematic basis for evaluating the potential operator errors within this simplified air traffic control task. Each component within the model is examined to assess the types of failure modes that are possible. Some of the failure modes associated with these components are shown in Figure 25.2, below.

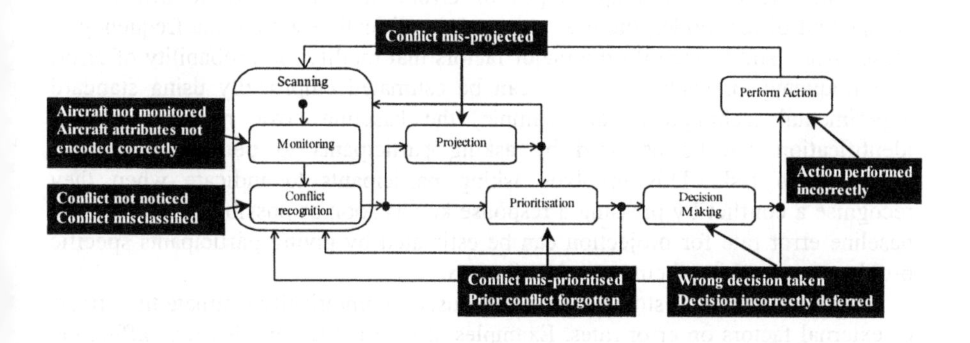

Figure 25.2 Potential failure modes for each component of the cognitive model

The principle types of errors associated with monitoring are likely to be a failure to monitor aircraft that are in potential conflict, and a failure to encode the attributes of the aircraft correctly. These types of errors are known to be a major problem for novice air traffic controllers, who are prone to so-called 'tunnel vision'. Tunnel vision occurs when controllers focus on one highly demanding problem, and fail to systematically scan for other potential problems. These types of errors may also occur under low workload conditions, if the participant has previously attended to the aircraft, but not noticed a conflict.

Conflict recognition can fail in at least two ways: the conflict may not be noticed, or it may be misclassified. These errors can be caused by failures in monitoring, but can also occur even when the participant is actually attending to the aircraft in conflict. For example, a participant may attend to two aircraft that are in potential conflict, but he or she may have seen a number of examples of aircraft in similar circumstances that were not in conflict. In this case, the prior examples in memory may cause the participant to misclassify the event as a 'near miss'.

The principle error associated with projection is mis-estimation. If a participant projects a pair of aircraft forward in time, they may estimate the separation between the aircraft incorrectly, or they may estimate the time at which the aircraft pass a specific point (e.g., the point of minimum separation) incorrectly.

Participants can make errors in prioritisation by assigning the wrong priority to the currently attended conflict, or by failing to retrieve one or more of the other current conflicts that are stored in short term memory. The result of this is that the participant either continues to work on the current conflict, when there is a more urgent conflict that needs attending to, or the participant switches to another conflict which should have lower priority.

The major errors associated with the later stages of the model include making the wrong decision (e.g., selecting the wrong speed), incorrectly deferring the decision (e.g., because the estimated time until separation is violated is incorrect), and incorrectly performing the action (e.g., by selecting the wrong aircraft with the mouse).

Having defined the major types of errors that are likely to affect each component of the model, the next step is to empirically estimate the frequency of these errors, and to identify the major factors that modify the probability of error. The frequency of each error type can be estimated empirically using standard experimental techniques. For example, the baseline error rate for conflict identification can be assessed by asking participants to perform a conflict identification task. This involves asking participants to indicate when they recognise a conflict by pressing a response key as soon as possible. Similarly, the baseline error rate for projection can be estimated by giving participants specific problems, and asking them to project forward.

These experimental studies can also be used to empirically estimate the effects of external factors on error rates. Examples of external factors that may affect the probability of error include:

- expertise;
- workload;
- memory load;
- fatigue; and

- HMI design.

HCI Redesign

Once developed, the error model can be used to systematically evaluate alternate HCI design options. By integrating the error model into the formal specification of the human-computer system, it is possible to identify the operator error modes that pose the greatest hazard to the safe operation of the system, and the operational conditions under which these errors are most likely to occur. Specific HCI designs can then be developed to address these problems, and the redesigns can be run through the model. For example, our preliminary testing of the ATC task suggests that conflict recognition represents a significant source of error for our participants. There are a range of design options that could be developed to address this problem. These include conflict detection probes that automatically alert participants to potential conflicts, and flight projection tools that allow participants to accurately estimate the time at which aircraft pass specific points. Each design option is evaluated by considering the effect that it should have on each error mode, and re-running the model to see if it significantly enhances the predicted safety score for the system (see Figure 25.3).

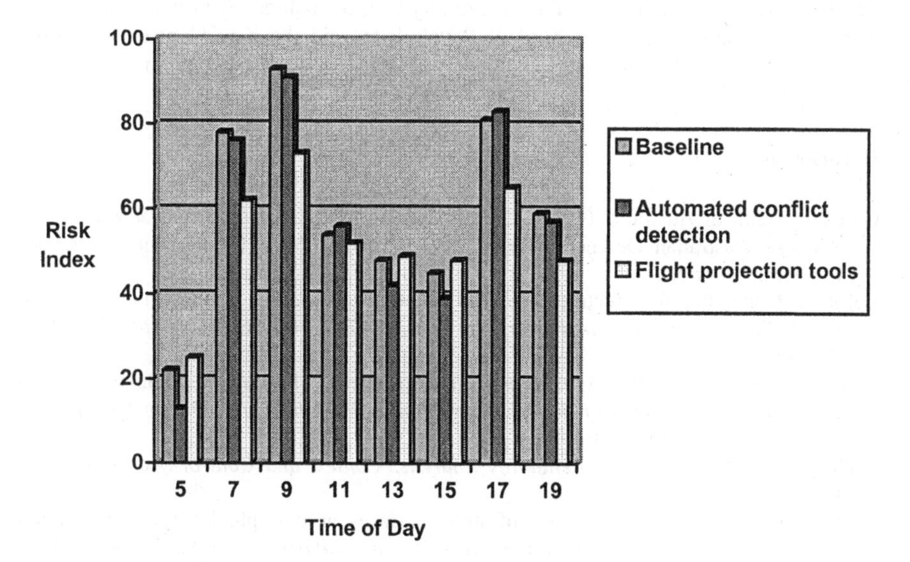

Figure 25.3 Illustrative example of a risk analysis for alternate HCI designs

Conclusion

In summary, the current paper has illustrated the ways in which a model of operator cognition can be used for evaluating the potential for human error within human-computer systems. This approach is promising, because it allows the user to simulate the operation of the system as a whole under a wide range of different conditions. This technique can be expanded to incorporate cost models in order to provide a comprehensive evaluation of the cost effectiveness of a range of interventions designed to enhance safety. In this manner, it is possible to compare the cost effectiveness of options, such as redesigning the HCI, changing staffing levels, or providing more training.

Acknowledgement

Michael Humphreys is Professor of Psychology, and the Director of the Key Centre for Human Factors and Applied Cognitive Psychology at The University of Queensland. Shayne Loft is a PhD candidate in the Key Centre for Human Factors and Applied Cognitive Psychology at The University of Queensland. Peter Lindsay is a Boeing Professor of Systems Engineering at The University of Queensland. Simon Connelly is a Research Officer in the School of Information Technology and Electrical Engineering at The University of Queensland. Andrew Neal holds a joint appointment with the School of Psychology and the Key Centre for Human Factors and Applied Cognitive Psychology at The University of Queensland.

References

Commonwealth of Australia. (1998). *The Procurement of Computer-based Safety Critical Systems* (Australian Defence Standard DEF [AUST] 5679). Department of Defence, Canberra, Australia.

European Committee for Electrotechnical Standardization. *Railway Applications: Software for Railway Control and Protection Systems* (European Standard prEN 50128). CENELEC.

Finklestein, A., & Dowell, J. (1996). A comedy of errors: The London ambulance service case study. *Proceedings of the 8th International Workshop on Software Specification and Design* (pp. 2-4). IEEE.

Hollnagel, E. (1993). *Human Reliability Analysis: Context and Control.* Academic Press Limited.

Hussey, A. (1998). Safety analysis of user-interfaces at multiple levels of interaction. *Proceedings of the 3rd Australian Workshop on Industrial Experience with Safety Critical Systems and Software* (pp. 41-57). Australian Computer Society.

Hussey, A., Leadbetter, D., Lindsay, P., Neal, A., & Humphreys, M. (2000). *A Method for Analyzing Hazards and Error Rates Related to Operator Activities* (Tech. Rep. No. 00-25). Software Verification Research Centre.

International Electrotechnical Commission. *Functional safety of electrical/electronic/ programmable electronic safety-related systems* (61508). IEC.

Kirwan, B. (1994). *A Guide to Practical Human Reliability Analysis.* Taylor and Francis.

Chapter 26

General Aviation Pilot Attitudes Towards GPS Use: Operational Implications

Michael D. Nendick, *The University of Newcastle, Australia*
Ross St. George, *Civil Aviation Authority, New Zealand*
Kurt M. Joseph, *SBC Technology Resources Inc., USA*
and
Kevin Williams, *FAA Civil Aeromedical Institute, USA*

Introduction

Surveys of General Aviation (GA) pilot attitudes toward Global Positioning System (GPS) receivers have been conducted over the last five years in Australia, New Zealand (NZ) and the United States (US; Nendick, 1994, 1997, 1998, 1999). Comparisons of the data indicate that there are commonalities between the way pilots view and use GPS receivers, as an example of a high-technology add-on, in a low technology GA environment. The converging evidence from these similar, but distinct, aviation cultures implies that some flight-safety issues associated with GPS receivers used by GA pilots can be addressed in a generic sense. Other issues appear to be related to the context, whether that is cultural or operational, such as regional regulatory differences or operator qualifications and experience. These issues may be best addressed on a specific basis.

Mitigation of some of these flight safety issues includes a review of the equipment design for new receiver models incorporating colour moving map displays, refinement of controls, and reduction in the complexity of GPS receiver operating logic (Heron, Krolak, & Coyle, 1997; Williams, 1998, 1999). Much of this approach is in the domain of the commercial manufacturers and is out of the immediate influence of the operators, Human Factors specialists, and indeed the regulatory authorities of the regions tasked with ensuring aviation safety.

An alternative approach for both design and operational issues is through improved education and training processes. The development of pilot licence syllabi and related training courses targeting specific critical competencies with regard to the human performance issues identified by the surveys is one avenue to be explored. Generic education initiatives such as flight safety seminars conducted by the regulatory authorities are another approach to raise pilot awareness of the Human Factors issues relating to the inappropriate operation of GPS.

An initial survey administered to 227 NZ pilots using GPS in 1994 indicated various issues of equipment design and use relating to the early adoption of the relatively new GPS technology for GA use under both Visual Flight Rules (VFR) and Instrument Flight Rules (IFR; Nendick, 1994). Modified versions of this survey were later administered as a follow up in the US, Australia, and again in NZ five years after the first survey (Nendick, 1997, 1998, 1999).

While some training syllabus guidelines have been published for IFR use of GPS in Australia, Canada, and NZ in the interim period as a result of GPS being an approved IFR navigation method, there are still no guidance or training requirements relating to the VFR use of GPS in these countries. While commercial organisations must conduct GPS IFR training in the US this is largely unspecified. There are no specific training, testing and recency requirements for GA IFR use of GPS in the US. The Practical Test Standards for an Instrument Rating in the US, requires that students demonstrate two kinds of non-precision approach, which may include a GPS approach only if the aircraft is equipped with a certified GPS receiver. Thus instrument rated pilots can legally fly IFR GPS approaches without necessarily being trained and tested on them. At the other extreme, GPS model specific regulations for initial endorsement and recency requirements that include ground courses and flight tests are in place in Australia and New Zealand for GPS IFR operations.

The previous survey research has shown general aviation VFR pilots typically learn to use GPS navigation as best they can 'on the job'. This can give rise to some hazardous attitudes and behaviours as they go about their flying business. Already accidents in the GA VFR environment have occurred over this period that implicate inappropriate use of GPS as a contributing factor to the accident (Heron & Nendick, 1999). For example, on May 30, 1996 in Quebec, Canada, a Piper Navajo PA-31 flight being conducted under VFR flew into terrain in adverse weather. The co-ordinates for the destination were entered incorrectly into the GPS. All indications are that the pilot did not know his actual position in relation to the crater (the destination) and evidently arrived at the crater one minute sooner than anticipated. It was determined that the pilot continued flight into reducing visibility and may have lost situation awareness due to incorrect information provided by the GPS (TSB, 1996). Research indicates that pilots using GPS rate 'continuing to the destination' as a less risky decision compared to pilots using traditional radio navigation aids (Henry, Young & Dismukes, 1999; Hughes & Nendick, 1997).

IFR flights are not immune to inappropriate use. On July 20, 1998 at Wagga Wagga, Australia, a Partenavia P68B was involved in a controlled flight into terrain (CFIT) accident in adverse weather on approach under IFR. The aircraft was equipped with an IFR certified GPS receiver and the pilot broadcast his intention to conduct the Albury-Wagga Wagga GPS Arrival procedure. The last altitude recorded on the GPS was 1,274 feet where the aircraft should have been flying not below 2,000 feet. No record was found of completion of the required ground course for GPS training or formal training on the GPS equipment installed in the aircraft. The GPS did not have a current data card and so was not approved for IFR use at the time (BASI, 1998).

Equally, information from aviation incidents and investigations has suggested that dependency or errors in GPS use are contributing factors for airspace incursions and navigation errors (St. George & Nendick, 1997). Anecdotal evidence suggests that 'GPS as sole-means navigation' has been occurring. One New Zealand incident involved a pilot arguing with a radar controller who advised the pilot that he was not heading for his destination (the aircraft was outside of the controlled environment – the assistance was just a helpful gesture). The aircraft flew to the incorrect airfield and then had to divert. The pilot was adamant that 'someone else' must have mis-set the GPS. The answer to the navigation error lay outside with the quite different terrain features, the absence of a plotted heading and standard VFR navigation procedures. (Nendick & St. George, 1995).

This paper reviews some findings from the international survey data, with emphasis on GA pilot attitudes to GPS training and use. The paper highlights generic or specific training strategies to minimise potential flight-safety implications as the use of satellite navigation becomes more pervasive.

Method

The 163-item GPS User Survey (Nendick, 1997, 1998, 1999) was administered to General Aviation pilots in the US, Australia and New Zealand over a three year period from 1997 to 1999. The survey included items examining the topic areas of: Receiver controls and displays, Operating logic, Receiver functions, Receiver operations, Operating procedures, Navigation performance, Pilot attitudes, and Training. Seven-point Likert-type rating scales and additional open-ended questions were utilised to measure pilots' responses to the survey questions. Pilot demographic data was also obtained to determine sample representativeness with the GA pilot population.

Two Federal Aviation Authority (FAA) regions were sampled in the US. Eighty-eight pilots from the Southwest (ASW) region completed the 35-minute survey at the Civil Aeromedical Institute (CAMI) in December 1997. Two hundred and twenty pilots from the Alaskan region (AAL) responded to a mail-out to 1,880 pilots in April 1998, a response rate of approximately 12 percent. Valid responses were received from 318 Australian pilots in October 1998 following a mail-out to 3,000 Departure and Approach Procedures (DAP) subscription holders through AirServices Australia, a response rate of approximately 11 percent. Valid responses were received from 88 NZ pilots in July 1999 following a mail-out through the NZ Civil Aviation Authority to 1,500 pilots, a response rate of approximately 6 percent. The relatively low response rates to the mail-out surveys are not ideal but are consistent with similar studies in the literature (e.g., see Driskill, Weissmuller, Quebe, Hand, & Hunter, 1997). Factors that may have contributed to the low rates of valid returns included a minority of returns with spoiled responses that did not include critical information and so were discarded, and inclusion with publication mail-outs from the regulatory authorities in Australia and NZ, which may have reduced the incentive for some pilots to participate. The low NZ response may also have been because it was a 'repeat' survey that was mailed to a random sample of 1,500 pilots across all licences, possibly missing the bulk of the GA GPS owner/users. The survey data from each response sample was combined into one data file for analysis.

Results

Pilot Demographics

The sampled groups from each region: American Alaska (AAL), American Southwest (ASW), Australia (AUS), and New Zealand (NZ) had distinctive demographic differences as shown in Tables 26.1 to 26.7. Overall the sample appears relevant to the general pilot population. Gender was 96 percent male that equates to the industry norm. AAL pilots were predominantly GPS owning, VFR,

Table 26.1 Total hours by region

Total hours	n	Mean	Median	SD	90th percentile	Range
AAL	214	1,883	988	2,919	4,200	90-20,000
ASW	88	1,999	957	2,887	4,000	30-16,000
AUS	315	4,406	2,000	4,993	12,500	170-22,000
NZ	83	4,078	1,400	6,257	14,000	80-30,000
All	700	3,293	1,400	4,570	10,000	30-30,000

Private Pilot Licence (PPL), single engine pilots with relatively low instrument time but high GPS hours. Their age was similar to both AUS and NZ. ASW pilots had similar total hours to the AAL group with higher instrument time and low GPS

Table 26.2 Instrument hours by region

Instrument hours	n	Mean	Median	SD	90th percentile	Range
AAL	188	80	10	202	200	0-1,500
ASW	86	387	85	1,180	800	0-10,000
AUS	308	673	210	1,464	1,500	0-12,000
NZ	80	577	55	1,456	1,275	0-8,000
All	662	456	92	1226	1,000	0-12,000

hours. They were the youngest group with an equal spread of PPL and Commercial Pilot Licence (CPL) holders, plus approximately 20 percent with Air Transport Pilot Licence (ATPL). There were a greater number of instrument and multi-engine ratings, and instructors, and they were least likely to own the GPS they were rating.

Australian and New Zealand pilots had the greatest total hours, and a similar spread between PPL, CPL, and ATPL holders to ASW pilots. There was a similar

Table 26.3 GPS hours by region

GPS hours	n	Mean	Median	SD	90th percentile	Range
AAL	211	414	240	620	1,000	0-6,000
ASW	87	212	40	450	600	1-2,500
AUS	295	730	400	925	1,600	0-5,000
NZ	81	637	200	992	2,000	0-6,000
All	674	553	250	820	1,500	0-6,000

GPS ownership level of around 60 percent. Australians were most likely to have an instrument and multi-engine rating. All groups had significant outliers with very high hours that skewed the means considerably in comparison to the medians. Small n variations here are the result of missing pilot demographic data.

Table 26.4 Flight ratings by region

Ratings %	n	Instrument	Instructor	Multi-eng
AAL	220	21%	2%	8%
ASW	88	65%	26%	47%
AUS	318	84%	25%	68%
NZ	88	40%	40%	43%
All	714	57%	20%	44%

Table 26.5 Age by region

Age	n	Mean	Median	SD	Range
AAL	220	49	49	10	24-87
ASW	88	41	40	14	18-70
AUS	318	45	46	13	18-75
NZ	88	47	50	13	21-74
All	714	46	47	12	18-87

Table 26.6 Licence by region

Licence %	n	PPL	CPL	ATPL
AAL	220	90%	2%	5%
ASW	88	38%	38%	20%
AUS	318	36%	36%	26%
NZ	88	45%	38%	17%
All	714	54%	26%	18%

Table 26.7 GPS ownership by region

Owner	n	Own GPS
AAL	220	91%
ASW	88	42%
AUS	318	57%
NZ	88	63%
All	714	66%

Factor Analysis

Underlying factor structures were extracted for pilot ratings within each of the survey topic areas by performing separate exploratory principal components factor analyses utilising a varimax rotation to obtain factor solutions. For this paper only the 'GPS Attitudes' and 'Training' topic areas will be considered. Responses relating to GA use only were included in the analysis.

An analysis of variance (ANOVA) was performed on each of the mean factor scores to determine differences in responses as a function of region (AAL, ASW, AUS, NZ).

Table 26.8 Results of the factor analysis: GPS attitudes (n=271)

Item/Factor	1	2	3	4	5	Communality
Confidence VFR with GPS	.594					.552
Confidence IFR with GPS	.520					.540
Confidence VFR without GPS	.755					.676
Confidence IFR without GPS	.786					.693
Reliance VFR		.781				.683
Reliance IFR		.835				.727
Confidence accuracy & reliability		.521				.586
Confidence basic functions			.838			.770
Confidence all functions			.772			.650
Complacency				.752		.720
Distraction				.735		.710
Current database					.784	.646
Database affordability					.737	.558
Eigenvalue	3.05	2.03	1.41	1.28	1.12	
Percent of Total Variance	21.8	14.5	10.0	9.1	8.0	63.5

Note: Extraction Method: Principal Component Analysis Rotation Method: Varimax with Kaiser Normalisation

GPS Attitudes

A total of 271 respondents completed all of the items relating to the GPS attitudes section of the survey. Five factors emerged from a factor analysis of the combined data for survey items relating to this topic area (see Table 26.8).

1. Confidence in navigation ability (VFR/IFR with/without GPS).
2. Dependence (reliance VFR/IFR and confidence in accuracy and reliability).
3. User Confidence (competency using basic and complex GPS functions).
4. Complacency/distraction.
5. Use of database (currency and affordability).

The factors are similar to the findings of Joseph, Jahns, Nendick, and St. George (1999). However complacency and distraction have moved from the dependence factor to a separate factor this time.

Cronbach's alpha coefficients of reliability were conducted for the items within each factor as estimates of internal consistency. For the 'GPS Attitudes' topic the standardised Alpha values were not high, with missing data from 'Not Applicable' question responses potentially reducing the overall results. Factor 1 (.65), Factor 2 (.58), Factor 3 (.72), Factor 4 (.37), Factor 5 (.40).

Table 26.9 Attitude factor 2 means for groups in homogeneous subsets

		Mean subsets for alpha = .05	
Region	**N**	**1**	**2**
NZ	38	4.5	
AUS	239	4.6	
AAL	189	5.9	
ASW	86	5.9	
Significance		**.723**	**.995**

Table 26.10 Attitude factor 3 means for groups in homogeneous subsets

		Mean subsets for alpha = .05	
Region	**N**	**1**	**2**
ASW	62	3.8	
AAL	47	4.1	
NZ	88		5.3
AUS	312		5.6
Significance		**.454**	**.621**

ANOVAs for the main effect of Region showed statistically significant differences at the $p<0.05$ level for the first three factors. A post-hoc comparison was performed using a Tukey's Honestly Significant Differences (HSD) comparison to explore these differences. There was a clear split between the US and Australasian regions for Factors 2 (Dependence), and 3 (User Confidence; see Tables 26.9 and 26.10) which are focussed on here.

Training

A total of 650 respondents completed all of the items relating to the GPS training section of the survey. Two factors emerged from the combined factor analysis of survey items relating to this topic area (see Table 26.11), consistent with the findings of Joseph et al. (1999).

1. Required knowledge and training (technical training, Human Factors training and GPS, knowledge needed).
2. Attained knowledge and training (user knowledge, training received).

Cronbach's alpha coefficients of reliability were conducted for the items within each factor as estimates of internal consistency. For the 'Training' topic the user manual item was removed leaving 6 items loading onto 2 factors, with a combined standardised Alpha (.71).

Table 26.11 Results of the factor analysis: Training (n=650)

Item/Factor	1	2	Communality
GPS knowledge needed	.830		.699
Tech training needed	.868		.764
HF training needed	.813		.674
Your knowledge		.883	.782
Training received		.769	.678
Eigenvalue	2.4	1.2	
Percent of Total Variance	47.9%	24.0%	71.9%

Note: Extraction Method: Principal Component Analysis Rotation Method: Varimax with Kaiser Normalisation

ANOVAs for the main effect of Region showed statistically significant differences at the $p<0.05$ level for both factors. A post-hoc comparison was performed using Tukey's HSD. There was a clear split between Australian pilots who scored significantly higher for this topic area compared to the other regions on both factors, 'Required knowledge and training', versus 'Attained knowledge and training' (see Tables 26.12 and 26.13).

Table 26.12 Training factor 1 means for groups in homogeneous subsets

Region	N	Mean subsets for alpha = .05	
		1	2
AAL	195	4.1	
NZ	88	4.1	
ASW	88	4.2	
AUS	302		4.6
Significance		**.802**	**1.000**

Table 26.13 Training factor 2 means for groups in homogeneous subsets

Region	N	Mean subsets for alpha = .05	
		1	2
NZ	88	3.7	
ASW	86	3.8	
AAL	209	3.9	
AUS	301		4.8
Significance		**.738**	**1.000**

Discussion

Four different regions were surveyed on pilot attitudes to using GPS, Alaska (AAL), American Southwest (ASW), Australia (AUS), and New Zealand (NZ). There are distinct cultural, contextual, and demographic similarities and differences between the regions and the pilot samples in each region. Pilots responding to the survey had some interesting regional differences. One contextual reason for these differences could be the availability of GPS education and the regulation of GPS training for pilots.

All regions recommend pilots educate themselves on the appropriate use of GPS, and offer some degree of educational material and guidance on a national level. It appears this is most comprehensive in Australia where Safety Seminars incorporating GPS information are periodically run throughout the main centres by the regulator. Australia has also produced educational material in the form of video, electronic, and hardcopy media available for general purchase. New Zealand is active with regulator sponsored educational seminars and flight safety magazine articles.

Generally VFR pilots are left to their own devices and there is no related theory syllabus or assessment requirement for GPS technical or Human Factors knowledge. For IFR operations regulations are in force. In Australia, pilots are required to complete a ground theory course and flight test for a specific GPS model, and to meet recency requirements to use GPS for primary means IFR navigation (St. George & Nendick, 2000). New Zealand pilots require a flight test certification using a specific GPS model to carry out an instrument approach procedure under IFR and to meet recency requirements (CAA, 1999). No specific ground training or flight test requirements are required in the US to use GPS for IFR operations.

Pilot Responses

From the factor analysis and ANOVAs for GPS Attitudes, US pilots responded significantly lower than Australasian pilots for GPS reliance and user confidence. The Australasians were the most confident in navigating and using GPS and the least dependent on it. However a possible confound in the current analysis may be as a consequence of missing data reducing the GPS Attitude sample size. Curiously, other sections of the survey were not as dramatically affected by missing data.

For Training Attitudes the Australians, who have a regulated exposure to educational and training processes in the IFR environment, responded significantly higher. They felt that they had more training and knowledge, and that greater knowledge and training is required to safely operate GPS. Interestingly in this analysis, the responses from the New Zealand sample, where IFR use of GPS is very similar to Australia, were grouped with the US samples, where there are far less directly regulated IFR GPS training requirements. Of course none of the regions surveyed mandate training for VFR GPS use.

Conclusion

Not withstanding that further analyses can be undertaken with this data set, first impressions suggest that less exposure to education and training is likely to be linked with greater dependence and lower confidence in pilots' ability to competently use GPS. Both these attitudes have been suggested to be causal factors in aircraft accidents and incidents. As in other areas of aviation there is a case for extending GPS education and competency-based training.

Civil aviation authorities in the dual function of regulation and safety promotion should actively pursue the development of GPS navigation education and regulation of GPS use. We believe that this involves the targeted development of coherent competency based education, assessment, and licensing programs relating to GPS application in aviation.

Acknowledgement and Disclaimer

The opinions expressed in this paper are those of the authors only, and do not represent policy of any regulatory body or authority. This research was supported in part by The University of Newcastle RMC Grant 45/299/643. The authors gratefully acknowledge the voluntary participation of the 714 pilots who completed the survey forms utilised in this paper. Thanks also to OMNI Corporation, the FAA Office of Aviation Medicine, the FAA Alaskan Region Flight 2000 Program, AirServices Australia, and New Zealand Civil Aviation Authority, for their support during survey distribution. Thanks also to Allen Lansdowne of The University of Newcastle for assistance with some of the initial statistical analysis.

References

BASI. (1998). *Air Safety Occurrence Report* (Report No. 199802757). Bureau of Air Safety Investigation. Retrieved on October 1, 2000 from http://www.basi.gov.au/occurs/asor9802757.htm

CAA. (1999). *Civil Aviation Rules: Part 19.* Wellington, NZ: Author.

Driskill, W. E., Weissmuller, J. J., Quebe, J. C., Hand, D. K., & Hunter, D. (1997). *The Use of Weather Information in Aeronautical Decision-making: II* (NTIS # DOT/FAA/AM-97/ 23). US Department of Transportation, Federal Aviation Administration, Office of Aviation Medicine, Washington, DC.

Henry, W. L., Young, G. E., & Dismukes, R. K. (1999). The influence of Global Positioning System technology on general aviation pilots' perception of risk during in-flight decision making. *Proceedings of the Tenth International Symposium on Aviation Psychology* (CD version). Columbus, OH.

Heron, R. M., Krolak, W., & Coyle, S. (1997). *A Human Factors Approach to the Use of GPS Receivers.* Vancouver, BC. Heron Ergonomics, Inc.

Heron, R. M., & Nendick, M. D. (1999). Lost in space: Warning, warning, satellite navigation. In D. O'Hare (Ed.), *Human Factors in General Aviation* (pp. 193-224). Aldershot, UK: Ashgate.

Hughes, K., & Nendick, M. (1997). Pre-flight risk taking with Visual Flight Rules (VFR) pilots using a Global Positioning System (GPS) receiver: An exploratory study. *Proceedings of the Third Australian Symposium on Satellite Navigation Technology* (pp. 150-157), Sydney, Australia.

Joseph, K. M., Jahns, D. W., Nendick, M. D., & St. George, R. (1999, October/November). An international usability survey of GPS avionics equipment. *Proceedings of the IEEE/AIAA 17th Digital Avionics Systems Conference*, Bellevue, WA, USA.

Nendick, M. (1994). *GPS Survey 1994.* Massey University, New Zealand.

Nendick, M. (1997). *GPS Survey 1997.* The University of Newcastle: Australia.

Nendick, M. (1998). *GPS Survey 1998.* The University of Newcastle: Australia.

Nendick, M. (1998). *GPS Survey 1999.* The University of Newcastle: Australia.

Nendick, M., & St. George, R. (1995). Human factors aspects of Global Positioning Systems (GPS) equipment: A study with New Zealand pilots. *Proceedings of the Eighth International Symposium on Aviation Psychology* (pp. 152-157). Columbus, OH.

St. George, R., & Nendick, M. (1997). GPS = 'Got Position Sussed': Some challenges for engineering and cognitive psychology in the general aviation environment. In D. Harris (Ed.), *Engineering Psychology and Cognitive Ergonomics: Vol. 1* (pp. 81-92). Aldershot: Ashgate.

St. George, R., & Nendick, M. (2000). GPS training for general aviation VFR pilots: To regulate or educate? In B. J. Hayward & A. R. Lowe (Eds.), *Aviation Resource Management: Vol. 2* (pp. 107-114). Aldershot, UK: Ashgate.

TSB. (1996). *Transportation Safety Board of Canada* (Aviation Occurrence Report A96Q0076). Retrieved on October 1, 2000 from http://www.tsb.gc.ca/eng/reports/air/1996/ea96q0076.html

Williams, K. W. (1998, October/November). GPS user-interface design problems. *Proceedings of the IEEE/AIAA 17th Digital Avionics Systems Conference*, Bellevue, WA, USA.

Williams, K. W. (1999). GPS user-interface design problems: II. *Proceedings of the Tenth International Symposium on Aviation Psychology* (CD version). Columbus, OH.

Chapter 27

Protecting the ATM System from Human Error: The JANUS Approach

Anne R. Isaac
Eurocontrol, Brussels, Belgium

Introduction

The Air Traffic Management (ATM) system in Europe is inherently safe, but the demands from the aviation industry will inevitably require changes in the way ATM conducts its business. These changes are already impacting on the operations environment in which procedures are changing to increase traffic. In harmony with these procedural changes are the increases in the technology to 'assist' the controllers. These advances in technology and procedures are not new in aviation, and much has been learned from the introduction of new technologies on the flight deck (Billings, 1997; O'Leary, 1999).

What is also well known is the fact that humans within an increasingly complex technological environment display an ever increasing number of errors, some of which are unpredictable. Past analyses of these human error contributions to incidents or occurrences have often been clear in their identification of 'what' went wrong but not so precise about 'why' it happened.

This paper deals not only with an approach to ATM incident analysis that is being developed for the European ATM system, but also how human error can be predicted and managed. This approach is known as the Human error in ATM (HERA) technique. These developments are intended to ensure that human error will continue to be understood and managed, both retrospectively and prospectively in the future ATM environment: The JANUS approach.

Background: Soft Systems

The majority of accidents in hazardous activities are caused by human error. This problem is not new, and a good deal of research, application, and development of practical techniques for the analysis, prediction, and reduction of human errors, or their negative effects, has occurred in a range of industries (Swain & Guttmann, 1983; Embrey et al., 1994; Reason, 1990, 1997; Helmreich & Merritt, 1998). The research mentioned has reported that human error contributions in nuclear power production is approximately 70-90 percent and in Medicine, 98 percent. Whilst human error within flight operations has for some time been the centre of exhaustive research and debate, a similar analysis within the field of Air Traffic Management (ATM) is not so comprehensive. As with other industries it is found that the human error contribution in ATM is in the order of 90 percent or more (Kinney, et al., 1977; FAA, 1990). However, it should be noted that the sole contribution by ATM to air accidents is very small and that the ATM environment poses some rather different problems, which characterise it from other environments. These problems include the highly dynamic and time critical elements that are considerations in most tightly coupled, low risk environments.

ATM is currently under pressure as traffic levels increase. Airspace in many parts of Europe is already complex and congested, and there is also pressure from the airlines, which are under strong competitive commercial constraints, to optimise routes and timings. These issues lead to complexity and time pressure on ATM operations that can subsequently lead to errors. Additionally, many ATM systems are currently being upgraded and developed into 'next generation' systems, which include computerised displays with new functionality, and computerised tools. There is also the prospect in the near future of the introduction of data-link technology, which will significantly impact upon the method of operation in ATM.

These major shifts in work practices will affect both controller and pilot performance, and new opportunities for error could arise, particularly in the 'transition period' during which new systems and practices are introduced. These developments suggest that the ATM system is at the beginning of a long period of significant change and evolution, a period that will possibly see increased error rates, and potentially new errors. This points to the need for the development of an approach to better understand errors, to monitor error trends, and to develop effective error reduction or tolerance strategies.

The majority of controller tasks are generally seen as cognitive, meaning a high reliance on mental processes. These cognitive skills include projection of aircraft movement in time and space, judgement, pattern recognition, maintenance of situation awareness, planning, rapid decision making, and rapid spoken communication. Such tasks are generally 'covert' and thus difficult to observe; thus, cognitive tasks must often be inferred from resulting behaviour. For example, whilst it is difficult to observe the cognitive processes involved in the judgement of separation, one can observe controller instructions, which attempt to maintain separation. This problem is the probable reason for the rather slow progress in error research in the ATM environment.

Development of the HERA System

The first objective of the HERA project was to identify a suitable and credible model of human performance in ATM, and from this model to derive a valid, robust, and usable technique for the identification of human error causes in this environment. The second objective was to use this system in the investigation of error management and prospective error issues within ATM. The development of HERA phase 1, occurred in the following five steps (see also Isaac et al., 2000).

1. Literature Review: academic and industrial research findings on human performance models and taxonomies of human error were reviewed. This review included models ranging from Reason's (1990) model of slips, lapses, and mistakes, to Rasmussen's (1981) skill, rule, and knowledge based behaviour framework, and Endsley's (1996) situation awareness approach. In total over 50 models and taxonomies were reviewed dating from 1948 to 1999.
2. Selection of the Chosen Conceptual framework: the literature review identified the most appropriate framework was from the human performance approach. Within this framework the information processing models from Marteniuk (1976) and Wickens (1992) were chosen. However there were some modifications made to more accurately reflect the controller's way of working.
3. Review of Current and Future ATM systems: a review of human performance aspects of ATM was conducted to ensure that all significant aspects of ATM activities were included. Different functional areas of ATC were considered, e.g., tower, approach, en-route, terminal area, and oceanic. This phase also considered future ATM systems, such as computerised conflict detection support tools, electronic flight strips, and data-link technology.
4. Adaptation of the Chosen Conceptual framework: the chosen model and framework was adapted in light of the ATM context. In addition the controller's mental 'picture', or representation of the ATM scenario, was added as this was seen as essential to the job of an air traffic controller, and the breakdown of the picture is often a precursor to incidents in ATM. In the resultant HERA model there was more focus on 'working memory' as a major resource for the controller's 'mental picture' and the controller's self-confidence, seen as necessary by controllers in order to control traffic effectively, particularly in a new technological environment.
5. Specification for the HERA Technique: also based on the literature review, two techniques were identified which would support the HERA approach. These techniques, were TRACE (Technique for the Retrospective Analysis of Cognitive Errors in ATM; Shorrock & Kirwan, 2000), which represented a flow-chart based taxonomy, and the Tryptich Pyramid model (Isaac, 1999). The Tryptich Pyramid model added the possibility to integrate the cognitive issues with the internal and external environmental factors that represent the contextual issues within the ATM system.

The literature review and the analysis of controller tasks led to the need for a number of high level 'cognitive domains', that were concordant with the information processing model, and allowed analysis of different types of failure at successively detailed levels. This hierarchical approach to error analysis was based partly on Rasmussen's Skill, Rule, Knowledge (SRK) classification system, and also followed other successful classification system developments (Taylor-Adams & Kirwan, 1995). The basic structure of these taxonomies reflect three types of understanding of the aetiology of an incident: what happened, how it happened (internally in terms of cognition), and why it happened (due to internal and/or external events, states, or pressures). The three perspectives in error analysis are given the titles of error type (ET), information processing level (IP: or internal mechanism of malfunction), and contextual conditions (CC: or environmental aspects). If all three are understood, then the reasons the error occurred will also be accurately understood, and hence effective error prevention can be developed. A final grouping of the error detail level, (ED) leant the HERA method its overall architecture and distinguished it from a simple classification list of error forms, into a true taxonomic system.

As has been mentioned, the notion of context was also incorporated into the HERA method. It was recognised from the outset that a contextual approach was needed, since errors do not occur in an abstract or generic environment (Dougherty, 1993). Therefore HERA identified aspects of the *controller task* (ATM function) or the behaviour at the time of the error (Fleishman & Quaintance, 1984), and the *equipment/information* being used (input devices, displays, etc.). These contextual classifications, including the Contextual Conditions themselves, should capture much of the actual context of the incident.

The HERA technique therefore resulted in seven main classifications, five coding the psychology of the error, and two coding the context.

The JANUS Approach – Looking Back: HERA Retrospective

The following section illustrates the HERA method, briefly analysing an actual incident, showing the analysis format, and giving extracts of the taxonomies that comprise the HERA technique.

Table 27.1 HERA incident analysis form

DETAILS OF INCIDENT			
Reference:	AIRPROX 24/96	**Date & Time:**	21/9/1996-1225 UTC
Country:	Europe		
Aircraft: position:	B767/B747 4nm South of BN	**Operators:**	Foreign Airlines
ALT/HT/FL:	FL310	**Airspace type:**	UAR – Class B
Reporter:	Upper Sector Controller		
Reported Separation:	1.3nm horizontal/300 feet vertical		
Recorded Separation:	1.1nm horizontal/400 feet vertical		
BRIEF DESCRIPTION OF INCIDENT			
A B747 was en route from Zurich to New York, cruising at FL310 on UAR UB4 via BN VOR to BP A B767 from Paris (Orly) to New York was routeing UB376, also via BN VOR at FL180. Both aircraft were under the control ETC			

The HERA technique firstly identifies, from the original event, the factual background data and also the sequence of events in a shortened format. This is recorded on a pre-defined form, as indicated below.

The HERA technique then guides the analyst through the detailed process of identifying each error and its causal routes. This process starts with the identification of the first erroneous act in the sequence. The analyst is asked to record how, if at all, the error was detected, how, if at all, the error was recovered and whether it could be considered (directly) causal, contributory (adding to the likelihood of an incident), compounding (making things worse), or non-contributory (adding nothing) to the overall sequence of events. The analyst then fills in the sections in the form which pertain to the task, the information and equipment which was associated with the event and then, in sequence, the Error Type (ETs), Error Detail (EDs), Error Mechanism (EMs), Information Processing level (IPs), and Contextual Conditions (CCs). The analyst is finally asked to record whether they had any assumptions during this analysis and if they wish to add any notes.

The example above had five error events, and although all errors were analysed, only the third identified error in the sequence will be described in detail with the HERA technique.

Table 27.2 Description of the error

DESCRIPTION OF ERROR # 3			
The Upper sector controller did not believe that the B767 was on frequency, because they did not remember its first call – hence, they did not take any action.			
How Detected:		N/A	
How Recovered:		Left avoiding action turn	
HERA CLASSIFICATIONS			
Task:		Control room communications	
Information/Equipment:		**Aircraft (on frequency)**	
ET:		**Omission**	
ED:		**Working Memory**	
EM:		**Forget previous actions**	
IP:		**Unknown**	
CC:		**Cross-cultural R/T difference, High/excessive R/T workload**	
Reporter's assumptions:	N/A	**Analyst's assumptions:**	N/A

The analyst identified Omission from the Quality and Selection section of the Error Type table, shown in Table 27.3.

Table 27.3 Error types

Omission	**A required action is not performed**
Action too much	Too much of the action is performed
Action too little	Too little action is performed
Action in wrong direction	An action is performed in the wrong direction
Wrong action on right object	The wrong action is performed on the correct object (e.g., aircraft)
Right action on wrong object	The correct action is performed but on the wrong object
Wrong action on wrong object	The wrong action is performed and on the wrong object
Extraneous act	An unnecessary act is performed

The analyst then identifies the Error Detail as associated with working memory.

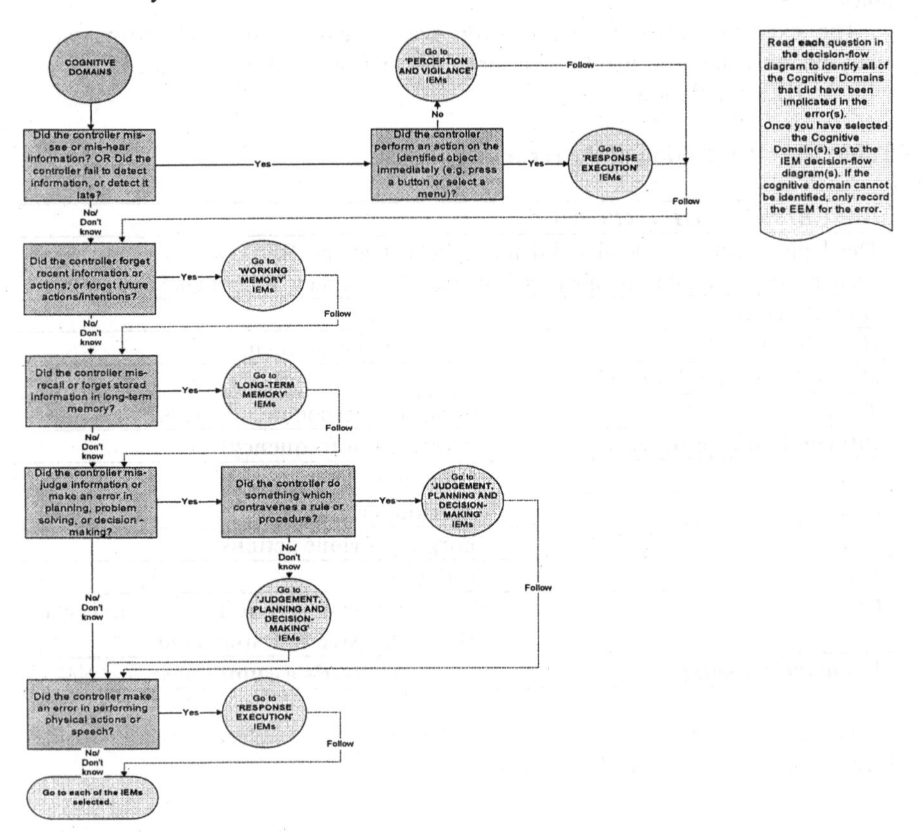

Figure 27.1 Error detail flow-chart

The analyst then uses the working memory flow-chart to identify the Error Mechanism (Figure 27.2) as associated with 'forget previous action'. The analyst fails to identify the Information Processing level. This can be due to the lack of evidence recorded in the original incident report.

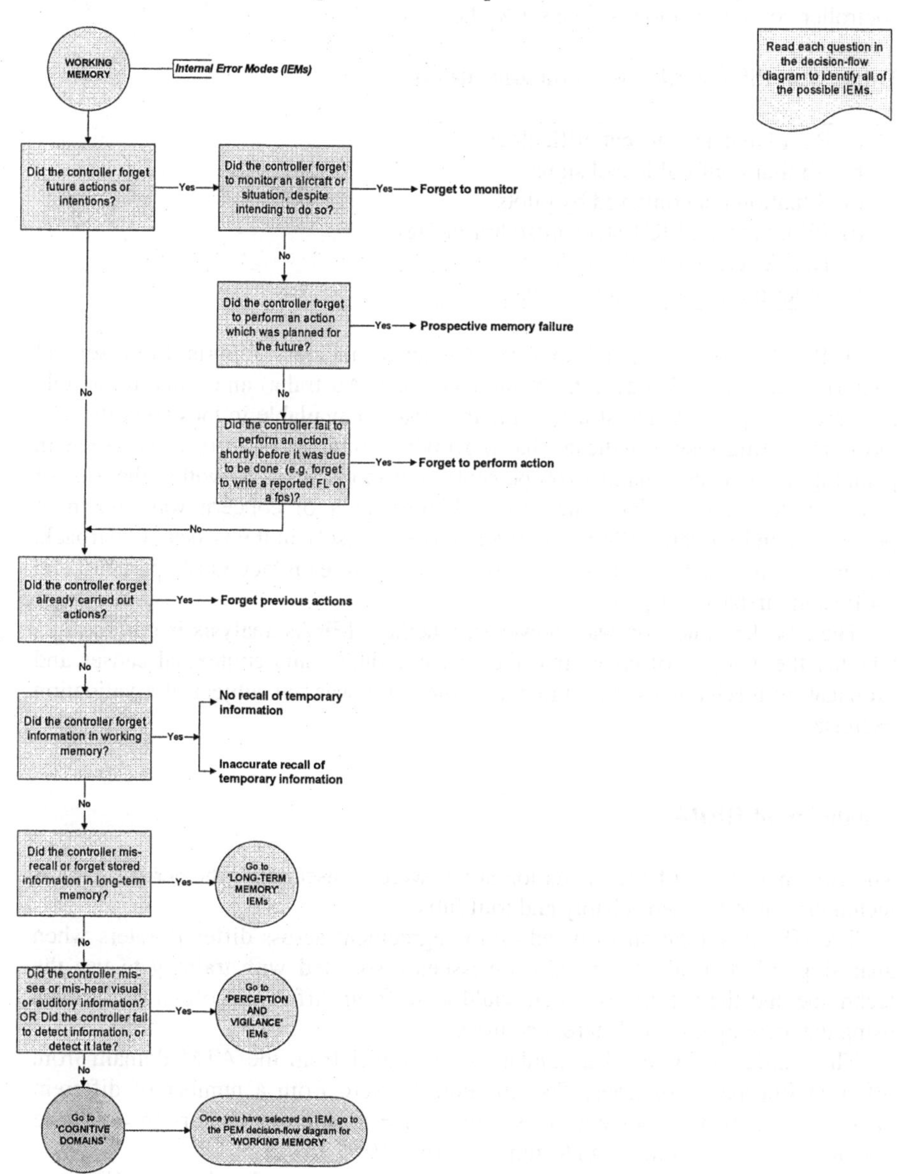

Figure 27.2 Error mechanism flow-chart

The analyst then identifies the associated contextual conditions (CCs) from the table which has nine sections associated with; communication with pilots, pilot actions, weather, personal and team issues, documentation and procedures, HMI, environment, and organisational issues. Only the section associated with pilot-controller communication will be shown here.

Table 27.4 Pilot–controller communications

a) Pilot language/ accent difficulties
b) Similar confusable call signs
c) Situation not conveyed by pilots
d) Pilot breach of R/T standards/phraseology
e) High/excessive R/T workload
f) Other pilot – controller problems

HERA has been applied to over 80 actual incident reports from several countries (e.g., UK, Sweden, the Netherlands, and Australia) and found to be able to interpret, represent, and structure the information available in these reports. The most interesting results indicate that the highest number of errors were seen in planning and decision making (60 percent), and more precisely within the area of incorrect decision or plan. The second highest area of concern was shown in perception and vigilance (26 percent) and more precisely in the issues of hearback. The main contextual conditions were associated with team factors (27 percent) and traffic and airspace (21 percent).

The next key question was, however, whether HERAs analysis is correct, i.e., whether the analysis of errors into their psychological and contextual causes and propagation mechanisms is accurate and true. This was the subject of a validation exercise.

Validation of HERA

The two main aims of the validation study were to ascertain whether the HERA technique could be used reliably and truthfully.

The HERA validation focussed on the agreement across different raters when analysing ATM incidents and also on issues associated with training to use the technique and the variations which could arise from different professional groups using the technique from different cultures.

The validation involved a number of personnel from the ATM domain from different European countries. The 26 subjects were from a number of different backgrounds, such as incident investigation (n = 8), Human Factors (n = 8), and operational or management ATM functions (n = 10).

The validation exercises occurred in three locations – Sweden, the UK, and Luxembourg. At each location the procedure and the protocols were identical. Subjects were given 1 1/2 days training which included practice with several incident analyses and in the remaining 1 1/2 days they analysed 8 test cases which were representative of incident reports in ATM. In total there were 26 error events to analyse and all reports and work was undertaken in English. All information was derived from real incident reports from several different countries.

Quantitative Results

The first validation issue concerned the question of whether HERA was used as intended. Table 27.5 shows agreement between the professional subjects. For the Information Processing and Error Mechanisms, most errors were classified into the two largest categories of Perception and Vigilance (PV) and Judgement, Planning, and Decision-making (JPDM). This table shows a range of agreement from 77 percent for Error Type to 46 percent for the Information Processing levels.

Table 27.5 Percentage agreement

Classification	Agreement
Task	63%
Information & Equipment	60%
Error Type	77%
Error Detail	73%
Error Mechanism /PV*	51%
Information Processing /PV*	46%
Error Mechanism /JPDM*	51%
Information Processing /JPDM*	46%
Contextual Conditions	60%

*PV = Perception & Vigilance; JPDM = Judgement, Planning & Decision-making

The principal objective measure of agreement applied in this study was the degree of agreement above chance – that is, a measure of agreement from which is subtracted the amount of agreement that would be obtained by chance alone.

The main question about the reliability of HERA concerns the ability of the subjects to attain a level of agreement of HERA classifications above chance which is above the conventional threshold of 40 percent (i.e., 40 percent above the chance level). The average agreement above chance for the error taxonomies was a kappa value of 0.46 which was acceptable.

Reliability, Training, and Nationality In terms of the different analyst groups, investigators had the best reliability, followed by Human Factors professionals, followed by other ATM personnel.

In terms of training for the HERA system, it appeared that those who were more experienced did perform better than those with the minimum training on HERA. In phase 2 of the HERA program of work, the appropriate training period will be further determined, but results here suggest that one or two day's training is not enough, and probably up to 5 days is desirable.

The validation was not able to statistically measure the influence of nationality or cultural differences.

In summary, HERA has shown a reasonable degree of validity, based on a range of different analysts and training levels, and therefore the system was thought to be worth pursuing and developing as a practical tool to be used in incident investigation. Evaluation of the technique in seven European States will take place in 2002.

The JANUS Approach – Looking Forward: HERA Prospective

The second phase of the HERA project – HERA 2 – was launched this year. The general objectives of this phase are to investigate those areas associated with the prediction, detection, and management of human error in ATM and to consider the application of these concepts at various levels of the Air Traffic Management system: such as safety training, safety management, and the application of human error vulnerability analysis within the ATM system. More specifically, there are three main activities which will be undertaken:

1. To develop an approach to investigate how human error can be detected and managed within a real-time simulated ATM environment.
2. To investigate the potential of the HERA classification as a predictive tool for use within the design stage of new ATM developments.
3. To develop an approach using the HERA classification tool for safety management within ATM.

This phase of the project will last approximately two years, from June 2000 to June 2002. At the end of this stage, HERA should be usable in the design stage (via usage in real-time simulations and through predictive analysis of future system designs), to identify and protect future systems from error. It should also be usable to learn from incidents and detect important trends, and show how these errors affect safety, enabling error management, and enhancing safety management itself as ATM progresses to its next level of automation and capacity. It is hoped that the HERA technique will also inform future ATM development in the difficult area of human error, and similarly help ATM avoid the dangers and disasters that have befallen other complex high-risk industries similarly dependent on human performance, and at risk from human error.

Discussion

The Human Error in ATM (HERA) project within Europe is an activity which deals with an approach to incident analysis in the air traffic management environment. The basis of the system and technique which has been developed, can be found in the human performance literature, but HERA has also had extensive development to ensure that the aspects which are particularly relevant and important in the ATM system are included. The resulting model, based on information processing principles, was modified to include aspects of the controllers' working practices, as well as the possible consequences of new planned technologies. The resulting HERA technique was changed through several iterations from expert discussion groups and was finally subject to a validation exercise. The results of this exercise indicated that, although the training of the technique was difficult to predict, the use of the technique was relatively robust. This work has led to the further development of the HERA system into the more predictive aspects of safety management within the ATM environment.

References

Billings, C. E. (1997). *Aviation Automation: The Search for a Human-centered Approach.* New Jersey: Lawrence Erlbaum Associates.

Dougherty, E. M. (1993). Context and human reliability analysis. *Reliability Engineering and System Safety, 41,* 25-47.

Embrey, D. E. (1986). *SHERPA: A systematic human error reduction and prediction approach.* Paper presented at the International Topical Meeting on Advances in Human Factors in Nuclear Power Plants, Knoxville, Tennessee.

Endsley, M. R. (1996). Design and evaluation for situation awareness enhancement. *Proceedings of the Human Factors Society 32nd Annual Meeting* (pp. 97-101). Santa Monica, CA: Human Factors Society.

Federal Aviation Administration. (1990). *Profile of Operational Errors in the National Airspace System in 1988.* Washington, DC.

Fleishman, E. A., & Quaintance, M. K. (1984). *Taxonomies of Human Performance: The Description of Human Tasks.* London: Academic Press Inc.

Helmreich, R. L., & Merritt, A. C. (1998). *Culture at Work in Aviation and Medicine.* Aldershot: Ashgate.

Isaac, A. R., Shorrock, S. T., & Kirwan, B. (in press). Human error in European air traffic management: The HERA Project. *Reliability Engineering and Systems Safety.*

Isaac, A. R., & Ruitenberg, B. (1990). *Air Traffic Control: The Human Performance Factors.* Aldershot: Ashgate.

Kinney, G. C., Spahn, M. J., & Amato, R. A. (1977). *The Human Element in Air Traffic Control: Observations and Analyses of the Performance of Controllers and Supervisors in Providing ATC Separation Services* (MTR-7655). METRIEK Division of the MITRE Corporation.

Martiniuk. R. G. (1976). *Information Processing in Motor Skills.* New York: Holt Rhinehart Winston.

O'Leary, M. (1999). *The British Airways human factors reporting programme.* Paper presented at the 3rd Human Error, Safety, and System Design Conference, Liege, Belgium.

Rasmussen, J. (1981). *Human Errors: A Taxonomy for Describing Human Malfunction in Industrial Installations.* Risø National Laboratory, DK-4000, Roskilde, Denmark.

Reason, J. (1990). *Human Error.* Cambridge, England: Cambridge University Press.

Reason, J. T. (1997). *Managing the Risks of Organisational Accidents.* Aldershot: Ashgate.

Shorrock, S. T., & Kirwan, B. (2000). The development of a human error identification technique for ATM. *Applied Ergonomics.* Manuscript submitted for publication.

Swain, A. D., & Guttmann, H. E. (1983). *A Handbook of Human Reliability Analysis with Emphasis on Nuclear Power Plant Applications* (NUREG/CR-1278). USNRC, Washington, DC 20555.

Taylor-Adams, S. E., & Kirwan, B. (1995). Human reliability data requirements. *International Journal of Quality and Reliability Management, 12*(1), 24-46.

Wickens, C. (1992). *Engineering Psychology and Human Performance* (2nd ed.). New York: Harper-Collins.

Chapter 28

Prospective Memory in Air Traffic Control

Shayne Loft, Michael Humphreys and Andrew Neal
Key Center for Human Factors & Applied Cognitive Psychology
The University of Queensland

Introduction

The air traffic control (ATC) environment is characterised not only by the necessity to remember past activities to support ongoing operations but also the requirement to remember to perform activities in the future (Neal, Griffin, Paterson, & Bordia, 1997). Controllers make frequent use of this type of memory; they frequently cannot execute a control action immediately either because the current situation does not allow it or their workload is too great (Neal, Griffin, Paterson, & Bordia, 1997). Successfully completing an intended action in the future depends on a type of remembering that has been labelled *prospective memory* (Harris, 1984). The objective of this paper is to overview the theoretical prospective memory literature, present a new experimental task for examining prospective memory, and outline potential general applications to en route air traffic control (ATC).

Prospective Memory and Air Traffic Control (ATC)

A recent survey of ATC related errors by the NASA Ames Research Center revealed that a high percentage of these errors involved failures to execute deferred actions (Freed & Remington, 1999). A specific incident that illustrates the importance of prospective memory occurred at Los Angeles International Airport in 1991. A controller cleared an aircraft to hold in takeoff position and shortly afterward directed another aircraft to land on the same runway, without clearing the first aircraft to takeoff beforehand (National Transportation Safety Board, 1991). On the surface, the accident was simple: the controller forgot about the action required for the plane that was on the runway because of intervening attention to other aircraft she was managing. This forgotten to-be-performed action represents a failure of prospective memory. Despite its importance this type of memory has received relatively little attention in the ATC literature.

The current paper reviews the current state of knowledge regarding prospective memory. Experimental studies have identified a range of general factors that affect the likelihood of prospective remembering. These include: (a) the characteristics of memory cues, (b) the length of the retention interval and the amount of mental rehearsal on the to-be-performed action, and (c) the nature and workload of concurrent activities (Mantyla, 1996).

Key Findings from the Prospective Memory Literature

One of the factors that affect prospective memory is the nature of the memory cue that is used. Information is stored in memory in the form of associations between events or objects. For example, when we have breakfast we store a memory trace that represents what we ate, together with information about where we ate it, the time of day, and any unique events that may have occurred. In order to retrieve information from memory, we need to use a retrieval cue that is associated with the information that we have stored. For example, when trying to remember what we had for breakfast, it is possible to use time of day and location to cue the retrieval of this information.

Prospective remembering is difficult because the environment has to spontaneously cue the retrieval of the intention to perform an action. Three types of memory cues have been identified by researchers: event-based, activity-based, and time-based cues (Einstein & McDaniel, 1990).

Event-based cues are located in the environment. Here people remember to perform an action when some external event occurs. For example, if I intend to invite a friend over for dinner next time I see him, this intention is cued when that friend walks into my office. Activity-based cues are also located in the environment, with individuals remembering to perform an action after finishing some previous activity. For example, if I intend to take some medication after dinner, this intention is cued when I finish my dinner. A time-based cue simply refers to the passage of time. Here the action is performed at a certain time or after a period of time has elapsed. For example, if I decide to take the garbage out at 8am in the morning I am relying on time to act as retrieval cue.

Research has shown that time-based cues are significantly less effective than event-based cues (Einstein & McDaniel, 1990; Einstein, McDaniel, Richardson, Guynn, & Cunfer, 1995). This is thought to be the case because time-based cued tasks rely on self-initiated processes and do not contain externally presented cues to signal the correct time for the initiation of the action (Einstein & McDaniel, 1990). Time-based cued tasks require participants to monitor elapsed time and initiate the prospective memory action on their own. No research to date has directly examined the effectiveness of activity-based cues.

A second factor affecting prospective memory is the retention interval. The retention interval refers to the amount of time between deciding to do something and the correct time for performing the intended action. Prospective memory performance is facilitated by relatively short retention intervals. Research suggests that the amount of rehearsal that an individual carries out during the retention interval can affect prospective memory (Koriat, Ben Zur, & Nussbaum, 1990; Vortac, Edwards, & Manning, 1995). Rehearsing the action during the retention interval in terms of both what has to be done and the correct time at which it should be performed enhances prospective memory.

The degree to which an event-based cue prompts a particular idea (something to be done) varies with its characteristics (Einstein & McDaniel, 1990; McDaniel & Einstein, 1993). A specific characteristic that facilitates prospective memory performance is the distinctiveness or salience of the memory cue in the environment. For example, if I intend to take medication before going to bed I am more likely to remember to do so if the medication box is a bright colour and/or placed on a bedside draw containing no other items.

The nature of concurrent activities being performed during the retention interval has been found to influence prospective memory (Brandimonte & Passolunghi, 1994; Ellis & Nimmo-Smith, 1993; Marsh & Hicks, 1998). In particular, concurrent tasks requiring high levels of attention, planning, and monitoring hinder prospective memory. Purely visual, phonological, and auditory tasks have been found to have little effect. The relationship between prospective memory and concurrent activities is considered to be particularly relevant to ATC. Controller's must plan and execute multiple tasks while attending to simultaneous changes in the environment. In addition, these tasks contain differential levels of workload. In general, high workloads have lead to higher decrements in prospective memory than low workloads (Marsh & Hicks, 1998).

Limitations

The prospective memory literature, to date, has used a number of different methodologies. A number of studies have been conducted in naturalistic settings using everyday tasks such as telephoning the experimenter (West, 1988) and mailing postcards to the experimenter at a certain data (Meacham & Leiman, 1982). These studies have been criticised for not allowing strict control or assessment of the memory strategies that people use nor any control over compliance. For this reason, recent research efforts have shifted to the laboratory. The dominant paradigm used was pioneered by Einstein and McDaniel (1990). This paradigm requires subjects to remember to respond (i.e., press a key) to particular words during a word recall task. Subjects are presented with a list of words to remember and recall at a later point in time. Before the commencement of the task they are given instructions to hit a key on each occasion they see the word TIGER presented to them. The measure of prospective memory is the number of times subjects remember to press the response key to this word.

This paradigm has led to substantial improvements in our understanding of prospective memory. However, it very difficult to generalise these results to complex applied environments such as ATC where operators must plan the execution of multiple concurrent tasks in the face of considerable uncertainty. Under these conditions no single mental operation determines behaviour. We believe that in order to understand prospective memory in domains such as ATC researchers must deal directly with the source of human error in a dynamic, multitasking environment. What is needed is an experimental paradigm that requires participants to perform a dynamic task, which involves competing demands on attention, while also allowing tight control over conditions. The use of such a task also allows the precise specification of the variables influencing error and the quantification of the probability of different error types under different conditions. In the following section we describe the development of one such task.

ATC Laboratory Task

The laboratory paradigm that we have developed is a simplified two-dimensional ATC task. It is designed so that naïve experimental participants can learn to perform the task to an adequate standard within 2 hours of practice. In order to

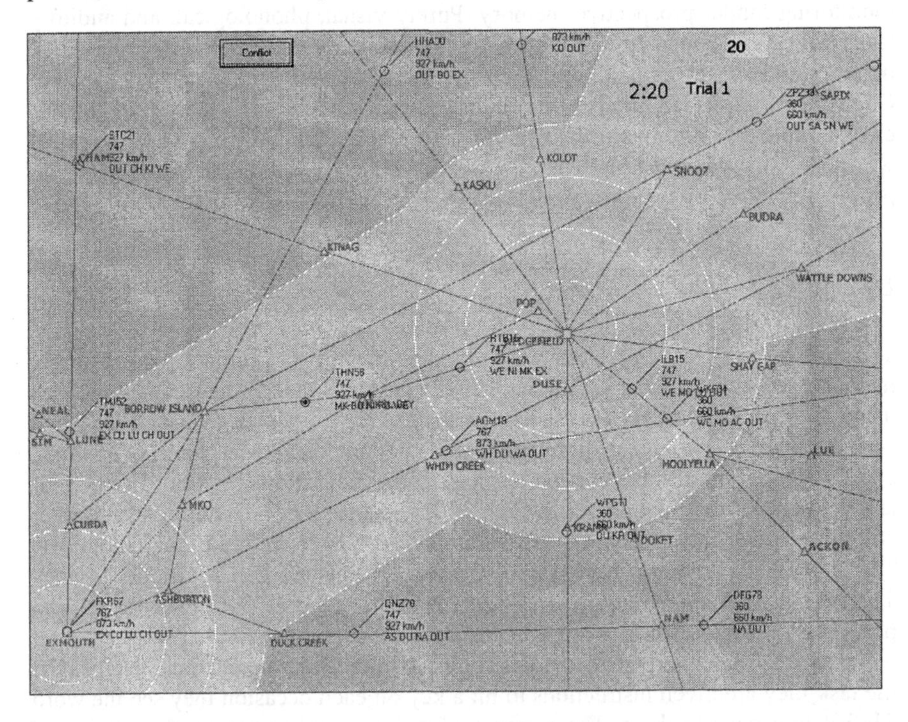

Figure 28.1 The ATC task. Small circles symbolise aircraft. Each aircraft has a light strip attached to it that displays the call sign of the aircraft, the type of aircraft (747, 767, 360), the speed of the aircraft and the designated route

meet this requirement, altitude has been eliminated from the display and aircraft fly on fixed flight paths. The participants have control over the speed of the aircraft and their job is to ensure that the aircraft do not violate a 5 nautical mile separation standard.

One of the difficulties of studying prospective memory in a natural setting is that the researcher can never be sure if and when the participant forms an intention to carry out an action, and does not know how the participant has encoded this intention in memory. In this task, we overcome these problems by embedding a prospective memory task in a primary ATC task. The primary ATC task requires participants to monitor evolving air traffic control scenarios on a radar-like screen in order to detect and prevent aircraft separation standard violations. The prospective memory instruction asks participants to remember to carry out a particular action in the future. Figure 28.1 presents the ATC task interface.

The factors reviewed in this paper, that have been found to affect prospective memory in word recall tasks, can be examined using the ATC task to determine their influence on prospective memory performance in a more complex dynamic setting. For example, if the participants are instructed to change the speed of a particular aircraft, we can manipulate the type of memory cue presented. We could ask to them to change the speed: (a) in five minutes (time-based cue), (b) when the aircraft reaches a nominated waypoint (event-based cue), or (c) after resolving a nominated conflict. In addition to the nature of the retrieval cue, it is also possible to manipulate the length of the retention interval, the opportunity for rehearsal (by representing the instructions), and the nature and workload of the intervening events (concurrent activities being performed).

Implications

One of the key features influencing the safety of computerised systems is the design of the Human-Computer-Interface (HCI). The HCI of a system consists of the physical mechanisms by which the operator interacts with the machine, as well as the procedures the operator follows. Findings from the literature have increased our knowledge of the psychological processes responsible for remembering to carry out previously planned actions. These findings give rise to some general applications for the design of human-computer-interfaces, applications that may help reduce the likelihood of prospective memory error. The new ATC task presented in this paper provides a sound setting for empirically examining these applications. Some starting points are listed below.

- Event-based cues are more effective reminders to perform an action. This is the case because time-based cues require the individual to monitor and initiate the action on their own. Human-computer systems should provide cues to signal the execution of to-be performed actions. These cues should be as distinctive as possible. Moreover, the system should draw operators' attention to events that will cue memory to perform actions.

- Prospective memory performance is enhanced with shorter retention intervals and when the individual rehearses the content and correct time for the intended action. The designers of human-computer systems could consider ways to encourage operators to rehearse information pertaining to the intended actions.
- The nature of concurrent tasks and workload appears to affect the likelihood of prospective memory error. Designers need to consider the likely variation in both the nature of tasks being performed and general workload in the task, paying particular attention to peaks and troughs in workload.

Acknowledgement

Michael Humphreys is Professor of Psychology, and the Director of the Key Centre for Human Factors and Applied Cognitive Psychology at The University of Queensland. Shayne Loft is a PhD candidate in the Key Centre for Human Factors and Applied Cognitive Psychology at The University of Queensland. Andrew Neal holds a joint appointment with the School of Psychology and the Key Centre for Human Factors and Applied Cognitive Psychology, at The University of Queensland.

References

Brandimonte, M. A., & Passolunghi, M. C. (1994). The effect of cue familiarity, cue-distinctiveness, and retention interval on prospective memory. *The Quarterly Journal of Experimental Psychology, 47*(a), 565-587.

Einstein, G. O., & McDaniel, M. A. (1990). Normal aging and prospective memory. *Journal of Experimental Psychology: Learning, Memory, and Cognition, 16,* 717-726.

Einstein, G. O., McDaniel, M. A., Richardson, S. L., Guynn, M. J., & Cunfer, A. R. (1995). Aging and prospective memory: Examining the influences of self-initiated retrieval processes. *Journal of Experimental Psychology: Learning, Memory, and Cognition, 21,* 996-1007.

Ellis, J. A., & Nimmo-Smith, I. (1993). Recollecting naturally-occurring intentions: A study of cognitive and affective factors. *Memory, 1*(2), 101-126.

Freed, M., & Remington, R. (1999). *A Conceptual Framework for Predicting Error in Complex Human-machine Environments* (Tech. Rep.). NASA.

Harris, J. E. (1984). Remembering to do things: A forgotten topic. In J. E. Harris & P. E. Morris (Eds.), *Everyday Memory, Actions, and Absent-mindedness* (pp. 71-92). New York: Academic Press.

Koriat, A., Ben Zur, H., & Nussbaum, A. (1990). Encoding information for future action: Memory for to-be-performed tasks versus memory or to-be-recalled tasks. *Memory & Cognition, 18*, 568-578.

Mantyla, T. (1996). Activating actions and interrupting intentions: Mechanisms of retrieval sensitization in prospective memory. In M. Brandimonte, G. O. Einstein, & M. A. McDaniel (Eds.), *Prospective Memory: Theory and Applications* (pp. 93-114). Hillsdale, NJ: Erlbaum.

Marsh, R. L., & Hicks, J. L. (1998). Event-based prospective memory and executive control of working memory. *Journal of Experimental Psychology: Learning, Memory, and Cognition, 24,* 336-349.

Meacham, J. A., & Leiman, B. (1982). Remembering to perform future actions. In U. Neisser (Ed.), *Memory Observed: Remembering in Natural Context* (pp.327-336). San Francisco: W.H. Freeman.

National Transportation Safety Board. (1991). *Aircraft Accident Report* (NTSB/AAR-91/08). Washington, DC: US Government Printing Office.

Neal, A., Griffin, M., Paterson, J., & Bordia, P. (1997). *Human Factors Issues: Performance Management in the Transition to a CNS/ATM Environment* (Final Report: Airservices Australia). Brisbane: The University of Queensland, Centre for Organisational Psychology.

Vortac, O. U., Edwards, M. B., & Manning, C. A. (1995). Functions of external cues in prospective memory. *Memory, 3*(2), 201-219.

West, R. L. (1988). Prospective memory and aging. In M. M. Gruneberg, P. E. Morris, & R. N. Sykes (Eds.), *Practical Aspects of Memory: Vol. 2* (pp. 119-125). Chichester, England: Wiley.

Chapter 29

Human Factors Issues in CPDLC

John Allin Brown
Boeing Commercial Airplanes, Air Traffic Management Services

Introduction

Controller-Pilot Data Link Communication (CPDLC) is recognised as one of the enabling technologies of the future, improved CNS/ATM1 environment. Anyone who has used High Frequency (HFq) radio communication in aviation or for any other purpose will recognise the shortcomings of the system. Its performance varies with the time of day, there is considerable background noise and reception fades and swells. In addition, in areas like the North Atlantic, where traffic is heavy, radio frequency congestion significantly hinders the transmission of operational data and adds to the other difficulties. Yet HFq is still the primary medium for Air Traffic Control communications in areas more than approximately 200 miles from the nearest Very High Frequency (VHFq) ground station. CPDLC, a communications method similar in principle to e-mail, represents a considerable improvement in quality of information transfer with enhanced message clarity and timeliness, and reductions in potential for errors.

In environments served by VHFq, communications clarity is good; however, frequency congestion is becoming a limiting factor, the chances of a missed or misheard message are significant, and transcription errors occur, especially when complex clearances are received. Digital communication, in the form of CPDLC, enables presentation of messages in a form that should be understandable without error and allows data included in complex clearances to be loaded directly into aircraft navigation systems, thus eliminating several stages of potential transcription error.

But this communications medium is not without its problems. Creation of messages simply by typing would increase workload unacceptably and so a standardised set of messages has been defined. To allow for use in as many scenarios as possible, the number of messages is large and the operator's ability quickly to select the correct message becomes a function of the design of the human/machine interface. While this is true for both pilot and air traffic controller; this paper limits itself to the Pilot's interface in Boeing aircraft. In addition, experience of use of CPDLC has shown that humans do not necessarily extract the same information from the written word as they do from similar auditory input.

Future Air Navigation System (FANS)

Data link is not new technology; it has been in use by the military for many years and for communication between pilot and airline for two decades. But it is new to air traffic services in civil aviation. This paper does not present the entire history of CPDLC in civil aviation since the purposes of the readers would not be well served. Instead, the description begins with the development of FANS by a group that became known as the FIT – the FANS Interoperability Team. The FIT consisted of air traffic service providers, data link service providers, operators, and aircraft and equipment manufacturers. The South Pacific routes between the US west coast and Australia/New Zealand were chosen because it was hoped that operational benefits would result in a strong return on investment and because, with only a limited number of operators, the environment was benign for new technologies and new operating procedures. All the operators in the FIT operated the Boeing 747-400 and so the first implementation was in this aircraft.

The chosen hardware for the Pilot interface was the multifunction control and display unit (MCDU) that the Flight Management System shares with other cockpit functions. The need for a relatively rapid and economical implementation and the general lack of space in which new control and display panels might be located, even in an aircraft the size of the 747, disallowed the use of a more exotic and user-friendly interface. However, despite the size of the display portion of the MCDU, the team managed to arrive at a design that minimised the number of accessible 'pages' without resorting to abbreviations in all but certain titles. Design philosophy was kept consistent with the control/display functions for other applications served by the MCDU (e.g., flight management computer).

FANS Functionality

In the 747, all FANS functions other than alerting are encompassed in the MCDU. Alerting is achieved by aural chimes from the aircraft warning system and by messages of various types appearing on the upper screen of the Engine Indicating and Crew Alerting System (EICAS) on the centre instrument panel. MCDU functions are menu-based with simple transitions between levels. Figure 29.1 shows the main menu which illustrates the operational communications functions served. Each menu item, which the Pilot selects with key adjacent to the end of the line on the screen (the line select key), leads to another menu, an example of which is seen in Figure 29.2. This is the most complex menu at this level and allows the Pilot to select all variations of requests to air traffic control.

The Pilot enters the required value of the parameter (e.g., ALTITUDE) by typing into the 'scratchpad' at the bottom of the screen and selecting it into the space provided using the adjacent line select key. The system recognises the difference between such pairings as altitudes and flight levels, and airspeed and Mach Number. The data entry procedure is consistent with all other MCDU functions and is thus well-known to the Pilot. Once a valid entry has been made, the Pilot selects the 'VERIFY' prompt allowing validation of the selection(s) prior to transmission. This feature is more important when requests are strung together, because each must be composed independently. The verification step allows the Pilot a final look at the message just as it will be presented to the controller, enhancing a shared understanding of the negotiation. As can be seen in Figure 29.3, the Pilot also has the opportunity to add a free text (hand typed) element to the message should any clarification be necessary. Pilots and controllers are encouraged to avoid use of Free Text because its use minimises the advantages of any automatic use of enclosed data while at the same time introducing the likelihood of considerable variation in formats, especially in abbreviations. Abbreviations in messaging are avoided almost completely.

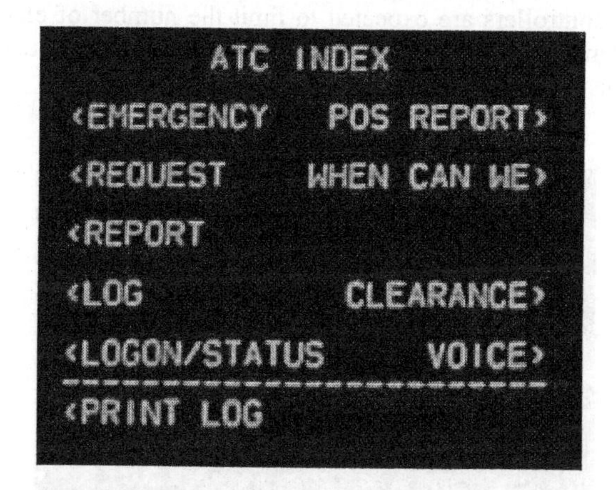

Figure 29.1 CPDLC main menu

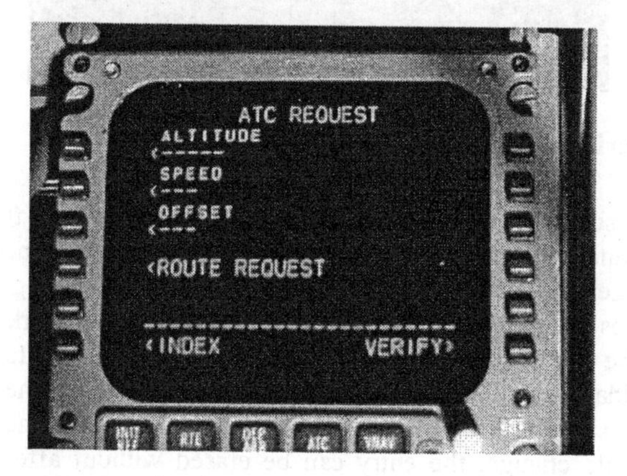

Figure 29.2 Altitude request page

Once the Pilot has validated the message, he can transmit it by pressing the line select key next to the 'Request SEND' prompt. The prompt changes to

'SENDING' after the key is pressed and to 'SENT' when the avionics receive a system acknowledgement, allowing the Pilot to monitor the message's progress.

When a message is received, the Pilot is alerted with a single chime and 'ATC MESSAGE' appears on the upper EICAS screen. The Pilot can gain access to the message simply by pressing the 'ATC' key on the MCDU. If there is more than one message awaiting the Pilot's attention, use of the ATC key displays a list of messages, allowing the Pilot to choose which message to deal with first. If no message is waiting, pilot access to the ATC MENU page is via the same ATC key, simplifying access to the application by reducing it to a single keystroke.

Each downlink or uplink message can consist of up to five message elements and each individual request or clearance represents a single element. However, both pilots and controllers are expected to limit the number of elements to those that are dependent upon one another, simplifying the operator's understanding of the final message and reducing interpretation errors.

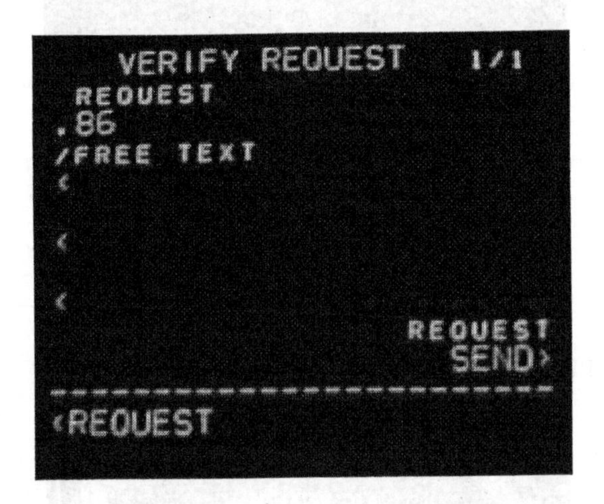

Figure 29.3 Verify request page

Figure 29.4 shows a 2-element uplink message presented to the Pilot. It also illustrates the function that allows the Pilot to load certain clearances directly into the flight management computer, reducing workload and reducing error potential. Route clearances in particular can be complex, and both workload and error potential can be significantly reduced by loading the data into the FMC for subsequent validation by the Pilot. Once the data are validated, the change can be executed in the same way as pilot-entered amendments, or, if the Pilot does not agree to the route change, the entry can be erased without affecting the active route. Consistency with other MCDU functions is thus preserved.

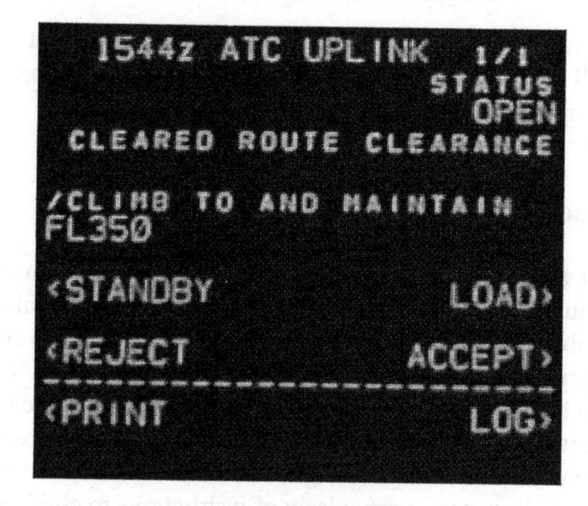

Figure 29.4 Two-element ATC uplink message

Figure 29.4 also shows the prompts that allow the Pilot to respond to ATC instructions. Acceptance and rejection result in different downlinks depending on the form of the uplink. If the uplink is an instruction, pilot selection of 'ACCEPT' results in a 'WILCO' and 'REJECT' in 'NEGATIVE'. Had the uplink been informational, only 'ROGER' could be downlinked by the Pilot and no rejection response would be possible. Only the messages appropriate to the uplink are made available for pilot response. The actual response is displayed to the Pilot along with the associated uplink following transmission.

FANS on South Pacific Routes

The first operations in which FANS was used in the South Pacific took no operational credit for the functions. First it was necessary to build experience and to ensure that the technology worked at a level of performance that was appropriate to the operation. The FIT had been disbanded following certification of the airborne element of the system, the stakeholders' relying on the usual processes of problem solution in an air traffic control environment. Unfortunately, few operators were involved and since there were no significant benefits, no one was motivated to report and correct the problems. System performance began to deteriorate and with it the confidence of the operators. The FIT was reformed so that problems could be identified and solutions found in a cooperative manner. The FITs activities continue to this day. The processes are now in place to maintain the operation and indeed to make increasing use of its capabilities.

As the system stabilised, operator confidence grew and it soon became clear that it was time to take operational credit for the functions and achieve real benefits. Some new procedures have become part of standard operations in the South Pacific while other benefits are beginning to be derived from CPDLC and other FANS functions. The use of FANS has now propagated beyond the South Pacific. The entire Pacific oceanic environment now has CPDLC capability as does western China, trials are under way in the Bay of Bengal and FANS CPDLC functionality will be added to North Atlantic and remote northern Canadian airspace quite soon.

Early Human Factors Issues

Unfortunately, at the time that FANS was developed, there was no experience in the use of data link for air traffic control and the format that evolved for uplinks placed unreasonable emphasis on the parameters included in the messages to the detriment of the Pilot's understanding the context of the message. This format was probably adopted because the parameters in a verbal message appear to the Pilot to be the most important elements. Validation processes used during development did not bring associated shortcomings to light. The result in service was occasional misinterpretation of clearances, particularly of instructions to climb at some time in the future or at some position down track. Such clearances are routinely used in the voice environment and no significant problems with comprehension have been reported. The standard message set allowed exactly the same wording to be used, but pilots sometimes derived a different meaning from the written word than they would have from the spoken word.

Figures 29.5 and 29.6 illustrate this type of problem message. The first page is free text, which was used inappropriately and can have done nothing to ease the Pilot's problems given the variations in data format included in it. The second page starts with the 'conditional' clearance. The '/AT' on the first line is small and easily overlooked. The positional information (latitude and longitude) on the second line is complex and only pushed the Pilot's attention to the 'CLIMB TO AND MAINTAIN...', a clearance that had been requested, was expected, and compliance was quick and inappropriate.

Figure 29.5 Problem uplink, page 1

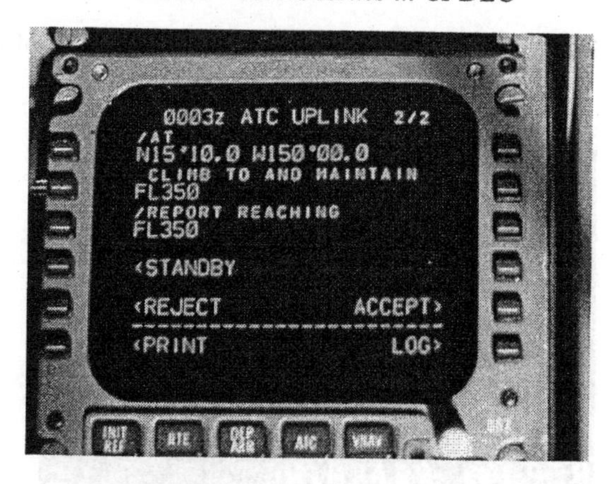

Figure 29.6 Problem uplink, page 2

Although this particular issue was partially attributed to a combination of training deficiency and lack of operational experience, Boeing recognised that improvements could be made to the uplink message format and a study was undertaken in support of FANS development for the 757 and 767 models. The changes to format that were arrived at and validated will be made available to 747-400 CPDLC operators toward the end of this year.

Figure 29.7 depicts the same message in the new format (the second page containing the response prompts is omitted from the paper). Major changes are the wrapping of words from one line to the next and the use of a uniform text size. The size provides good line separation while the use of as much of the display line as possible allows a larger percentage of messages to be displayed on a single display page. It should be noted that this represents an unusually long instruction. As a supporting measure, operators have changed procedures to make best use of the cross-checking abilities of the crew members as well as the difference in format provided by the hard copy printer.

Second Generation CPDLC Interface

The Boeing 777 design features a more sophisticated control and display system than other Boeing models, and the designers chose to make full use of the multi-function display (MFD) and cursor-control device (CCD) to provide a superior CPDLC function. All CPDLC tasks, for both uplinks and downlinks, can be achieved using the touch-pad CCD on a display page that can appear on either pilot's inboard screen or on the lower centre screen, giving full recognition to the importance of the communications task. Alpha-numeric input to the function is made through the MCDU; the Pilot types into the scratchpad and then transfers the text to the appropriate data field using the CCD and cursor. Figure 29.8 is an example of an MFD ATC uplink page containing a message similar in content to the example above. Note here again, however, that free text has been used, again inappropriately. Nevertheless, the superior format and legibility of the display are evident.

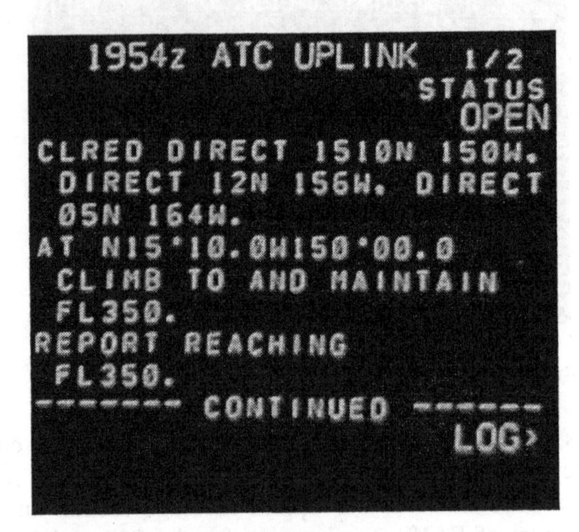

Figure 29.7 Problem uplink, new format

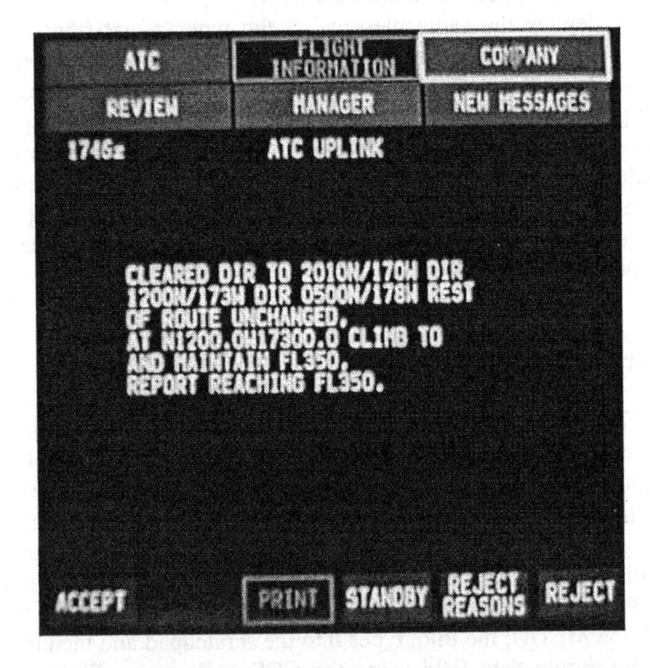

Figure 29.8 Second generation data link, example uplink message

While the touchpad is similar to devices common on personal computers, it has been possible to modify its functions further to simplify the Pilot's communications task. A prime example is the provision of 'hot corners'. If the Pilot taps close to any of the four corners of the pad, the cursor jumps to the associated corner of the MFD. The most commonly used control function command boxes are located at these corners and so pilot access to these functions is optimised. Workload is further reduced by the choice of default positions for the cursor following MFD page changes and function activations within the communications application. For example, when the communications menu is displayed prior to logging on to the system, the cursor defaults to the LOGON/STATUS page select box allowing selection of the page by a single button push at the side of the touchpad. There are many similar examples of such optimisation.

Most uplink messages are displayed, without pilot action, on the upper centre display along with response options available to the Pilot. Simple responses (acceptance or rejection of the instruction) can be achieved through dedicated glareshield-mounted buttons. The end result is that, for the vast majority of ATC instructions (messages occupying less than 5 lines of 30 characters), the Pilot has no reason to gain access to the communications page on the MFD. Workload is thus reduced and pilot response time, an important part of the overall delay performance of the communications function, is also reduced. An example of an ATC uplink displayed in what is known as the 'data block' is shown in Figure 29.9.

This depiction also shows the '□ATC' communications message which replaces the 'ATC MESSAGE' in the 747. In addition, the coloration of '350' in the clearance illustrates the 'dial' function; the numerals turn green only when the correct altitude is dialled into the altitude window of the aircraft's mode control panel. A number of other parameters including altimeter settings and radio frequencies similarly give feedback to the Pilot. The lowest line in the data block displays available responses and, like the 'SEND/SENDING/SENT' sequence in the 747, allow the Pilot to monitor the message's progress.

In trials conducted with the FAA Technical Centre, the 777 interface showed itself to be more than capable of supporting the use of CPDLC in domestic airspace, including in the high workload terminal area, using current ATC procedures (e.g., vectoring). It is Boeing's belief that procedures could be developed that would take advantage of more of the capabilities of the FMC in guiding the aircraft along a predictable flight path, thus reducing communications workload and allowing a simpler CPDLC interface to be used in domestic operations.

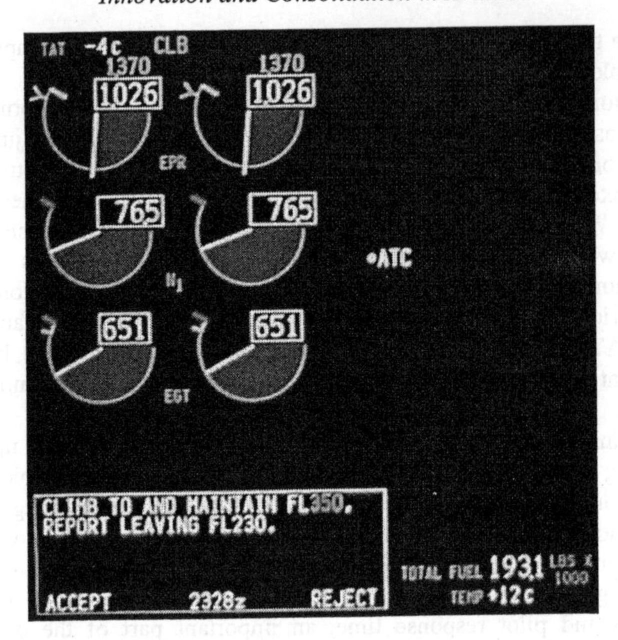

Figure 29.9 'Data Block' and ATC communications message

Second Generation Data Link System

FANS was developed for use in oceanic and remote areas. End-to-end system performance and all current pilot interfaces are adequate for the communications tasks in these environments. However, the system will not perform well enough to meet the stated requirements for an ATC data link system which can be used in busy, domestic airspace. A new system called the Aeronautical Telecommunications Network (ATN) is being developed to serve this high demand environment. Among other things, ATN will have quicker message delivery and more robust security than FANS. It will also have a slightly different message set although many of the messages are common. The combination of similarities and differences between the systems promises to pose some interesting Human Factors problems given that FANS and ATN are likely to operate in parallel for more than 10 years, in some cases in the same aircraft.

Boeing feels that it is very important that the Pilot be able to achieve seamless operation when transitioning between FANS and ATN environments. Procedures should be as common as the environments allow and, above all, the cockpit interface should minimise errors while maximising training benefit. To this end, the Pilot should be required to carry out the same action(s) to display CPDLC control pages in ATN and FANS applications. The high level menus should be common as, where possible, should the selection functions which the Pilot must use to construct a downlink message. Only where additional messages are available should an obviously different page format be adopted. If additional messages are to be included in existing page formats, it will be necessary to hide the additions when the FANS function is in operation; it will be unacceptable to display functions that are not available to the Pilot. However, having similar pages, some with the additional messages and some without, might also confuse the operator when transitioning from FANS to ATN environments. It might be possible for the additional messages to be muted but still visible on the display page when FANS is in use to help to build and maintain the Pilot's knowledge of selection locations in ATN operations.

A similar issue will be experienced as a result of the incremental implementation of messages in US domestic CPDLC Build 1, 1A, and 2. Designing separate software applications for each of the Builds would be expensive and would reduce flexibility in the evolution of the system. As a result, it is likely that pilots will be presented with the far greater number of message options used in Build 1A even when they are operating in Build 1 environments. Similar issues are expected to be encountered in aircraft carrying out ATN operations in both Europe and the USA, perhaps during the same flight, at least in early implementations. In addition, in Europe, different countries are adopting different mixes of procedures and, therefore, messages. If a Pilot downlinks a message that is not supported by the ground facility, the result will be a 'SERVICE NOT AVAILABLE' message in reply, hardly a reaction to engender pilot confidence in the system. Perhaps the most egregious case is the likelihood that an Emergency downlink report will result in a SERVICE NOT AVAILABLE message in one of the environments.

ATN system designers' expectations of timeliness of response by pilots will be partly met by the high performance of the communications system and partly by the reactions of the pilots themselves. It is clear that, for some of the operations for which CPDLC will be used in the domestic environment, a stand-alone uplink message display will be required if pilot response time and workload are to be acceptable. Boeing believes that the 777 'data block' will serve this function and the company's long-term aim is to provide such functionality, including the multi-function display and cursor control device interface, in all models. However, Boeing also recognises that such a solution may not be cost effective for in-service aircraft, and a retrofit solution must also be designed.

The data block function could be replaced by a dedicated uplink display, most likely one for each pilot given the constraints of placing such a device in an already crowded cockpit. Dedicated response switches, similar to those mounted on the glareshield in the 777, would facilitate rapid pilot response, would minimise workload and would prevent the CPDLC function's denying the Pilot use of other MCDU functions while carrying out the communications task. As in the 777, the combination of dedicated display, response buttons and both visual and aural alerting functions would mean that the Pilot need not use the MCDU for most responses to uplink messages. The MCDU must still be used in composing and sending downlink messages like requests and reports.

Conclusions

The introduction of data link for air traffic services communications in air transport aircraft has proved to be a bigger Human Factors challenge than had been imagined and despite the use of accepted methods in validation of the design, some Human Factors deficiencies made their way into the production version. The major lessons that can be learned from technological and procedural evolution can be summarised as follows:

1. It is very difficult to arrive at an adequate design without full knowledge of the procedures and operating environment in which the function will be utilised.
2. Adding such a function to a control and display unit with a well-established design philosophy increases the degree of difficulty.
3. What may appear to be the most important parts of an aural message may not be the most important in a visual communications medium.
4. It is much easier to design such functions as integrated parts of a new aircraft design.

Chapter 30

Alertness and Awareness of Long Haul Aircrews: The Contribution of a New Interface Concept as an Effective Fatigue Countermeasure

Jean-Jacques Speyer, Adrian Elsey
Airbus Industries, Flight Operations Support, France
Philippe Cabon, Regis Mollard, Samira Bougeois-Bougrinne
Université René Descartes, Paris V, France
Nicolas Parriaux and Marc Perrinet
Dead-PHAROS

Introduction

Aircrew fatigue during long-haul flights is recognised as one of the major Human Factors that can impair performance and situational awareness (Bourgeois-Bougrine, Gounelle, Cabon, Mollard, & Coblentz, 2000). Using physiological recordings on 156 commercial airline flights, previous work has shown that reductions in alertness are frequent during flights, including the descent phase. Most decreases in alertness occur during the monotonous part of the cruise and were often observed simultaneously on both pilots in two person crews (Cabon, Mollard, & Coblentz, 1995; Cabon, Mollard, Coblentz, Fouillot, & Speyer, 1995). Based on these results specific operational recommendations have been designed. Further studies have shown the positive effects of these recommendations (Cabon et al., 1995; Cabon, Mollard, Coblentz, Fouillot et al., 1995). Practical recommendations were gathered into a booklet for the use of long-haul aircrews. Currently, this booklet is available in French, English, and Chinese (Speyer, Cabon, Mollard, Bougrine, & Coblentz, 1995; Mollard, Coblentz, Cabon, & Bougrine, 1995). One of the main recommendations promoted in these guidelines is based on the alternation of crew rest and activities, including cockpit napping also called controlled rest on the flight deck. The efficiency of cockpit napping was first emphasised by Graeber, Rosekind, Connell, & Dinges (1990). However, one of the main drawbacks of cockpit napping is that it could contribute to increase the monotony inside the cockpit (lower light intensity, reduced communications...) and thus could decrease the alertness and awareness of the Pilot remaining at the controls. Therefore fail-safe monitoring of the non-napping pilot should avoid simultaneous sleepiness of the two pilots. The Electronic Pilot Activity and Alertness Monitor (EPAM) is in fact intended to provide this type of support warning both crew members by means of activity and alertness measures. Moreover, the EPAM also provides useful feedback on wakefulness monitoring when cockpit napping is not allowed.

Fatigue Countermeasure Systems

Fatigue countermeasure systems are vehicle-mounted devices designed to monitor operator performance to provide alert signals if operator impairment reaches a predetermined level as validated by a pre-established mathematical model.

Automobile and Railway

Numerous studies and developments have been conducted and several systems have been developed for private cars, lorries, public transport, and railways. Although driving an automobile is clearly different from piloting an aircraft, these safety systems aim to detect when activity and alertness levels have reached a critical level either based on performance in the main task:

1. Safety Driver Adviser (reduction in steering wheel amplitude and speed of movement).
2. Stay Awake Alarm (reduction in trajectory corrections versus speed).
3. Driver Alertness Aid (reduction in frequency of steering wheel movements).
4. The Peugeot System (pressure on the accelerator, steering wheel movements, speed).

or based on the evaluation of the status of the operator using physiological indications:

5. Dozer's Alarm (based on posture by monitoring the inclination of the head).
6. Dormalert (based on cutaneous resistance, but interference with ambient temperature).
7. Onguard device (closing of the eyelids by means of an infrared transceiver).

The French railway uses a driver's safety and vigilance device called VACMA. This system was initially designed to stop the train in dead-man cases and not to act in cases of reductions in vigilance. The type of VACMA pedal used will track complete absence of reaction but cannot reveal vigilance decrement. Laboratory research has shown that a dozing driver continues to automatically actuate the pedal. The sound restores the driver's pressure for a brief period of time before the complete stop but does not reactivate his vigilance.

Aeronautics

Before anything is reviewed it is essential to mention that these systems are engineered to provide non-intrusive and confidential feedback to the pilots. In other words, any event recording similar to the Cockpit Voice or Flight Data Recorders is specifically banned from the concept, as this would alter the intended use. That is for pilots:

1. to be informed of reduced activity and/or alertness;
2. take this into consideration when the flight is strategically stable, e.g., in cruise;
3. to practice controlled rest on the flight deck or recuperative rest in dedicated crew rest areas for augmented crews; and
4. to systematically organise this for all crewmembers in order to enable them to operate at their best in the more crucial phases of flight, whether the crew is augmented or not.

Pilot Guard System (PGS)

Originally, the PGS was designed as an interface to collect indicating signals from pilots' interactions with the aircraft interfaces to retrigger their activities by means of visual and aural warnings if no action occurred within a given time. Before each action, the time remaining before the warning is given is displayed by means of a strip of light-emitting diodes.

On completion of the complete 'warning' initialisation cycle, the system will reset itself at the initially selected basic period. The process thus described is supposed to ensure that the pilots become aware of their decreasing activities in order to resolutely adopt the counter-measures required to increase their levels of activity and attention.

However, it is just a pure activity-monitoring device, which in no way guarantees alertness decrements would be intercepted. An evaluation (Speyer & Elsey, 1995) was performed on a Sabena A310-300 for activities pertaining to:

1. the FMS CDU panel;
2. the ECAM control panel;
3. the FCU panel;
4. the VHF/HF, VOR/DME or ADF selection panels; and
5. the PGS itself.

The PGS was deemed non-intrusive to the pilots enabling them to organise their flight watch in cruise with alternating Pilot/Co-pilot active/passive vigilance to facilitate and improve resource management.

Crew Alertness Monitor (CAM)

Based on the same principle as the PGS, Boeing developed the Crew Alertness Monitor to install it on the Boeing 767-200/300, 747-400, and 777-200/300 aircraft. The CAM is a function of the FMS and its purpose is to trigger a warning if – during a predetermined period – it detects no activity:

1. on the FMS and on the EFIS/EICAS display control units;
2. on the control panel of the auto-pilot and flight director; or
3. on the VHF/HF transmissions.

Apart from the resetting operations, the pilots have no control over the CAM. They are aware of its presence only when a warning is generated: after a sleep episode, there is no way of ensuring that they will not immediately again succumb to this phenomenon. The PGS is more of a vigilance and workload management tool that goes along with sound crew resource management. Should the two pilots doze off simultaneously, it ensures they will not immediately go back to sleep after having been awakened.

Monitoring Activity and Alertness

The concept evolved from an interface principally related to the exogenous aspects, i.e., task activities to one also including information directly related to the endogenous state of wakefulness. There is evidence that pilot inactivity would not be sufficient to detect efficiently all decreases in alertness, pilots still having ample interactions with their flight systems during low alertness phases. The device is hence based on both pilot activity monitoring and eyes' camera recording.

Pilot Activity Monitoring

The activity part includes two modes. In the first mode (EPAM), pilots' interactions within the flight deck are continuously monitored. It is based on the assumption that a pilot who is dozing off will have fewer interactions with the aircraft. The device is to be connected to different systems of the aircraft (Flight Monitor System, Electronic Centralised Aircraft Monitor, etc.). When the EPAM mode is switched on, and if no activity on one of these systems is detected after some pre-set period (5, 10, or 25 minutes depending on the flight phase or on the pilot's choice) a visual alert is released. Subsequently, after another 1-minute inactivity period, an auditory alarm is released for 15 seconds. The second mode (TIMER) can be considered as an alarm clock. The Pilot who plans to nap activates this mode. The delay before the alert sequence is released is 45 minutes. Hence the EPAM should help the crew manage their in-flight rest-activity cycles.

Eyes' Camera Recording

Adding continuous physiological monitoring to activity monitoring, the second part of the device is based on camera recording of pilots' eyes during flight. Using specialised image processing software, various parameters can be automatically analysed such as eye movements and eyelid closures. Initial studies in car driving did show that just a few criteria were sufficient to efficiently detect some low alertness episodes (Perinet-Marquet & Parriaux, 1999). The nature of these criteria depends on the depth of the alertness decrements. Early drowsiness is associated with strabism and long eye fixations while outright stage 1 sleepiness is associated with increased eyelid closures and slow eye movements.

Operational Evaluation

Commercial Airline Flights The device was put in an operational evaluation stage during long-range airline flights carried out on Airbus A330/340 aircraft operated on commercial flights at Sabena. Some 10 flights have been performed on the Brussels-New York-Brussels route and rosters were chosen for the significant fatigue they are known to induce: outbound leaving in the evening, inbound flights returning in the early morning hours with 24 hour layovers.

Airbus Industrie Test Flights Airbus Industrie flight trials were carried out in February 2000 during a round the world FANS flight evaluation campaign consisting of some 5 very long-range flight legs up to 15 hours that demonstrated the operational interest of the concept to test pilots. The device was mounted on the factory A340 and effectively integrated to the flight testing avionics.

Data collected

Data collected are the same as those gathered during previous studies. Physiological parameters such as the electro-encephalogram (EEG), the electro-oculogram (EOG), and the Heart Rate Variability (HRV) are continuously recorded to evaluate the impact of EPAM both in terms of its sensitivity to fatigue and of its ability to maintain alertness. Simultaneously, an observation of the crew is carried out in order to describe and analyse their activity by means of dedicated Aircrew Data Logging software (ADL) in correlation with measured changes in physiological parameters.

General Results

In a first stage, data processing focussed on: sleep quantity and quality during in-flight naps, in-flight alertness decrements, and EPAM alert warning occurrences. Figure 30.1 presents an example of results for a New York – Brussels leg with 3 types of data:

1. *the hypnogram* during scheduled in-flight nap. This graph represents the occurrence of sleep stages 1 to 4 (no Rapid Eyes Movement sleep was observed during these flights). However, when the Pilot is supposed to be alert, some stage 1 sleep can occur which corresponds to 'microsleeps';
2. *the inactive time* which would have results in EPAM warnings for the different selectable periods: 5, 10, 15, 20, or 25 min; and
3. *the alpha/delta ratio* from the electro-encephalogram. When the Pilot is supposed to be alert, an increase of this ratio represents an alertness decrement (i.e., an increase of alpha power). On the contrary, during in-flight naps, a decrease of this ratio corresponds to deeper sleep (i.e., an increase of delta power), increases of the ratio meaning lighter sleep.

Figure 30.1 shows that potential alerts would have occurred around micro-sleeps after at least 15 minutes of inactivity. The very first micro-sleep is not related to any significant increase of inactivity time. This confirms the need for another source of information related to the internal state of the Pilot, which can best be traced by means of ocular activity.

Overall, it appears that inactivity precedes increases of the alpha/delta ratio, typical of alertness decrements linked to monotony but does not predict the onset of micro-sleeps typical of sleep pressure and performance degradation. Pilot activity often goes along with decreases of the alpha/delta ratio typical of higher alertness states.

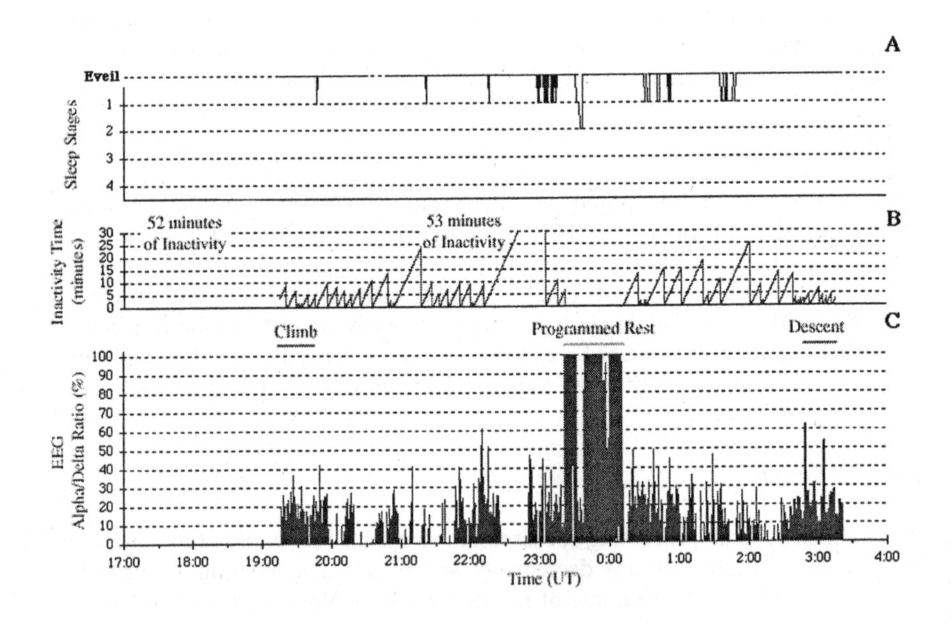

Figure 30.1 Activity graphs, EEG, and sleep stages

Figure 30.2 shows an example of video analyses with 2 parameters, i.e., the duration of eye closures and the duration of eye blinking. First results suggest that an increase of these two parameters could efficiently predict micro-sleep occurrences. Further analyses are currently conducted on other parameters such as eye fixations and strabism to select parameters and thresholds for the best algorithm. At this stage, the study confirms that high quality recordings are possible in flight and that aircrews indicate the acceptability of this recording device.

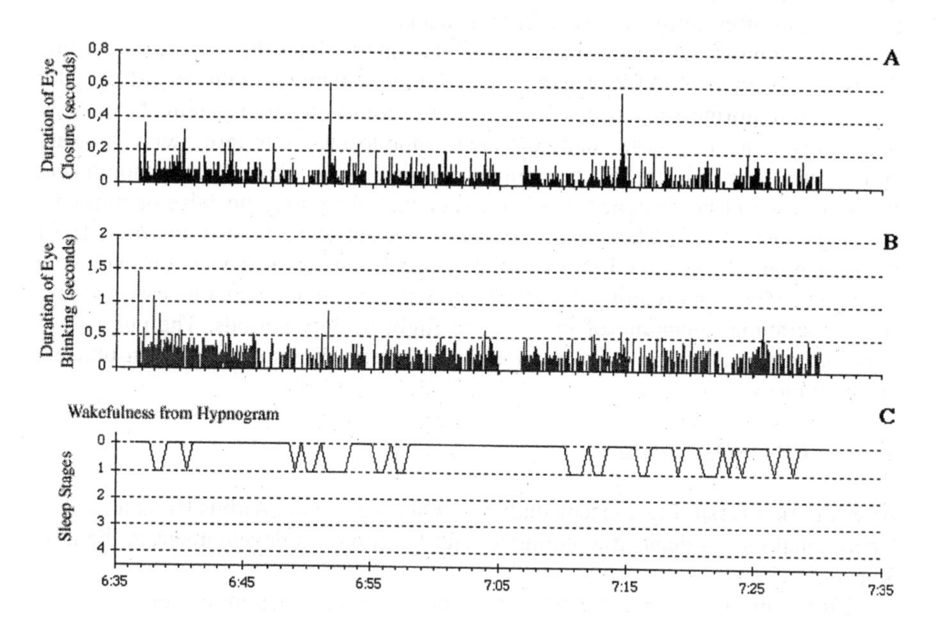

Figure 30.2 Eye closures, blinking, and sleep stages

Implications for System Specification and Design

Provisions from ALPA and IFALPA

The anticipated use of cameras may intrude cockpit privacy and raises ethical questions that certainly have to be correctly addressed before anything else as stated earlier in the paper. Literally, the mere fact of being continuously observed by tele-monitoring might 'at first glance' be potentially disturbing and distracting. There is however operational evidence this is not the case. The US Airline Pilots Association (ALPA) supports the use of cameras for its value in accident investigation but with the caveat that no parts of the pilots, including their hands, be visible. Cameras would typically record large segments of the cockpit including instruments, displays, as well as specific switch and selector positions. In this context, Cockpit Voice Recorders (CVDR) are an equivalent tool containing information that mandates cautious treatment. Pilots and safety officials agree that privacy legislation similar to current voice recorder laws would be needed to keep CVDR footage protected from court cases and Freedom of Information Act requests. It is also fair to say that the increased practice of Flight Operations Quality Analysis (FOQA) is going much further beyond anything seen so far. FOQA has largely expanded the use of Digital Flight Data Recorders to make inferences about quality assurance using sophisticated benchmark techniques that are meant to offer non-jeopardy safety feedback.

In the context of fatigue management the International Federation of Airline Pilots Association (IFALPA) has issued policy statements with regard to pilot activity monitoring. It states that devices should be transparent in normal operations. In no case should such monitoring interfere with normal communications within, or transmitted from, the cockpit. Such monitoring functions should be designed in such a way that absolutely no false or nuisance warnings are issued. Manual or electronic means should keep no records of warning events, nor should the data-link transmission of such warning events be permitted. Most importantly pilot activity monitoring should never be used for pre-flight operational planning of increases in flight or duty periods. The purpose of pilot activity monitoring is to provide an additional margin of safety in unusual circumstances.

Steps in EPAM Development

After the first feasibility concept study in 1999 (Speyer) the Airbus Human Factors Operational Group decided to continue with the project's development in the next 3 years.

Efforts in the year 2000/2001 are focussing on cockpit cameras and on algorithm design:

1. tests are now being carried out with double cameras to secure the field of view to better cover the range of pilot head movements and therefore include the fine-tuning of camera cockpit positions;
2. experimental work also concentrates on improving image processing in parallel with EEG recordings, including laboratory tests to calibrate a full scope images and EEG samples covering the complete spectrum of cerebral activation states;
3. use of in-flight and laboratory recordings for algorithm development and fine-tuning; and
4. design of a real time image processing device not to store images of cockpit scenes (as performed for algorithm development) but to temporarily keep factual data on pilot alertness states in a memory buffer after initial processing by the algorithm.

Efforts in the year 2001/2002 will mainly concentrate on algorithm evaluation and device validation:

1. initial evaluation of the prototype real time image processing device in airline flights with real-time in-flight generation of under-alertness sequences and immediate comparison with EEG to envisage various sequences of visual, auditory, and tactile EPAM alerts under the direct control and supervision of flight observers;
2. use of in-flight and lab recordings for further algorithm fine-tuning to minimise false alerts and omissions and further adapt the real time processing computer to its needs;
3. timely evaluate certification requirements and engage into participatory design with the Pilot community making sure the issue of image or alert signal storage is clarified; and
4. gradually – and as the project matures – airline flights will no longer see volunteer pilots being equipped with EEG monitors and real-time in-flight generation of under-alertness sequences will produce direct alerts however under observer supervision who will still control the under-activity sequence as this will still not be part of avionics.

The total EPAM package will then be handed over for future application and integration whether as an avionics retrofit or as a package for new aircraft such as the A3XX.

Conclusions and Prospects

Initial results presented in this paper confirm the feasibility of the EPAM concept. Results also show that reductions of pilots' interactions with cockpit interfaces are often related to decreased alertness detected by physiological means. But results also suggest that the sole measurement of these interactions is not sufficient to predict alertness decrements and that these could be intercepted by non-invasive individual measures such as those stemming from eye movements. The data confirm the efficiency of in-flight napping strategies on the specific roster selected for airline flights as it induced a high level of monotony and fatigue. In adopting this strategy however, pilots preferably have to be supported by a device such as the EPAM. Moreover, to this day no flight deck means exists to enable detecting simultaneous sleep onsets for both pilots.

The Sabena commercial flights demonstrated that:

1. activity decrements would tend to be attributed to monotony as they correlated with heart rate variability decrements related to reduced mental effort; and
2. eye closures and eye-blinking would tend to be attributed to the stage 1 onset of micro-sleeps typical of sleep pressure and performance degradation.

The Airbus test flights on the other hand proceeded with augmented crews relaying on each other to practice recuperative rests in between intensive cockpit test activities. The FANS test activities could not be used as reliable predictors since all crew members were less subject to both monotony and sleep loss. However, the flights confirm that inactivity is indeed related to decreased alertness, albeit less frequently because of crew augmentation.

Further work is underway to fine-tune the concept, as it is essential to avoid both false positives and false negatives. Whether positive or negative, mixing activity and alertness signals will be performed by bringing time buffers into play to wait for ample signal confirmation before triggering alerts. Further work is also underway to develop specific procedures intended to increase pilot's alertness and situational awareness after EPAM alerts. Initial research on this issue (Speyer) documented that after getting visual then auditory alerts pilots who perform a specified situational awareness reactivation procedure had seen their alertness rebound for at least 10 minutes if the procedure was being performed within a minute or two. Those who performed the procedure less vigorously over a longer time did not see alertness rebound. This finding has to be further verified.

After certification, the EPAM could be implemented on future Airbus aircraft, first with activity monitoring and in a later stage with eye video recording with the clear understanding that any traces should be erased after feedback signal processing. It should also be emphasised that this concept is not a panacea to do cheaply away with drowsiness and wakefulness problems. Responsible rostering and effective education for all actors in the process should in the first place address pilot fatigue problems. Existing knowledge in this field has turned fatigue and alertness management into a corpus of solid, common sense, and scientifically validated recommendations. The Reason model tells us only too well that a technological device alone cannot be relied upon to prevent latent and organisationally related issues. It is about time the industry remembers that about one in five accidents has fatigue at its origin either as a causal or as a circumstantial factor. The EPAM is nothing but an added layer of protection like a last line of defence vis-à-vis otherwise perfectly normal human phenomena such as fatigue, sleep pressure, and consequently normal variations of alertness. Let's not forget it, if we want safety so much.

References

Bourgeois-Bougrine, S., Gounelle, C., Cabon, P., Mollard, R., & Coblentz, A. (2000, March). Fatigue in aeronautics: Point of view of pilots. *Proceedings of the 4th International Conference on Fatigue and Transportation 'Coping with the 24 Hour Society'* (pp. 2), Fremantle, Australia.

Cabon, P., Mollard, R., Coblentz, A., Fouillot, J-P., & Speyer, J-J. (1995). Recommandations pour le maintien du niveau d'éveil et la gestion du sommeil des pilotes d'avions long-courriers. *Médecine Aéronautique et Spatiale, XXXIV, 134*, 119-128.

Cabon, P., Mollard, R., & Coblentz, A. (1995, April). Prevention of decreases of vigilance of aircrew during long haul flights: Practical recommendations. *Proceedings of the 8th International Symposium on Aviation Psychology* (pp. 914-920), Colombus, OH, USA.

Graeber, C., Rosekind, M. R., Connell, J. J., & Dinges, D. F. (1990). Cockpit napping. *ICAO Journal*, 5-10.

Mollard, R., Coblentz, A., Cabon, P., & Bougrine, S. (1995, October). Vols long-courriers. Sommeil et vigilance des équipages. *Guide de recommandations: Volume I et II*(pp. 202), DGAC ed. Paris.

Perrinet-Marquet, M., & Parriaux, N. (1999, September). Activités oculaires et vigilance des conducteurs. *Journées d'études 'Facteurs Humains et Sécurité sur Autoroutes'* (pp. 10-1/10-9), Paris.

Speyer, J-J., Cabon, P., Mollard, R., Bougrine, S., & Coblentz, A. (1995, November). *Coping with Long Range Flying: Recommendations for Crew Rest and Alertness* (pp. 215), Airbus Industrie ed. Blagnac.

Speyer, J-J., Elsey A. (1995). *Towards the Integration of Pilot Guard Systems for Monitoring Crew Activity in Flight: A Cooperative Evaluation Conducted by Airbus* (Airbus Industrie Document AI/ST-F 472.6797/95), Toulouse, France.

Speyer, J-J. (1999). Specification, Conception et Mise au Point d'un Système de Support pour le Maintien de l'Attention, de l'Activité et de la Vigilance en Vol, *Mémoire de DU. Bases Facteurs Humains pour la Conception de Systèmes Homme-Machine en Aéronautique.* Université René Descartes, Paris.

Chapter 31

Inadvertent Slide Activation

Dieter Reisinger
Lauda-Air Luftfahrt AG, Airport Vienna, Austria

Introduction

Airlines traditionally have a big interest in keeping the number of inadvertent slide activations to a minimum. It is not only a costly issue, as slides have to be replaced and revenue is lost when passengers have to be offloaded, the deployment of a slide can also pose a significant hazard to ground personnel who approach the aircraft door from the outside. With less numbers of inadvertent slide deployments, ground safety can be improved significantly. Statistics show that not only cabin crew are involved – quite often catering and maintenance personnel is engaged in such incidents. A recent investigation indicated that most airlines have an established procedure to limit such incidents as most of them had bad experience in this area. The investigation also showed that there seems to be a correlation between aircraft type, and thus door design, and frequency of inadvertent activation. This paper aims to identify the causal factors that lead to inadvertent deployments and makes suggestions as to how to improve the situation. In this context 'inadvertent slide activation' means a non-intended deployment of a slide/raft when the door is opened and the door is in automatic position (i.e., the slide is armed).

Causal Factors

When looking at incidences, the causal factors leading to an inadvertent slide activation most often cited are: Fatigue/Distraction of cabin crew, Out-of-flow situations, Time of Pressure, Non-adherence to procedures, Unfamiliarity with the tasks, Environmental factors, and Poor man-machine interface among others.

Typically, slide deployment (as incidences in general) is a combination of latent and active conditions, such as distraction combined with inexperience or deviation from procedures combined with out-of-flow situations, etc.

Fatigue of Cabin Crew

Naturally, one expects a performance degradation when cabin crews suffer from fatigue or disturbed sleep patterns, which is the case on long-haul flights but also on short range flights, where duty times can easily match those achieved on long haul duties. In the author's opinion, the influence of fatigue is over-estimated. Statistics show that slides are deployed even after short duty periods (see also Table 31.1).

Distraction of Cabin Crew

Normally, when the command is given to 'disarm slides' cabin crews will immediately perform that task and report back via cabin interphone. When distracted while the command is given, e.g., by passengers or fellow crew members, there is a high likelihood that the disarming-task is not performed. When a cross-check procedure is established, the mistake should be detected by the other cabin crew. Distractions can also lead to out-of-sequence-events, for example when the request is made to open the doors again in attempt to ventilate the cabin (see also 'out-of-flow-situations').

Out-of-flow Situations

These are situations where something unusual happens that distracts the crew member responsible for the door. A typical example could be a last minute catering car approaching the aircraft while the slides are already armed and the ground staff, unsure whether he can enter the aircraft at this stage or not knocks on the door in attempt to get in. As a natural reaction to someone 'knocking on the door' the crew opens without first disarming the slide. A good defence strategy here is to establish the procedure that slides are only armed after the movement (push-back) has started.

Time Pressure

This is almost always a factor and does not need an explanation. Experienced crews and inexperienced crews can run into the time-trap. Also, maintenance personnel are affected.

Non-adherence to Procedures

An on-purpose deviation (violation) from established procedures is rare in aviation (see IATA statistics on H1). Unintentional deviations from procedures are not. This could be a cabin crew member leaving the assigned duty area without notifying the remaining crew. In one case a cabin crew member decided to spend the final minutes of taxi on the flight deck, forgetting to delegate his door responsibility to another crew member. The slide was deployed when the door was opened by an inexperienced crew member when catering personnel knocked.

Table 31.1 Error producing conditions

Condition	Factor	Condition	Factor
Unfamiliarity with task	17	Misperception of risk	4
Time shortage	11	Poor feedback from system	4
Poor signal-to-noise ratio	10	Inexperience (not lack of)	3
Poor interface design	8	Inadequate checking	3
Designer-user mismatch	8	Educational mismatch/Person	2
Irreversibility of errors	8	Disturbed sleep patterns	1,6
Information overload	6	Hostile environment	1,2
Negative transfer between	5	Monotony and boredom	1,1

Unfamiliarity with a Task

Research has shown that personnel unfamiliar with a task will have a much higher error rate (factor 17). Those who lack experience will make more errors by factor 3 (see Table 31.1). This of course implies that people with a lack of training or insufficient practice should not be given responsibility to operate a door unless supervised. Aviation is an industry with heavy training requirements, both initial and recurrent, and high quality standards with respect to checking. Cabin crew typically fly more than one type of aircraft and depending on the rostering, one type may not be flown at regular intervals. However, both experienced and inexperienced crews have been involved in slide incidences in the past.

Environmental

Cabin Crew reported that when condensed water within the door frame froze, it caused stuck girt bars. The only way to assure that slides are disarmed is a detailed look at the floor area, i.e., have a second person take a look at the position of the girt bar while opening the door carefully.

Man-machine Interface

The author conducted a survey and found out that certain aircraft types have a worse slide-history than others. The Boeing 767 and the Airbus A320 rank among those aircraft with a high number of reports.

Poor interface design ranks high among the error producing conditions. According to Table 31.1 it shows a factor of 8.

Other

There also have been some curious cases, such as:

1. A passenger, opening an aircraft door in an attempt to practice the newly acquired knowledge on Emergency Procedures, after the safety briefing.
2. An in-flight slide deployment on a B747-400. The reporter cites that the causal factor could never be determined but that there were three possible scenarios: (a) a child would be able to put its hand in the bustle of door 3 and could have activated the firing lanyard, (b) a possibly leak inflation cylinder, and (c) lanyard installed with a hair-trigger situation during packing. For (a) an engineering order was issued to place a plastic cover over the inspection hole port in the bustle. As a result of possibility (b) extensive tests were carried out that would proof a leakage in the cylinder. The 747-400 crew deflated the slide with a Swiss army knife.

While these causal factors are rare, the reports show that they have happened. Ferry flights pose a certain hazard because they are usually conducted with a non-standard crew composition. Due to unfamiliarity with the task the risk of deploying a slide inadvertently is higher. We suggest a two-step procedure to overcome the problem:

1. the Commander makes a person responsible for the doors (pre-flight briefing with whoever is in the cabin, such as maintenance staff); and
2. always have the First Officer rather than the senior cabin crew (Purser) do the 'cabin crew disarm slides and cross-check' announcement. Then the FO will be accustomed to making the announcement, no matter who travels in the cabin.

Door Lever Design

In this section, several door handle locations and slide selectors are discussed. Different ways to indicate that a slide is armed are also discussed. Unlike in other areas of aircraft design, standardisation among door handle mechanisms is not very advanced.

Boeing 737

The B 737 features a girt bar that has to be manually put into girt bar brackets after the door is closed. While this is an old design, it proved to be very efficient – there are hardly any reports about inadvertent slide deployments. The arming of a B737 slide is in fact a 'process' rather than a simple act. Because the arming/disarming process is completely different to the opening of the door (which starts with a simple rotation of the door handle in direction of the arrow) it is very unlikely that a cabin crew confuses the two actions. The design can therefore be considered to be very error tolerant – should the girt bar remain in the brackets when the door is opened, flight attendants will:

1. Feel a much higher resistance when pushing the door open as there is no powered assistance.
2. When looking down the cabin crews can actually verify that the slide has not been disarmed.

As a result the B737 has a good history with respect to inadvertent slide deployment despite the fact that it is a short range aircraft with an accordingly higher frequency of door operations. These designs, although very error-tolerant have been superseded by the advance of technology with much more complex systems.

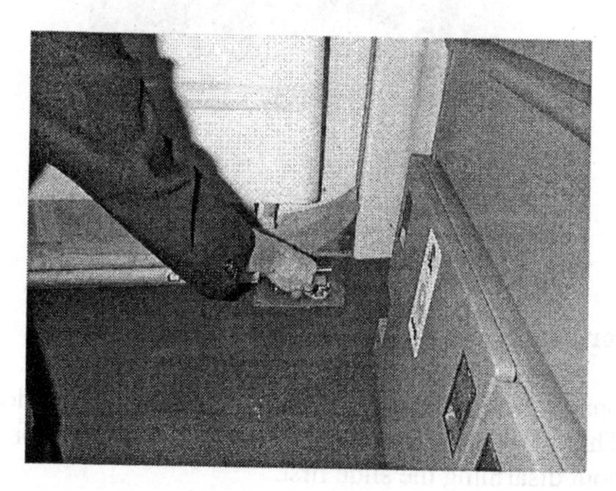

Figure 31.1 Engaging girt bar in floor brackets: Cabin crew has to reach down and physically place girt bar into brackets

Boeing 767

The Boeing 767 is a long-range aircraft where doors are opened typically once every 10 hours. Despite this low frequency this type has had a very unfortunate history with respect to inadvertent slide activation. The door handle is in immediate vicinity to the slide selector.

To arm a slide, the following steps need to be performed:

1. A flap needs to be lifted.
2. A push button needs to be depressed.
3. The slide selector needs to be pushed outboard.

To disarm a slide the following steps need to be performed:

1. The flap needs to be lifted.
2. The slide selector needs to be pulled towards you – no depressing of the button.

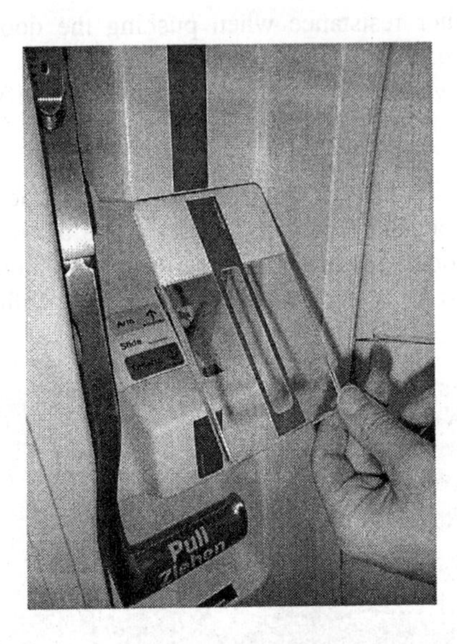

Figure 31.2 Door handle and slide selector on a Boeing 767

As an additional indication a 'yellow flag' covers part of the door handle if the slide is armed. This shall warn the Flight Attendant in case he or she reaches the door handle without disarming the slide first.

Now lets take a look at the intrinsic features of this man-machine interface.

1. Are the handles separated? NO – the handles for opening the door and for arming/disarming the slides are in close proximity.
2. Are the movements conspicuous? NO – when the slide is disarmed, the lever has to be moved towards the body. So has the door lever when the door is to be opened. The slide lever is protected by a flap. Does this protect from inadvertently grabbing the wrong handle? Because lifting the protection is a natural thing to be done every time the door has to be opened, this 'protection' is worthless. A protection that is disabled every time the door is opened will not help prevent inadvertent activation.
3. Does the push button on the side actually safeguard against inadvertent activation. NO – because the danger lies in grabbing the door handle and pulling it towards the body rather than grabbing the slide arm/disarm handle.
4. Is the yellow flag a good protection? NO – because the flag is there for most of the time – average 10 hours trip time – one gets used to the look of it. A flag is worthless when it is present during normal operation for most of the

time. In addition, the flags either tend to break away or they do not show due to mechanical faults within the system.

5. Is the system error tolerant? NO – because once the door handle is lifted, the door will be pulled open by a strong spring in the ceiling. The process cannot be stopped. This pulling force of the spring is not so evident in normal operation when the slide is disarmed, as the slide remains within the door structure so the weight of the slide adds to that of the door and counteracts the upward movement. When the door is opened and the slide is armed the weight of the slide will not be in the door.
6. Also note that the labelling of the door handle says PULL – perhaps an 'invitation' to pull. Obviously, the label below 'Pull Handle to Open' is not eye-catching enough. On other designs, the PULL and OPEN are more in the line of sight. On the B767 the 'PULL-OPEN context' definitely is not as evident.

One word about girt bar indicator windows – in some incidences the cabin crew thought that the slide was disarmed – they carried out all actions in proper sequence – when in fact the girt bar was still engaged in the brackets because of a mechanical fault in the disarming system. In order to avoid this, designers have implemented a means that lets cabin crew verify whether the girt bar is still engaged or not. On Boeing 767 and 777 so-called 'girt bar viewing windows' are installed. This window shows a yellow flag when the girt bar is engaged.

The advantage of this system is its simple mechanical design with little probability for failure. The disadvantages are:

1. that the viewing windows might get covered with dirt due to the close proximity to the ground;
2. flight attendants have to bend down to see the windows, an exercise that is not always carried out although it is procedure; and
3. The frame around the girt bar indicator can get detached. Some yellow may then remain visible when it should not.

Cabin crews reported on the B777 that the yellow flag comes into view when the door handle is moved to the open position, despite prior disarming. It stays yellow while the door is opened. During initial training flight attendants were instructed that this is 'normal' for a B777-door. In the author's opinion this is either a maintenance or a design flaw and should be eliminated. The yellow flag-issue caused some confusion when the type was introduced, according to one operator.

Airbus A320

The A319/320/321 also features a history of significant numbers of inadvertent slide deployments. First, it can be argued that the A320 is a short/medium range aircraft and due to the higher number of cycles of such aircraft the likelihood of mishandling a door is fairly high. Secondly, the A319/320/321 is a very popular model with airlines so the absolute numbers will be high. Therefore, to be more accurate one would have to look into deployments per cycle. However, the man-machine interface design is worth a few words. Intrinsic to the design of an A320's door lever and slide selector are:

1. The handles are comparatively close together. However, different tactile feedback is achieved by the shape of the grip.
2. To open a door, the door lever has to be pulled upwards. To disarm a slide, the slide selector has to be pulled upwards as well.
3. Both opening a door and disarm a slide is a (physically) easy task that does not require extraordinary strength.
4. Some crews reported that the green 'armed' indicator next to the 'disarm'-label caused confusion.

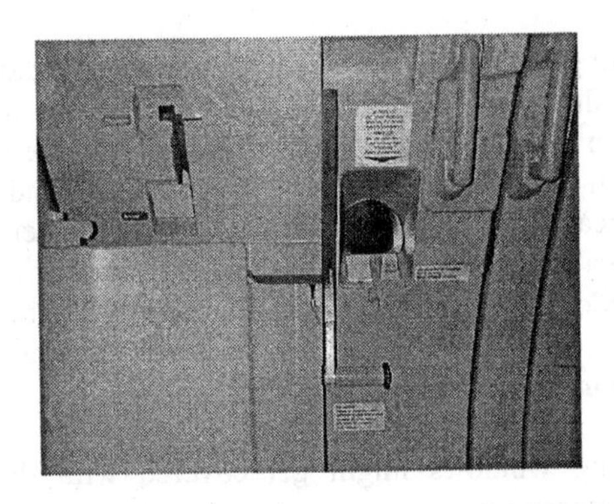

Figure 31.3 Door handle A320

When the door handle is grabbed with the slide still armed, a buzzing sound will be triggered to warn the Flight Attendant. Obviously, because the risk of confusing the levers has also been detected by the design engineers, such a warning was implemented. This buzz was then later removed by request of the authorities. There is no such warning on Boeing aircraft – the buzz could be a distraction in case of an evacuation. One Flight Attendant, however, reported that when she was distracted (both the cockpit crew and a passenger had a request at the same time while she wanted to disarm the slide) the warning helped her avoid an inadvertent deployment. One airline reported never to have had any problems with inadvertent slide activation until the A320 was introduced. Before, this airline operated other short/medium range aircraft with slides, so the FAs were used to the arm/disarm procedure. Once the A320 was introduced at this airline, they had five deployments within one year. As mentioned before, the absolute numbers need to be related to number of sectors critical with respect to inadvertent slide deployments. However, five is a high number in this case irrespective of the cycles because before the number was nil.

Airbus 340

The A340 features a different door operation system compared to the A320. All doors on the A330/A340 are basically operated in the same way – the overwing exits look slightly different but the operation of the door is still identical to all others.

The door handle and the slide selector are very dissimilar with respect to location and operation. The door handle is pulled upwards/downwards whereas the slide selector has to be deflected sideways. It is hidden behind a flap. The condition of all slides can be verified on a central Flight Attendant panel, located next to door 1L. Unlike Boeing 767 or 777 aircraft, there are no girt bar indicator windows. The Airbus 340 has had only a small number of inadvertent deployments so far.

Boeing 777

With the B777 the door handle and the slide selector are dissimilar in size and shape. To open the door, the big lever has to be rotated approximately through 180 degrees. To arm the slide, a flap has to be lifted and a much smaller lever, located well above the door handle, has to be rotated through approximately 50 degrees. As mentioned before, the B777 features girt bar indicator windows below each door. There is an indication on the Flight Attendant control panel next to Door 1L that shows whether a door is closed or not. It does not, however, show the status of the slides (a feature implemented in the A 340 FA panel). This information is displayed on the 'door page' on a LCD-tube in the flight deck, only.

Defence Strategies

There are numerous methods that can help to reduce the number of inadvertent slide activations.

Training

Door training is required by JAR-OPS 1 for all pilots and flight attendants. Good training will reduce the 'unfamiliarity with a task'-type of causal factor. It will emphasise the dangers of accidentally opening an aircraft door while slides are armed and will consist of a sufficiently high number of practice runs to give flight attendants enough experience. Training is preferably conducted by means of a door trainer: some aircraft doors, such as Boeing 737 show additional resistance when opened while the slide is armed. It is this slight difference in feel that can 'warn' a Flight Attendant should the door be entered while the slides are armed. Training-wise it is essential not only to cover flight and cabin crews. Catering personnel or security staff (air marshals) could be among those opening an aircraft door, as well as maintenance staff. All these either need to be trained or they must not operate an aircraft door. In some companies it is procedure that air marshals receive training (able-bodied passenger introduction), maintenance crews receive training, and hold certificates that make them eligible to operate a door while catering is not allowed to open a door from inside the aircraft.

Cross-check procedure

Before the door is opened, have the crew member of the opposite door check your door. This procedure is intended to detect mistakes of a perhaps distracted crew-member that forgot to disarm the door or who thought that the door is already disarmed. A typical command is 'Cabin Crew disarm slides and crosscheck'. Is this cross-check procedure sufficient in all circumstances? NO. This procedure will not be of any assistance in the following two situations:

1. When the door handle is grabbed and pulled instead of the arm/disarm lever.
2. When aircraft is operated with reduced number of cabin crew and one person is responsible for two doors.

A cross-check procedure that would be efficient in all situations where two crew members are in close proximity is that followed by pilots in case they have to handle memory items during an emergency: While one pilot (the non-flying, PNF) points out the action he is going to do (e.g., 'left thrust lever to idle') the other pilot (usually the flying pilot, PF) verifies this by looking at the lever that the PNF grabbed (in this example the left and not the right thrust lever). Only after the PF confirms, will the PNF pull the lever to idle. A similar procedure could be set-up in case of a cabin door:

Table 31.2 Example, assignment of door responsibility for B777 with two different numbers of cabin crew (FA – Flight Attendant, AbP – Able-bodied Passenger)

Number of Cabin Crew: **8-12**		Number of Cabin Crew: **7**	
FA-Position	Door Location	FA-Position	Door Location
FA1	Door1L	FA1	Door1L
FA3	Door 1R	**AbP**	Door 1R
FA4	Door 2L	FA4	Door 2L
FA5	Door 2R	FA5	Door 2R
FA7	Door 3L	FA7	Door 3L
FA8	Door 3R	FA8	Door 3R
FA6	Door 4L	FA6	Door 4L
FA9	Door 4R	FA9	Door 4R

Again, it takes two crew members to do such a cross-check.

Clear Assignment of Duties

Typically, a higher number of flight attendants than there are doors are available on the aircraft. In such cases persons need to be made responsible for the door. Otherwise, a crew member might think that the job has been done already and opens the door without checking. A typical door-responsibility is shown in the following Table 31.2.

Reduce Time Pressure

When the aircraft approaches the final parking position the likelihood that a catastrophic situation arises and immediate disembarkation via the escape slides is necessary becomes less likely. At this stage the energy levels are very small as the aircraft travels at small pace. When the disarm lever and the door handle are operated under time pressure, the chance of grabbing the wrong lever is high. With the implementation of a procedure, where slides are disarmed when the final parking position is reached rather than waiting for the aircraft to come to a complete stop at the final parking position, the actions 'disarm slides' and 'open door' are separated. Hence, there is less time pressure. This procedure, however, implies that someone has to determine, when the final parking position is reached. The best-qualified person on board is the First Officer as he or she usually does not have any other tasks at this stage of flight except monitoring the Captain's performance during the docking process. As an alternative, a senior cabin crew will be able to determine final parking position by taking a look outside the cockpit window. This however requires some training, as it might not be always as obvious that the parking position is reached. Also, on some aircraft types the Jump Seat would have to be removed first to gain access to the cockpit door.

Naturally, this procedure tends to raise discussion among safety experts whether indeed it is safe to disarm the slides before passenger stairs have been attached to the aircraft.

Wording and Labelling

Across the industry different wordings are in use for one and the same thing:

1. Slides armed/disarmed.
2. Doors in automatic/manual.
3. Doors in Park/Flight position.

Preferably, the labelling and the wording used in a procedure should be uniform, i.e., when giving the command to disarm slides, there should be a label next to the disarm handle that says 'disarm' rather than something else. Let us take a look at the different wordings. Arm/Disarm seems to be quite common in the industry. The drawback to the author's opinion is the fact the words arm/disarm are not typically part of school English. It is an expression common in military and aviation but has hardly any use in everyday English. Therefore, flight attendants coming from school will not be familiar with both these words. They do not relate to flight phase directly, as does 'park' and 'flight' which are similar except for the 'dis'.

In aeronautical communication the phrase 'affirmative' was changed to 'affirm' several years ago when it was obvious that there is a risk of confusing 'affirmative' with 'negative', meaning the exact opposite, especially when the beginning of the word was blocked out with keying the mike button. Here, we have a similar situation.

The advantage of 'park' and 'flight' is obvious: both are short and conspicuous, they are part of everyday English, they relate to the phase of flight. The only drawback that needs to be addressed is the fact that when passengers have to be evacuated on the runway after the aircraft aborted a takeoff will also 'be parked' on the runway.

Checklist for Door Operation

Why not stick a checklist next to the door? Pilots are used to working with checklists even for the simplest tasks. Why should checklists not be useful to cabin crews?

Additional Procedures

Some companies rely on the fact that slides will be disarmed automatically when the door is opened from the outside. It could therefore be made a procedure that cabin crews are not allowed to open the doors from inside except in an emergency. Doors are always opened from personnel on the outside of the aircraft. When establishing such a procedure consider that there might be a group of people in the aircraft, such as air marshals who need to be made aware of such a procedure, too, and that some doors, such as B 737 or DC 9 will not disengage the slides, when the door is opened from outside. It could be contra-productive when a Flight Attendant who flies mixed fleet has to disarm on one type of aircraft and leave the door to itself on another.

Reduce Distractions

1. Maintain cabin crew discipline when approaching final parking position – similar to the 'silent cockpit' – concept below 10.000 ft for flight crews. A 'silent cabin' – procedure will reduce communication to the minimum required and hence reduce distractions.
2. Implement clear commands also for cabin crew to minimise asking back.
3. PA announcements that passengers shall remain seated after the aircraft has landed, so to avoid instructing passengers when flight attendants have to concentrate on the doors. In this context, give clear advice how passengers should be addressed if they get out of their seats early. Typically, a passenger cannot tell whether the aircraft is at the final parking position or not – the

aircraft could have stopped on a taxiway to let other traffic pass. Here it is important to point out the Fasten Seat Belt Sign – only when this has been turned off, the final parking position has been reached.

4. Suggest that cabin crews should not wait for the Fasten Seat Belt-sign to disarm slides. If the flight crew forgets to turn it off, the trigger to disarm slides is not there.
5. Organise the cabin work such that coats are not handed over to passengers shortly before doors have to be opened (creates time pressure).

Door Design

The recommendations given so far concentrated on procedures within an airline – the ideal way for any design, however, would be a system that is failure tolerant in every respect so that neither training, nor a procedure would be required. This ideal can hardly ever be achieved.

1. Typically, it is not sufficient to produce a piece of equipment that just fulfils the basic engineering requirements (working system). The mismatch between this basic design and an ideal error-proof design has to be overcome. In a first step this is done by INTERFACE ENGINEERING, where an interface is added to the basic design, that makes it more 'user friendly'.
2. When the design of the interface shows weaknesses, then in a next step TRAINING to overcome the flaws is required.
3. If still some degree of miss-use is left over after the training, then in a last step PROCEDURES are established to make sure that a system is operated by the human operator in a certain (and only in a certain) way.

An aircraft door is a piece of engineering: there is training required and there are procedures implemented.

Summary

Inadvertent slide deployments are costly incidences and dangerous to ground staff in the vicinity of aircraft doors. The article has addressed several causal factors and gave an insight into the intrinsic problems with interface design. As such it does not only address airline safety officers but also aircraft door system design engineers. The previous section showed that the tasks for operation of the doors/slides is basically similar for all aircraft, however there are intrinsic differences with respect to the means by which the task is achieved. Cabin Crew typically operate more than one type. In other areas of aircraft operation, a much higher level of standardisation has been achieved. From a design viewpoint the disarm/arm action needs to be completely different to the door opening sequence, both in tactile feedback and work pattern. From the airlines point of view it is probably best to establish clear procedures as to which cabin crew is responsible for the door, include cross-check philosophies, improve training. Because this is a complex problem, simply taking disciplinary action to deter personnel from misuse of door mechanisms is not a solution.